OXFORD MONOGRAPHS ON GEOLOGY AND GEOPHYSICS NO. 26

Series editors

OXFORD MONOGRAPHS ON GEOLOGY AND GEOPHYSICS

1. DeVerle P. Harris: *Mineral resources appraisal: mineral endowment, resources, and potential supply: concepts, methods, and cases*

2. J. J. Veevers (ed.): *Phanerozoic earth history of Australia*

3. Yang Zunyi, Cheng Yuqi, and Wang Hongznen (eds.): *The geology of China*

4. Lin-gun Liu and William A. Bassett: *Elements, oxides, and silicates: high-pressure phases with implications for the Earth's interior*

5. Antoni Hoffman and Matthew H. Nitecki (eds.): *Problematic fossil taxa*

6. S. Mahmood Naqvi and John J. W. Rogers: *Precambrian geology of India*

7. Chih-Pei Chang and T. N. Krishnamurti (eds.): *Monsoon meterology*

8. Zvi Ben-Avraham (ed.): *The evolution of the Pacific Ocean margins*

9. Ian Mc Dougall and T. Mark Harrison: *Geochronology and thermochronology by the $^{40}Ar/^{39}Ar$ method*

10. Walter C. Sweet: *The conodonta: morphology, taxonomy, paleoecology, and evolutionary history of a long-extinct animal phylum*

11. H. J. Melosh: *Impact cratering: a geologic process*

12. J. W. Cowie and M. D. Brasier (eds.): *The Precambrian-Cambrian boundary*

13. C. S. Hutchison: *Geological evolution of south-east Asia*

14. Anthony J. Naldrett: *Magmatic sulfide deposits*

15. D. R. Prothero and R. M. Schoch (eds.): *The evolution of perissodactyls*

16. M. Menzies (ed.): *Continental mantle*

17. R. J. Tingey (ed.): *Geology of the Antarctic*

18. Thomas J. Crowley and Gerald R. North: *Paleoclimatology*

19. Gregory J. Retallack: *Miocene paleosols and ape habitats in Pakistan and Kenya*

20. Kuo-Nan Liou: *Radiation and Cloud Processes in the Atmosphere*

21. Brian Bayly: *Chemical Change in Deforming Materials*

22. Allen K. Gibbs and Christopher N. Barron: *The Geology of the Guiana Shield*

23. Peter J. Ortoleva: *Geochemical self organization*

24. Robert G. Coleman: *Geological evolution of the Red Sea*

25. Richard W. Spinrad, Terry L. Carder, and Mary Jane Perry: *Ocean Optics*

26. Clinton M. Case: *Physical principles of flow in unsaturated porous media*

Physical Principles of Flow in Unsaturated Porous Media

Clinton M. Case

OXFORD UNIVERSITY PRESS   New York
CLARENDON PRESS   Oxford
1994

Oxford University Press

Oxford   New York   Toronto
Delhi   Bombay   Calcutta   Madras   Karachi
Kuala Lumpur   Singapore   Hong Kong   Tokyo
Nairobi   Dar es Salaam   Cape Town
Melbourne   Auckland   Madrid

and associated companies in
Berlin   Ibadan

Library of Congress Cataloging-in-Publication Data
Case, C. (Clinton)
Physical principles of flow in unsaturated porous media / Clinton M. Case.
p.   cm.   (Oxford monographs on geology and geophysics ; no. 26)
Includes bibliographical references and index.
ISBN 0-19-504622-6 (alk. paper)
1. Groundwater flow.   I. Title.   II. Series.
GB1197.7.C36   1993   551.49–dc20   92–40046

9   8   7   6   5   4   3   2   1

Printed in the United States of America
on acid-free paper

PREFACE

The study of the flow of fluids through unsaturated porous media, commonly called unsaturated flow, is both ongoing and timely. Applications range from irrigated agriculture to nuclear waste disposal. The concepts and experimental methodology have evolved primarily from the domain of soil science, with contributions from petroleum engineering and work generic to porous media pictured as being composed of collections of capillary tubes or spherical grains. As the concepts and methodology of unsaturated flow have become increasingly widespread, there has occurred a corresponding decline in the efforts among practitioners to continue to examine the theoretical underpinnings of the field, and to accept theoretical understanding at a certain level as being generally sufficient.

The material presented here is aimed at dealing primarily with the theoretical underpinnings of the study of flow in unsaturated porous media and extending the understanding of these in a number of respects.

The fundamental point of view taken here is that phenomena at a given level are explained by considering phenomena at a more microscopic level. Phenomena are discussed at a number of levels with generally an attempt to relate happenings at the macroscopic (Darcy's law, say) level to causes at the microscopic (pore/capillary) level.

The example porous medium considered throughout is soil in the vadose zone. The transport of liquids containing impurities as well as gases is considered, and a unified view of these flows, as well as the possible movement of the porous medium itself, is presented.

It is hoped that the physical and mathematical concepts presented here along with the perspective they represent will help to stimulate further work, leading to ever-improved descriptions of flow in the many forms of unsaturated porous media.

A number of people have made this work possible; Ms. Jan Walker, who did the illustrations, Dr. Mike Whitbeck, for essential computer support, the personnel at Oxford University Press, in particular Mr Stanley George, Ms. Anita Lekhwani, Dr. Jacki Hartt, and above all, Ms. Joyce Berry, for their steadfastness and good spirits, my parents, who encouraged scholarship, my brother and his family for constant positive support, and finally, my wife Marie, for creating a home environment conducive to external achievement, and my two daughters, Hillary and Annette, for constant, unwavering faith in the success of this work.

Reno   C.M.C.
July, 1993

CONTENTS

INTRODUCTION

This work is intended for researchers, students, and applied workers interested in acquiring, from a unique point of view, a knowledge of the fundamental physical basis of flow in unsaturated porous media.

The point of view adopted here is that macroscale effects are best understood in terms of phenomena on the relative microscale.

A number of results are presented using the example of flow in capillary tubes, even though this is not an accurate physical picture of a porous medium. The reason for this is that a number of exact results of considerable expository value are available for this case.

In addition to the capillary level of detail, phenomena are discussed at the macroscopic, that is, Darcy's law level as well. The transition between the microscopic and macroscopic points of view has been considered from various points of view by a number of authors, probably the earliest of whom were Green and Ampt in 1911, who employed the bundle of capillaries picture.

Predicting the exponential dependence of unsaturated hydraulic conductivity on the degree of saturation in the case of water in soil as well as the hysteretic behavior of the relation between matric potential and degree of soil saturation when a laboratory sample undergoes repeated wetting and drying remains elusive. Because of this, no attempt, except phenomenologically at the Darcy's law level, is made in this work to deal with hysteresis. Some further justification for this omission may be found in the fact that in neither practical field applications nor in numerical modeling of unsaturated flow phenomena is hysteresis considered.

The statistical analysis of data is often honored more in the breach than the observance. However, the practical importance of statistical analysis, for assigning levels of significance to field and laboratory observations as well as its use in the environmental monitoring of hazardous waste disposal sites, which are generally loacated in the vadose zone, have made the inclusion of a chapter on basic statistical techniques both necessary and desirable.

The origin of the coupled (for multispecies transport) partial differential equations developed in this work to be solved to describe multispecies flow in unsaturated media is worthy of note. For the past few years, the view among workers interested in multispecies transport in porous media increasingly has been that more attention should be paid to the inclusion of momentum terms in the partial differential equations to be solved to describe the coupled flows. Given the fact that the partial differential equations, which are the starting point(s) for the study of the mechanics of deformable media, of which fluid mechanics and elasticity are special cases, are derived using conservation of momentum as the starting point, this perspective is seen to be not only reasonable but correct. This point of view is adopted here and leads to new equations of motion, new parameters to be measured to describe the results of experiments, and new experiments corresponding to the new parameters as well. Unfortunately, the coupled partial differential equations that result are nonlinear and as such do not possess either standard form analytic solutions or a standard solution methodology. Thus, though physically and mathematically correct, and long used in such contexts as numerical fluid mechanics and earth motion calculations carried out in support of the underground nuclear weapons testing program, this formulation applied to describe multispecies flow in porous media leads to more questions than answers at this stage.

Chapter 1 contains an overview of unsaturated flow phenomenology and definitions of some fundamental parameters. Chapter 2 deals with the consequences of the interfacial energy between two materials in contact, including concepts of curvature. These concepts are fundamental to the understanding of the matric potential of liquids that wet the unsaturated porous media containing them and so are discussed at some length. A connection is then made between the microscopic and macroscopic pictures of flow in porous media via a bundle of capillaries model which includes an explicit solution for the height of capillary rise in a gravitational field as a function of time. Chapter 3 has as its subjects equilibrium thermodynamics and an introduction to statistical mechanics. Thermodynamics describes physical phenomena from a macroscopic, phenomenological point of view. The connection between the microscopic world of atoms and molecules and the macroscopic view of thermodynamics is provided by statistical mechanics. In Chapter 4 the evolution and transport of vapor through a porous medium are considered from various points of view. Aspects of the kinetic theory of gases are developed at a fundamental, though useful, level and, somewhat anticipating the material in Chapter 5, partial differential equations derived from the conservation of momentum point of view are given. Chapter 5 treats the porous medium and the liquid in it as interpenetrating, possibly viscoelastic, continuae. The viscoelastic properties are represented using fractional calculus. This mode of representation, which is by no means universal, was considered by rheologists of the past generation, most notably Scott-Blair. However, the fundamental mathematical result needed to complete the construction of equations of motion similar to those arising in fluid mechanics and elasticity, namely the analog in fractional calculus of the chain rule for differentiation, did not exist until 1970. Some of the analytic approximations made in the course of the development given here are fairly drastic, however, and an area for future work is the further development of aspects of this point of view.

The departure in viewpoint of the material given here from the traditional treatment of equations of motion for liquids and gases in porous media is the following. Traditionally, the equations taken to describe the movement of liquids and gases in porous media have been derived by combining the equation of continuity of mass with a suitable flux law resulting in some form of a second-order, parabolic differential equation known colloquially as the diffusion equation. Mathematical difficulties exist with the solutions of this form of equation when this particular application is made, however, and so the treatment given here is based on the more fundamental principle of conservation of momentum. This methodology is used and accepted as the correct basis for the existing treatments of fluid mechanics and elasticity, that is, continuum mechanics as that discipline is currently understood. The analytic solutions of the equations so derived are, somewhat unfortunately, still areas for future work. Tensor notation is used for most of the discussion. This provides a level of precision and generality that is absent when vector and Cartesian tensor notation and concepts are used as the starting point for equation development.

Chapter 6 gives an overview of nonequilibrium thermodynamics, which contains, in principle, the complete coupled equation set for describing the simultaneous movement of heat, water, water vapor, and impurities, in unsaturated porous media.

Chapter 7 deals with the principles of operation of a number of instruments and methodologies useful in the measurement of various physical properties and transport parameters relevant to unsaturated flow. Chapter 8 contains selected techniques of statistical analysis of data. Nonparametric statistics are discussed first, and extended tables used in the nonparametric analysis of data have been generated as well. Standard techniques of analysis of data are then discussed, along with a brief discussion of extreme-value theory. This last is useful in estimating the return period of the largest or smallest event of a series.

1

OVERVIEW OF UNSATURATED FLOW PHENOMENOLOGY

1.1   MATRIC POTENTIAL

The matric potential, $^{(i)}\Psi$, of the ith species of liquid in a porous medium, is written as

$$^{(i)}\Psi = {}^{(i)}\bar{p}_l - {}^{(i)}p_{ol} = {}^{(i)}\Psi_o \ln\left(\frac{^{(i)}\bar{p}_v(r)}{^{(i)}p_{ov}}\right) \qquad (1.1a)$$

where $^{(i)}\bar{p}_l$ is the average pressure of the ith species of liquid in the medium at the location at which $^{(i)}\Psi$ is measured, $^{(i)}p_{ol}$ is the pressure just beneath a flat liquid surface having the same chemistry and temperature as that of the liquid in the porous medium,

$$^{(i)}\Psi_o = \frac{^{(i)}\rho kT}{^{(i)}m} \qquad (1.1b)$$

in which $^{(i)}\rho$ is the mass density of the ith species of liquid in the porous medium, k is Boltzmann's constant, 1.38054×10^{-16} erg/K, T is the absolute temperature in K, and $^{(i)}m$ is the mass of a molecule of liquid of species i. For water, $^{(i)}m$ is approximately 3×10^{-23} g. The quantity $^{(i)}\bar{p}_v(r)$ is the average vapor pressure of the ith species of liquid in the porous medium. It is the weighted average of the vapor pressures above the liquid–air interfaces in menisci of varying average effective radii of which r is the largest one containing water. The quantity $^{(i)}p_{ov}$ is the vapor pressure of the ith species of liquid when it is at the same temperature and contains the same chemical impurities as the liquid in the porous medium and its surface is of infinite radius of curvature (flat). Thus, in a certain sense, $^{(i)}\bar{p}_v(r)/{}^{(i)}p_{ov}$ may be regarded as the relative humidity, divided by 100%, of the ith species of

3

liquid in the porous medium. Logarithms to the base $e \simeq 2.7183$, natural logarithms, are denoted by $\ln()$, and since $^{(i)}\bar{p}_v(r)/^{(i)}p_{ov}$ is always less than one for liquids that "wet" the material of the pore walls, as will be seen shortly, $^{(i)}\Psi$ is always negative (note that $^{(i)}\Psi_o$ is always positive) in that case. The dimensions of $^{(i)}\Psi_o$ are pressure and for water at 300 K,

$$^{(i)}\Psi_o \simeq 10^3 \text{ bars} = 10^3 \times 10^6 \frac{\text{dyn}}{\text{cm}^2} \tag{1.1c}$$

For porous media nearly saturated with liquid that "wets" its surfaces, water in soil, for example, $^{(i)}\Psi < 0$, and as the porous medium dries, $^{(i)}\Psi$ becomes increasingly negative. For example, for arid region soil surfaces in the summer, such as at the Nevada Test Site, values of matric potential of up to -300 bars have been reported. Special instrumentation exists to measure laboratory samples having values of degree of saturation (fraction of the pore space filled with liquid) corresponding to matric potentials of up to -1000 bars. A theoretical expression for $^{(i)}\Psi$ can be developed for a porous medium if a geometric model of it is assumed. If the geometric model is assumed to be a bundle of parallel capillary tubes having circular cross sections, then the Kelvin relation, derived in Chapter 2, which gives the vapor pressure over a curved liquid surface of mean radius of curvature r_a in terms of the vapor pressure over a flat (infinite radius of curvature) surface of the same liquid containing the same impurities and at the same temperature, yields

$$^{(i)}p_v(\xi) = {}^{(i)}p_{ov}e^{-2\,{}^{(i)}\gamma\,{}^{(i)}m\,\cos({}^{(i)}\varphi)/\,{}^{(i)}\rho kT\xi} \tag{1.2}$$

where $^{(i)}\gamma$ is the air–liquid interfacial energy for the ith species of liquid, commonly called the surface tension of the liquid, $^{(i)}\varphi$ is the angle of capillarity between the liquid and the capillary walls, $1/r_a = \cos(\,^{(i)}\varphi)/\xi$, ξ is the radius of the capillary, which is assumed to be of circular cross section, and the rest of the symbols are as previously defined. The average vapor pressure, $^{(i)}\bar{p}_v(r)$ for the ith species of liquid in the porous medium, is computed for this model as

$$^{(i)}\bar{p}_v(r) = \int_{r_m}^{r} f(\xi)\,{}^{(i)}p_v(\xi)\,d\xi \tag{1.3}$$

where r_m is the smallest pore/capillary radius of interest, $f(\xi)$ is the frequency distribution of average effective pore/capillary radii normalized to one on the interval $r_m \leq \xi \leq r$, where r is the average effective radius of the largest pore/capillary of interest, and $^{(i)}\Psi$ is computed by using Eqs. (1.2) and (1.3) in (1.1). Equation (1.3) is one of the connections between the largest pore/capillary of interest, and $(i)\Psi$ is computed by using Eqs. (1.2) and (1.3) in (1.1). Equation (1.3) is one of the connections between the micro- and macroaspects of unsaturated flow. Another such connection is given by

$$^{(i)}\theta = \frac{\int_{r_m}^{r} D(\xi)\,d\xi}{\int_{r_m}^{R} D(\xi)\,d\xi} \tag{1.4}$$

where $D(\xi)$ is the pore volume distribution parameterized on average effective radius, $^{(i)}\theta$ is the degree of saturation of liquid of species i, the fraction of the pore space of the porous medium containing liquid of species i, and R is the largest pore/capillary radius in the sample.

The assumption made here is that the pores/capillaries are filled in order of increasing size and that all pores/capillaries that are accessible to the fluid are filled up to radius r. $D(\xi)$ is related to $f(\xi)$ via

$$f(\xi)\,d\xi = \frac{[D(\xi)/v(\xi)]\,d\xi}{\int_{r_m}^{R}[D(\xi)/v(\xi)]\,d\xi} \tag{1.5}$$

where $v(\xi)$ is the volume of a pore/capillary of average effective radius ξ and may assume different functional forms in different ranges of ξ. As will be demonstrated later on, for a single liquid of species i, subject to the sequential filling assumption, $^{(i)}\theta$ can be expressed as

$$^{(i)}\theta = \frac{\int_{r_m}^{r} v(\xi)f(\xi)\,d\xi}{\int_{r_m}^{R} v(\xi)f(\xi)\,d\xi} \tag{1.6}$$

Knowledge of the moisture characteristic of the liquid–porous–medium system of interest, namely $^{(i)}\Psi(\,^{(i)}\theta)$, matric potential of the ith species as a function of degree of its saturation, and $v(\xi)$ enables, in principle, Eqs. (1.1a) and (1.6) to be solved simultaneously for $f(\xi)$. The moisture characteristics for water in most, and perhaps, in principle, all, earth materials display hysteresis. In this context hysteresis means that the $^{(i)}\Psi - {}^{(i)}\theta$ relations for the wetting of a porous medium sample hold for larger (less negative) values of $^{(i)}\Psi$ than those for which the $^{(i)}\Psi - {}^{(i)}\theta$ relation represents drying of the sample. The physical reason for this behavior is the following: After liquid initially wets an initially dry porous medium sample, it is drawn, both by direct capillary action and vapor evolution and recondensation, into smaller pores/capillaries than those into which it initially moved. This process is called *spontaneous redistribution* and leaves the porous–medium–liquid system in a lower energy state than it had immediately following the initial wetting. The initial wetting, of course, yields a system having a lower energy state than the liquid and dry porous medium separately. Both the initial wetting and the spontaneous redistribution produce heat, which is conducted away, to leave the net energy of the system lower than before the process occurred. Concomitantly, the entropy of the soil–water system is increased. That this is so can be seen qualitatively when it is remembered that mathematically entropy is the negative of information (Shannon, 1948) and that both the spreading of liquid through the porous medium on initial wetting and the later spontaneous redistribution of the liquid reduce the information available as to the location of the liquid, thus increasing the entropy of the liquid–porous–medium system. More specifically, it is shown in thermodynamics that the condition for equilibrium for any system, the ultimate goal of any natural process, is that its internal energy be a minimum (energy minimum principle) and correspondingly that its entropy be a maximum (entropy maximum principle) (Callen, 1961, p. 86).

1.2 EQUILIBRIUM THERMODYNAMICS

Equilibrium thermodynamics proves to be very useful in understanding many aspects of moisture movement in arid regions. A first example is the work term associated with the combined first and second laws of thermodynamics, which can be written as

$$dU = T\,dS - \sum_i {}^{(i)}\Psi\,d\,{}^{(i)}\theta \tag{1.7}$$

where U is the internal energy per unit bulk volume of the soil–water system, T is the absolute temperature, as noted previously, S is the entropy per unit bulk volume of the soil–water system, and ${}^{(i)}\Psi$ and ${}^{(i)}\theta$ are as previously defined. If it is presumed that $-{}^{(i)}\Psi\,d\,{}^{(i)}\theta$ is the only work term describing the thermodynamic system, and that hysteresis can be neglected, a number of relations defining such quantities as the heat capacity per unit bulk volume at constant ${}^{(i)}\Psi$, the heat capacity of the soil–water system per unit bulk volume at constant ${}^{(i)}\theta$, and U can be written down. An example of thermodynamic manipulations at this level is offered by writing the usual identity

$$\left(\frac{\partial x}{\partial y}\right)_z \left(\frac{\partial y}{\partial z}\right)_x \left(\frac{\partial z}{\partial x}\right)_y = -1 \tag{1.8}$$

and setting for a single species, i, $x = {}^{(i)}\Psi$, $y = {}^{(i)}\theta$, and $z = T$ to yield, after rearrangement,

$$\left(\frac{\partial\,{}^{(i)}\Psi}{\partial\,{}^{(i)}\theta}\right)_T = -\left(\frac{\partial\,{}^{(i)}\Psi}{\partial T}\right)_{{}^{(i)}\theta} \left(\frac{\partial\,{}^{(i)}\theta}{\partial T}\right)_{{}^{(i)}\Psi} \tag{1.9}$$

Equation (1.9) expresses the relation between the slope of the moisture characteristic, $\left(\partial^{(i)}\Psi/\partial^{(i)}\theta\right)_T$ and the change of $^{(i)}\Psi$ with temperature as computed from the results of a set of moisture characteristic measurements, each made at a different constant temperature. Since both $\left(\partial^{(i)}\Psi/\partial^{(i)}\theta\right)_T$ and $\left(\partial^{(i)}\Psi/\partial T\right)_{(i)\theta}$ are positive, the rate of change of $^{(i)}\theta$ with T at constant $^{(i)}\Psi$ is negative. This means, for example, that as the temperature of a liquid–porous–medium system increases, the degree of saturation must decrease to allow $^{(i)}\Psi$ to remain constant. This can seen in another way as follows. As the temperature of a block of partly saturated porous medium, soil for example, partly saturated with liquid water containing dissolved impurities, is increased, both the medium and the liquid in it expand (water expands more readily than the soil containing it by a factor of about 8–10). Thus, the liquid expands to fill larger menisci. This increases the vapor pressure of liquid vapor in the sample via Eqs. (1.2) and (1.3) and concomitantly increases (makes less negative) its matric potential as well. Thus, the degree of saturation must decrease sufficiently for the remaining liquid meniscus surfaces to be somewhat less than their pre-expansion average radii, since the vapor pressure of the liquid increases with temperature as well, so that the average vapor pressure in the porous medium at the higher temperature can be maintained at its lower temperature value. Hence, for constant $^{(i)}\Psi$, the final value of $^{(i)}\theta$ must be less than the initial value, and hence

$$\left(\frac{\partial^{(i)}\theta}{\partial T}\right)_{(i)\Psi} < 0$$

Another example of the use of equilibrium thermodynamics in describing an aspect of flow in unsaturated porous media is the well-known difference in concentration of impurities in the surface layer of a liquid from that of the bulk, thus affecting $^{(i)}\gamma$ and r of Eq. (1.2). The first effect may be quantified for a dilute solution of a single component i via

$$^{(i)}C = -\frac{^{(i)}n}{R_g T}\frac{\partial^{(i)}\gamma}{\partial^{(i)}n} \tag{1.10}$$

where $^{(i)}C$ is the excess or deficit in concentration of the ith impurity in solution in the surface layer over (under) that in the bulk of the solution, $^{(i)}n$ is the number of moles of solute per unit volume of solution of the ith species of impurity, R_g is the universal gas constant, and the rest of the quantities are as previously defined. The minus sign indicates that the concentration of impurities in the surface layer of water that lower the interfacial energy that exists between the water–impurity surface and its surroundings is increased over its value in the bulk solution, whereas the impurities that would increase the interfacial energy in the surface layer are reduced below their concentrations in the bulk solution.

The variation of concentration of impurities in a column of water or in a partly saturated soil column with height is due to the existence of the ambient gravitational field and arises from the condition that the change in the Gibbs free energy of a column containing impurities in the water at equilibrium is zero and that one of the potentials acting on the system is the gravitational potential. The change of $^{(i)}n$ with column height, l, is given for a dilute solution of a single solute species by

$$\frac{d\,^{(i)}n}{dl} = \frac{^{(i)}n}{R_g T} \left(^{(i)}\theta \frac{\partial\,^{(i)}\Psi}{\partial l} + \,_s^{(i)}v\,_s^{(i)}\rho g - \,^{(i)}\mathcal{W} \right) \qquad (1.11)$$

where $_s^{(i)}v$ is the volume per mole of species i of the solute, $_s^{(i)}\rho$ is the mass density of the solution, $^{(i)}\mathcal{W}$ is the mass per mol of the solute of species i, and the other symbols are as previously defined (compare Lewis and Randall, 1923, p. 244). Integration of Eq. (1.11) shows that the relation between $^{(i)}n$ and l is that of molar density in an external field, examples of which are the density of an isothermal atmosphere, the density of macroscopic particles suspended in a liquid, and the change of concentration along a tube of solution in a rotating centrifuge. This last has direct application to the question of the possible change in concentration of impurities in water in a soil sample being centrifuged to remove the water from it. The qualitative conclusion of such an examination is that the degree to which a net concentration change occurs depends on the average effective pore/capillary radii at a given location in the soil sample–the smaller the radii, the larger the surface–to–volume ratio and the larger the centrifugal force required to remove the water and hence the more pronounced the concentration variation of the soil solution becomes. In soil samples having a high proportion of clay, the surface adsorption forces can easily dominate the adsorptive–bulk capillary effects, thus requiring much higher centrifugal forces to remove the water than for other soils, and the concentration change effect would be expected to be quite pronounced. Comparison of Eqs. (1.10) and (1.11) shows that $^{(i)}C$, and hence $^{(i)}\varphi$, the angle of capillarity, and $^{(i)}\Psi$ will vary with height in a soil column or with the centrifuge speed of a soil sample being dried. The point of importance here is that centrifugation is not a viable means of removing water from soil samples if the measurement of groundwater impurity chemistry is the object.

1.3 NONEQUILIBRIUM THERMODYNAMICS

Nonequilibrium thermodynamics deals explicitly with irreversible processes, which all processes in nature actually are, and with the coupling of fluxes of heat, liquid, charge, etc., with generalized forces, related to the usual gradients of thermodynamic potentials. For example, a generalized force related to the gradient of temperature causes not only a flux of heat, a direct effect, but also a flux of water and water vapor and coupled effects as well. For example, a more thermodynamically complete form of Darcy's law than the one usually discussed would include terms proportional to the gradients of temperature, chemical potential, etc. (the "etc." depending on the system being studied) as well as the usual gradient of matric potential head. This formalism will be explored in more detail later on.

1.4 DARCY'S LAW (SIMPLE FORM)

The isothermal, isochemical, etc. (the "etc." depending on the system being studied) form of Darcy's law for unsaturated flow in which all coupled effects, in the sense of nonequilibrium thermodynamics, are neglected, may be written in vector notation as

$$^{(i)}\vec{J} = -\,^{(i)}\overleftrightarrow{K}(^{(i)}\theta) \cdot \vec{\nabla}\,^{(i)}H \tag{1.12a}$$

where $^{(i)}\vec{J}$ is the Darcy velocity, or volume flux (volume per unit area per unit time) of the ith liquid species moving in the porous medium, $^{(i)}\overleftrightarrow{K}(\,^{(i)}\theta)$ is the unsaturated hydraulic conductivity, a second-rank symmetric tensor, the · indicates the inner product of $^{(i)}\overleftrightarrow{K}(\,^{(i)}\theta)$ and $\vec{\nabla}\,^{(i)}H$, the gradient of hydraulic head, which, for unsaturated flow, is given by

$$\vec{\nabla}\,^{(i)}H = \vec{\nabla}\left(\frac{^{(i)}\Psi}{^{(i)}\rho g} + gz\right) \tag{1.12b}$$

where g is the acceleration due to gravity, z is the coordinate parallel to the direction of g, $\vec{\nabla}$ is the gradient operator, and $^{(i)}\Psi$ and $^{(i)}\rho$ are as previously defined. The unsaturated hydraulic conductivity can be parameterized on either matric potential $^{(i)}\Psi$, in which case $^{(i)}\overleftrightarrow{K}$ displays definite hysteresis, or on $^{(i)}\theta$, in which case the hysteretic behavior is reported in the literature to be almost absent to within experimental accuracy.

In unsaturated flow, the tensorial nature of $^{(i)}\overleftrightarrow{K}(^{(i)}\theta)$ has not yet been taken into account, principally due to a lack of experimental data [at the moment no direct method exists for measuring $^{(i)}\overleftrightarrow{K}(^{(i)}\theta)$ in situ, and cores of undisturbed material for laboratory analysis are difficult to obtain and in any case lead to a measurement of a projection of $^{(i)}\overleftrightarrow{K}(^{(i)}\theta)$ along an arbitrary direction] so that $^{(i)}\overleftrightarrow{K}(^{(i)}\theta)$ is generally replaced by $^{(i)}K(^{(i)}\theta)$, which depends essentially exponentially on $^{(i)}\theta$, and Darcy's law for unsaturated flow thus becomes

$$^{(i)}\vec{J} = - \,^{(i)}K(^{(i)}\theta)\vec{\nabla}\,^{(i)}H \tag{1.13}$$

where now $^{(i)}\vec{J}$ and $\vec{\nabla}\,^{(i)}H$ are in the same direction.

1.5   EQUATIONS OF CONTINUITY AND MOTION

The equation of continuity formalizing the principle of conservation of mass written for a fixed location (the limit to which a volume element shrinks as its volume becomes small macroscopically while retaining its identity as a volume of porous medium microscopically) for unsaturated flow is

$$\frac{\partial\left(^{(i)}\rho_b\right)}{\partial t} = -\vec{\nabla}\cdot\left(^{(i)}\rho_b\,^{(i)}\vec{v}\right) \tag{1.14}$$

where $^{(i)}\rho_b$ is the bulk density of the fluid of species i, $^{(i)}\vec{v}$ is the velocity of the material of species i at the location at which (1.14) is taken to hold, and the rest of the symbols are as previously defined. In order to develop (1.14) further, we now explore the concept of bulk density in more detail, beginning with volumetric saturation and a broadened view of porosity. The volumetric saturation of the ith species, $^{(i)}\phi$, of $n+1$ interpenetrating continuae, one of which is the porous medium, containing material of the ith species no matter how dispersed through the elementary volume the species may be, is the fraction of the elementary volume containing the material of that species. Thus,

$$\sum_{i=0}^{n}\,^{(i)}\phi = 1 \tag{1.15}$$

Similarly, an enlarged view of the concept of effective porosity, $^{(i)}\epsilon$, of each of $n+1$ interpenetrating continuae yields

$$^{(i)}\epsilon = 1 - \,^{(i)}\phi \tag{1.16}$$

so that

$$\sum_{i=0}^{n}\,^{(i)}\epsilon = 0 \tag{1.17}$$

If ΔV is the elementary volume containing the $n + 1$ interpenetrating continuae of interest, then,

$$^{(i)}\rho_b = \frac{\Delta \,^{(i)}M_v}{\Delta V} = \frac{^{(i)}{}'\rho \,^{(i)}\phi \,\Delta V}{\Delta V} = {}^{(i)}\rho \,^{(i)}\phi \tag{1.18}$$

where $\Delta \,^{(i)}M_v$ is the mass of species i contained in ΔV and the rest of the quantities are as previously defined. If $i = 0$ is taken to refer to the porous medium, then

$$^{(o)}\epsilon \,^{(i)}\theta = {}^{(i)}\phi \tag{1.19}$$

The velocity of fluid in the porous medium, measured relative to the medium, called the *pore velocity*, $^{(i)}\vec{v}$, is given by

$$^{(i)}\vec{v} = \frac{^{(i)}\vec{J}}{^{(o)}\epsilon \,^{(i)}\theta} \tag{1.20}$$

which is seen by writing the volumetric flux, $^{(i)}\vec{J}$, as

$$^{(i)}\vec{J} = \frac{^{(i)}\vec{v} \,^{(o)}\epsilon \,^{(i)}\theta \,\Delta A}{\Delta A} = \frac{\Delta \,^{(i)}d \,\Delta A \,^{(o)}\epsilon \,^{(i)}\theta}{\Delta t \,\Delta A} \tag{1.21}$$

where ΔA is the elementary area through which the volume of fluid ΔV ($\Delta V = \Delta \,^{(i)}d \,\Delta A \,^{(o)}\epsilon \,^{(i)}\theta$) of species i passes in time Δt, where $\Delta \,^{(i)}d$ is the linear distance perpendicular to ΔA that species i travels in time Δt so that $\Delta \,^{(i)}d/\Delta t = {}^{(i)}v$. The corresponding mass flux, mass per unit area per unit time, of species i, is

$$^{(i)}\vec{j} = {}^{(i)}\rho_b \,^{(i)}\vec{v} \tag{1.22}$$

which is seen by writing

$$^{(i)}j = \frac{\Delta \,^{(i)}M_v}{\Delta t \,\Delta A} = {}^{(i)}\rho_b \frac{\Delta \,^{(i)}d \,\Delta A}{\Delta t \,\Delta A} = {}^{(i)}\rho_b \,^{(i)}v \tag{1.23}$$

where $\Delta \,^{(i)}m_v = {}^{(i)}\rho_b \,\Delta \,^{(i)}d \,\Delta A$ is the mass of species i traversing a distance $\Delta \,^{(i)}d$ perpendicular to ΔA in time Δt. Thus, from Eqs. (1.18)–(1.20) and (1.22) or by direct reasoning we obtain the result

$$^{(i)}\vec{j} = {}^{(i)}\rho \,^{(i)}\vec{J} \tag{1.24}$$

Assuming that the variation in $^{(i)}\rho$ is small compared to variations in the other quantities appearing in (1.14) yields

$$\frac{\partial \left({}^{(o)}\epsilon \,^{(i)}\theta \right)}{\partial t} = -\vec{\nabla} \cdot {}^{(i)}\vec{j} \tag{1.25}$$

which is the usual form of the continuity equation as it is encountered in unsaturated flow. A source or sink term, $^{(i)}W$, say, for the ith species, may be added to (1.25) to yield

$$\frac{\partial \left(^{(o)}\epsilon \, ^{(i)}\theta \right)}{\partial t} = -\vec{\nabla} \cdot \, ^{(i)}\vec{j} + \, ^{(i)}W \tag{1.26}$$

The equation of motion for species i corresponding the (1.14), the Eulerian representation, is

$$\frac{\partial \left(^{(i)}\rho_b \, ^{(i)}\vec{v} \right)}{\partial t} = -^{(i)}\vec{v} \, \vec{\nabla} \cdot ^{(i)}\vec{j} - ^{(i)}\rho_b \left(\vec{\nabla} \, ^{(i)}\vec{v} \right) \cdot ^{(i)}\vec{v} + \vec{\nabla} \cdot ^{(i)} \overleftrightarrow{\tau} + ^{(i)}\rho_b \, ^{(i)}\vec{F} \tag{1.27}$$

where the (second rank, symmetric) stress tensor for a liquid of species i, $^{(i)} \overleftrightarrow{\tau}$, is given in tensor notation in Eqs. (5.148a) and (5.148b), $^{(i)}\vec{F}$ is the sum of possible body forces per unit mass acting on the fluid at the location in question, and the rest of the quantities are as previously defined. The quantity $\vec{\nabla}\vec{v}$ is a second-rank tensor, and hence (1.26) is most directly and easily written in Cartesian coordinates when vector notation is used. This is not suitable for all problems, however. Hence, in Chapter 5, where a fuller discussion of the equations of motion is given, and it is seen that the Lagrangian and not the Eulerian form of the equation of motion is the most appropriate descriptor of flow in porous media, tensor notation is used throughout. The advantages of this over the more familiar vector notation used here are simplicity and precision of expression, generality, and greatly enhanced clarity in the expression of fundamental concepts. Equation (1.27) is derived as an application of the principle of conservation of momentum. An equation of state for each fluid, that is, a relation between the pressure, density, and temperature of the fluid, is required to complete the equation set. For an isothermal liquid the equation of state takes the form

$$\frac{\partial \, ^{(i)}p}{\partial t} = -\frac{1}{^{(i)}\kappa} \vec{\nabla} \cdot \, ^{(i)}\vec{v} \tag{1.28}$$

where $^{(i)}p$ is the pressure of the ith species, $^{(i)}\kappa$ is the compressibility of the ith species, and the rest of the quantities are as previously defined. For gases, the ideal gas law, in the form

$$^{(i)}P = \frac{^{(i)}\rho k T}{^{(i)}m} \tag{1.29}$$

where $^{(i)}P$ is the pressure of the ith species of gas, k is Boltzmann's constant, or gas constant per molecule, T is the absolute temperature, and $^{(i)}m$ is the mass of a gas molecule of species i, may be used as a first approximation of the equation of state.

1.6 VAPOR TRANSPORT

Vapor transport is a more important mechanism in arid zone unsaturated flow than in applications involving a higher degree of saturation such as irrigated agriculture. The Kelvin relation, given in Eq. (1.2), shows that for a region of porous medium at constant temperature and given concentration of impurities in the water, the vapor pressure over a given water–air interface, meniscus in this case, is lower the smaller its mean (in the sense of differential geometry) radius of curvature. This means that the evolution of vapor occurs more readily from "flatter" menisci and that the soil–water system will have as one mechanism for reaching equilibrium the transfer of vapor from larger to smaller menisci and condensation there. Under isothermal, isochemical conditions, a net transfer of vapor from smaller menisci to larger ones cannot occur and so the general trend from both the capillary movement point of view, and the vapor transport point of view, is for water to migrate to and remain in the smallest pores/capillaries available. If temperature differences exist between one location of the porous medium and the next, then vapor can be transported from smaller to larger pores/capillaries, and if the temperature differences are cyclic, as is the case for near-surface diurnal fluctuations, vapor can be transported back and forth along a given path in the porous medium. In addition, when temperature gradients exist, the porous medium can act as a molecular sieve as follows. From the kinetic theory of gases, it is known that the kinetic energy E_k of a molecule of mass m in thermal equilibrium with its surroundings at temperature T is

$$E_k = \frac{3}{2}kT = \frac{1}{2}mv_{\mathrm{rms}}^2 \qquad (1.30)$$

where v_{rms} is the rms (root-mean-square or square root of the average of the square of the magnitude of the particle's velocity; the magnitude of the particle's velocity is its speed) speed of the molecule, and the other quantities are as previously defined. Since the kinetic energy of the molecules depends on the ambient temperature, for a given temperature the kinetic energies of two particles of masses m_1 and m_2, which have rms speeds $v_{1\mathrm{rms}}$ and $v_{2\mathrm{rms}}$, respectively are equal so that

$$\frac{1}{2}m_1(v_{1\mathrm{rms}})^2 = \frac{1}{2}m_2(v_{2\mathrm{rms}})^2 \qquad (1.31a)$$

or

$$\frac{v_{1\mathrm{rms}}}{v_{2\mathrm{rms}}} = \left(\frac{m_2}{m_1}\right)^{\frac{1}{2}} \qquad (1.31b)$$

from which we see that for molecules in equilibrium at a given temperature those of lower mass (than others) have higher speeds. If a given region of an unsaturated porous medium is at a higher temperature than its surroundings, vapor will tend to diffuse from the region of higher temperature to the surroundings at lower temperature. However, the less massive molecules in the higher–temperature region will diffuse away more rapidly than those of larger mass. Since air is composed principally of nitrogen, 78% by volume, oxygen, 20% by volume, and argon, .93% by volume, and since oxygen and nitrogen occur as diatomic molecules, with atomic mass numbers of 32 and 28 respectively, as compared to 18 for a water molecule, the water molecules move away preferentially from warmer to cooler regions, since their velocities are higher, as shown by Eq. (1.31b), thus depleting the atmosphere in the warmer regions of water vapor. Since increasing the temperature in a region of an unsaturated porous medium increases the vapor pressure over the menisci, as shown by the Kelvin relation [Eq. (1.2)], both by the increase of T explicitly and by the simultaneous decrease of $^{(i)}\gamma$, water vapor is evolved from the meniscii to be diffused toward the cooler regions of the porous medium there to be condensed. The diffusion process for water vapor in a porous medium is one of adsorption onto the pore/capillary walls and re-emission therefrom, rather than one of direct flight down the capillaries (de Boer, 1968). Since the re-emission of the water molecules is essentially isotropic and varies in intensity as the cosine of the angle the velocity vector of the emitted molecule makes with the normal to the emitting surface (Knudsen's cosine law), the process of diffusion of water vapor in a porous medium is slowed considerably over what it would be in a large open volume. The equations of motion to be applied to the transport of gases in porous media at the macroscopic level are essentially those of the Lagrangian form of Eq. (1.27) and will be developed in more detail in Chapter 4.

SELECTED REFERENCES

de Boer, J. H., *The Dynamical Character of Adsorption*, Chapters VI, X, Oxford University Press, London, 2nd ed., 1968.

Callen, H. B., *Thermodynamics*, John Wiley & Sons, N.Y., 1961.

Lewis, G. N., and M. Randall, *Thermodynamics*, McGraw-Hill Book Co., Inc., N.Y., 1923.

Shannon, C. E., "A Mathematical Theory of Communication", *Bell System Technical Journal*, Vol. XXVII, July, 1948.

PROBLEMS

1.1 Following a heavy rain, soil moisture contents are measured as a function of depth below land surface. The values obtained are shown in the table.

Saturation θ_s	Depth (cm)
1.0	0.0
0.9	-1.0
0.8	-2.0
0.7	-3.0
0.6	-4.0
0.5	-5.0
0.4	-6.0
0.4	10.0
0.4	-20.0

(a) Plot the tabulated data (degree of saturation, θ_s, as a function of depth below land surface). Linear axes are suggested for this plot.

(b) Compute $\vec{\nabla}\theta_s$ in the depth interval $-5 \leq z \leq -1.0$ cm assuming no lateral variation in θ_s. If depths below land surface are measured in cm, what are the units of $\vec{\nabla}\theta_s$?

(c) Assume that the soil is sufficiently sandy that the dependence of the matric potential on the degree of soil saturation can be represented for wetting by the following empirical fit to laboratory data for Adelaide Dune sand: $\Psi = -e^{-f_1(\theta_s)}$ bars, where $f_1(\theta_s) = 18.3017\,\theta_s^3 - 33.4885\,\theta_s^2 + 20.6245\,\theta_s - 1.66472$ for the moisture range $0.4 \leq \theta_s \leq 0.95$. Derive an expression for the gradient of the matric potential head in the depth range $-5 \leq z \leq -1$ cm. [Recall the chain rule for the differentiation of a function of a function, which in this case takes the form $\vec{\nabla}\Psi = (\partial\Psi/\partial\theta_s)\vec{\nabla}\theta_s$.]

(d) Assuming the validity of the empirical fit to laboratory data for unsaturated hydraulic conductivity for the wetting of Adelaide Dune sand, $K(\theta_s) = e^{-f_2(\theta_s)}$ cm/sec, where $f_2(\theta_s) = 3.97029\,\theta_s^2 - 10.2179\,\theta_s + 12.2313$, compute the instantaneous volume flux of water infiltrating at -5 cm using Darcy's law and other quantities as appropriate.

(e) Assuming that $\epsilon = 0.29$, compute the value of v_p that corresponds to the value of J computed in (d).

(f) Develop a semiempirical formula for $J(\theta_s)$ for the depth range $-6 \le -1$ cm in cgs (centimeter–gram–second) units using Darcy's law, the results of (c) and (d), and any other relations you may require.

1.2 Compute (a) the rms speed (in metric units) of a water molecule that is in thermal equilibrium with a heat reservoir (as envisioned in equilibrium thermodynamics) at a temperature of $23°C$, (b) its kinetic energy, and (c) its momentum (the momentum, p, of a particle of mass m is $p = m v_{\text{rms}}$).

1.3 Assume that $f(\xi)$ of eq.(1.3) equals $\delta(\xi - r)$, where $\delta(\xi - r)$ is the Dirac delta function and has the property (among others) that $\int f(x)\,\delta(x - b)\,dx = f(b)$ for $a \le b \le c$ or $= 0$ if b is outside the interval $a \le x \le c$.

(a) Discuss the physical significance of this form of $f(\xi)$.

(b) Derive the functional form of Ψ of Eq. (1.3) that follows from assuming this form of $f(\xi)$ and discuss the physical significance of the result.

1.4 Using the values $\rho = 1$ g/cm^3, $k = 1.38 \times 10^{-16}$ ergs/K, $T = 293.16$ K, $\gamma = 72.75$ dyne/cm, $m = 3 \times 10^{-23}$ g, $\xi = 10^{-5}$ cm, and $\varphi = 0$, compute $p_v(\xi)/p_{ov}$ of Eq. (1.2).

2

PHYSICAL BASIS OF
THE MATRIC POTENTIAL

2.1   BASIC PHENOMENOLOGY OF THE MATRIC POTENTIAL

The matric potential is the energy per unit bulk volume of porous medium required to remove liquid from it isothermally against the capillary forces holding it in place. It depends on temperature, impurity concentration(s) in the liquid, the air–porous-medium and porous-medium liquid interfacial tensions, and the internal geometry of the porous medium itself.

The interfacial tension between two media is the work per unit area or force per unit length required to increase the interfacial area by a unit amount. The term *surface tension* refers to the interfacial energy of a pure substance with its own vapor. The energy input required to maintain a surface at constant temperature as its area is increased and the concept of total or "internal," in the sense of equilibrium thermodynamics, interfacial surface energy per unit area will be discussed in Chapter 3, as will the changes in interfacial energy and tension with dissolved impurities and temperature. The change in the air–water interfacial tension at a given temperature depends on whatever impurities are in the surface layer, and in many cases of practical interest, the concentration of impurities in the surface layer of water in contact with air or soil is sufficient to cause it to differ significantly from the pure water value.

We begin our discussion of the basic phenomenology of liquid, water, for example, in unsaturated porous media by deriving the Laplace equation for the pressure difference across a curved interface between two media due to the interfacial tension that the media in contact possess.

2.2   PRESSURE CHANGE ACROSS AN INTERFACE

We consider the curved element of an interface having principal radii of curvature r_1 and r_2 and interfacial tension γ shown in Figures 2.1(a) and 2.1 (b).

The interfacial tension in a homogeneous, isotropic interface is expressed as a force per unit length. This force acts at right angles to any imaginary line "drawn" in the interface. Thus, in Figure 2.1(a) the lines ds_2 "drawn" as two edges of the element of the curved interface are maintained in position against the tendency of the surface to contract due to the "surface tension" forces acting perpendicular to them, namely the $\gamma\, ds_2$, by some unspecified external agency. The curve connecting these two edges of the surface element has radius of curvature r_1 and, since the forces are tangent to the surface element at its edges, the vectors representing them are perpendicular to ds_2 at each edge and make an angle α_1 with each other, the same angle that is subtended at O_1 by the radii of length r_1 drawn from O_1 to the two edges of the area element having lengths ds_2. Thus, the magnitude of the resultant force normal to the surface element due to $\gamma\, ds_2$ (directed upwards in the figure) is represented by the line segment OP_1 in the force triangle OP_1A, where angle $OAP_1 = \alpha_1/2$, as seen by considering the isosceles triangle AP_1Q, and the hypotenuse AP_1 represents $\gamma\, ds_2$. Thus,

$$OP_1 = \gamma\, ds_2 \sin\left(\frac{\alpha_1}{2}\right) \tag{2.1}$$

[note in passing that the component of $\gamma\, ds_2$ in the plane of the element of the surface at its edge is $\gamma\, ds_2 \cos(\alpha_1/2)$]. The magnitude of the total force on the two ds_2 edges of the surface element due to "surface tension" is twice OP_1, or

$$F_1 = 2\gamma \sin\left(\frac{\alpha_1}{2}\right) ds_2 \tag{2.2a}$$

Similarly, we have for the force components along OP_2 in Figure 2.1(b) due to the forces $\gamma\, ds_1$ acting perpendicularly to the ds_1 edges of the element of surface

$$F_2 = 2\gamma \sin\left(\frac{\alpha_2}{2}\right) ds_1 \tag{2.2b}$$

The magnitude of the total force acting perpendicularly to the surface element due to surface tension ["up" in Figures 2.1(a) and 2.1(b)] is

$$F = F_1 + F_2 \tag{2.2c}$$

and balances the force of the upper medium on the lower one, also of magnitude F, which is directed downward in the figure.

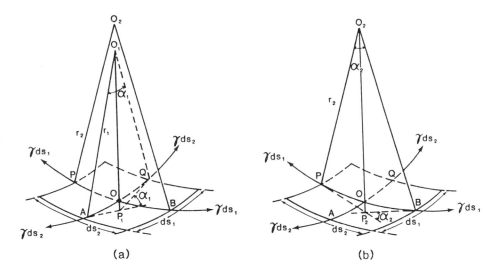

(a)   (b)

Figure 2.1: An element of surface area at a curved interface between two media where a) α_1 is the angle subtended by the element of interface of length ds_1 and b) shows the angle α_2 subtended by the element of interface of length ds_2.

Since, in radians,

$$\alpha_1 = \frac{ds_1}{r_1} \tag{2.3a}$$

and

$$\alpha_2 = \frac{ds_2}{r_2} \tag{2.3b}$$

we have on substituting these relations and Eqs. 2.2(a) and 2.2(b) into Eq. 2.2(c)

$$F = 2\gamma \sin\left(\frac{ds_1}{2r_1}\right) ds_2 + 2\gamma \sin\left(\frac{ds_2}{2r_2}\right) ds_1 \tag{2.4a}$$

Since for angles of 5^0 or less, $\sin(\theta) \approx \theta$, where θ is in radians, Eq. 2.4(a) becomes,

$$F \approx \gamma \, ds_1 \, ds_2 \left(\frac{1}{r_1} + \frac{1}{r_2}\right) \tag{2.4b}$$

Pressure is defined as force per unit area, where the area in this case is that of the surface element, given approximately by $ds_1 \, ds_2$, so that the difference in pressure across the curved interface between the two media is

$$\Delta p = \frac{F}{ds_1 \, ds_2} = \gamma \left(\frac{1}{r_1} + \frac{1}{r_2} \right) \tag{2.4c}$$

or,

$$p_1 - p_2 = \gamma \left(\frac{1}{r_1} + \frac{1}{r_2} \right) \tag{2.5}$$

where r_1 and r_2 are the principal radii of curvature of the surface at the point O in Figure 2.1, and are taken to be positive, p_1 is the pressure exerted on the interface by the material of medium 1, located above the interface in Figure 2.1, and p_2 is the (smaller) pressure exerted on the interface by medium 2, located below the interface in Figure 2.1. In equilibrium this pressure difference is balanced at each point of the surface by the surface tension force per unit area given on the right-hand side of Eq. (2.5). Unless the interface is a section of a hemisphere, r_1 and r_2 vary from place to place on the interface, with the result that the pressure difference varies from place to place as well. Familiar examples of this pressure difference are an air bubble in a liquid, a drop of mercury in air, in which the pressure inside the bubble and drop are greater than those of the surrounding liquid and air, respectively, and the meniscus surface of a capillary tube in air filled with water for which the pressure exerted by the water (medium 2) on the interface is less than that exerted by the air (medium 1). We now examine the notion of curvature in more detail.

2.3　CURVATURE IN THE PLANE

We consider a plane curve being traversed from left to right and a position vector $\vec{s}$ drawn from the origin of a Cartesian coordinate system to a point $P(x, y)$ on the curve as shown in Figure 2.2. If

$$\vec{\zeta} = \hat{i} x + \hat{j} y \tag{2.6a}$$

where $\hat{i}$ and $\hat{j}$ are unit vectors along the x and y axes respectively, then, formally,

$$\frac{d\vec{\zeta}}{ds} = \hat{i} \frac{dx}{ds} + \hat{j} \frac{dy}{ds} \tag{2.6b}$$

where Δs is an element of arc length along the curve. From Figure 2.2 we see that

$$\lim_{\Delta s \to 0} \frac{\Delta \vec{\zeta}}{\Delta s} = \frac{d\vec{\zeta}}{ds} = \vec{T} \tag{2.7}$$

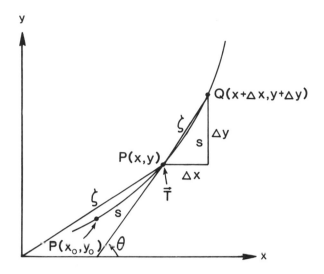

Figure 2.2: Illustration of the change in arc length, Δs, as $P(x,y)$ moves along the curve from $P(x_o, y_o)$ and Q moves toward $P(x, y)$, this latter yielding the tangent, $\vec{T}$, to the curve at $P(x, y)$ in the limit as $\Delta s \to 0$.

where the limit $\Delta s \to 0$ is taken to mean $\Delta x \to 0$ so that in the limit the point Q merges with the point P and $\vec{T}$ is the tangent vector to the curve at the point P. From Eqs. (2.6b) and (2.7) we have

$$\vec{T} = \hat{i}\frac{dx}{ds} + \hat{j}\frac{dy}{ds} \qquad (2.8a)$$

and if x and y are parameterized on some common variable, say t, then

$$\vec{T} = \hat{i}\frac{dx}{dt}\frac{dt}{ds} + \hat{j}\frac{dy}{dt}\frac{dt}{ds} \qquad (2.8b)$$

where the relation between ds and dt can be found from the expression for the square of the differential of arc length (see Figure 2.2)

$$(ds)^2 = (dx)^2 + (dy)^2 \qquad (2.9)$$

where dx and dy are the total differentials of the x and y components of $\vec{\zeta}$. The direction of $\vec{T}$ may be specified by the angle θ that $\vec{T}$ makes with the x axis. This angle changes as the point P with which $\vec{T}$ is associated moves along the curve, and we define the curvature, κ, of the curve at the point x_0 as

$$\kappa = \frac{d\theta}{ds}\bigg|_{x=x_0} \tag{2.10a}$$

where

$$\tan(\theta)|_{x=x_0} = \frac{dy}{dx}\bigg|_{x=x_0} \tag{2.10b}$$

or, at a general point, (x, y),

$$\theta = \tan^{-1}\left(\frac{dy}{dx}\right) \tag{2.10c}$$

Thus,

$$\kappa = \frac{d\theta}{ds} = \frac{d\theta}{dx}\frac{dx}{ds} = \frac{d\theta/dx}{ds/dx} \tag{2.11}$$

where, from Eq. (2.10c),

$$\frac{d\theta}{dx} = \frac{\frac{d^2y}{dx^2}}{1 + \left(\frac{dy}{dx}\right)^2} \tag{2.12a}$$

and, from Eq. (2.9)

$$\frac{ds}{dx} = \left[1 + \left(\frac{dy}{dx}\right)^2\right]^{1/2} \tag{2.12b}$$

so that from Eqs. (2.10a), (2.12a), and (2.12b) we have for the curvature

$$\kappa = \frac{\frac{d^2y}{dx^2}}{\left[1 + \left(\frac{dy}{dx}\right)^2\right]^{3/2}} = \frac{d}{dx}\left\{\frac{dy}{dx} \bigg/ \left[1 + \left(\frac{dy}{dx}\right)^2\right]^{1/2}\right\} \tag{2.13}$$

We now define the radius of curvature r_a of the curve $y(x)$ having curvature κ given by Eq. (2.13) as

$$r_a = \frac{1}{\kappa} \tag{2.14}$$

The circle of radius r_a tangent to the curve at P, as shown in Figure 2.3, is termed the *circle of curvature*, and its center is termed the *center of curvature* of the curve at P. We see from Figure 2.3 that if (x, y) are the coordinates of the general point $P(x, y)$, then x_c and y_c, the coordinates of the center of curvature there, are given by

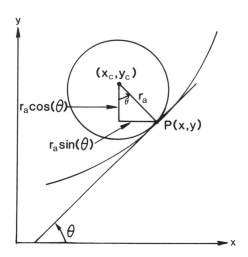

Figure 2.3: The circle of curvature is tangent to the curve at the point $P(x, y)$, with radius r_a, and (x_c, y_c) are the coordinates of its center.

$$x_c = x - r_a \sin(\theta) \tag{2.15a}$$

and

$$y_c = y + r_a \cos(\theta) \tag{2.15b}$$

where θ is the angle that $\vec{T}$ makes the x axis so that $y' = dy/dx = \tan(\theta)$ or, as in Eq. (2.10b), $\theta = \tan^{-1}(dy/dx)$, so that

$$\sin(\theta) = \frac{y'}{\sqrt{1 + y'^2}} \tag{2.16b}$$

and

$$\cos(\theta) = \frac{1}{\sqrt{1 + y'^2}} \tag{2.16c}$$

Combining Eqs. (2.16b) and (2.16c) with equations (2.15a) and (2.15b) yields

$$x_c = x - \frac{y'(1 + y')^2}{y''} \tag{2.17a}$$

and

$$y_c = y + \frac{1 + y'^2}{y''} \tag{2.17b}$$

We see that, as the point $P(x, y)$ moves along the curve, the center of curvature moves also. The curve generated by the moving center of curvature is called the *evolute* of the original curve, and, conversely, the original curve is called the *involute* of the second curve. We now consider as a next step the more general case, a surface of finite extent, rather than a curve.

2.4 DIFFERENTIAL GEOMETRY AND CURVATURE

A surface has two principal radii of curvature at each point. The circles of curvature associated with these radii of curvature are at right angles to each other at the point in the surface at which its curvature is being evaluated. These radii of curvature are not equal in general. We shall call the curvatures corresponding to the radii of curvature r_1 and r_2, κ_1 and κ_2 respectively. We consider a vector $\vec{s}$ in three dimensions defining a surface as a function of two independent parameters, u and v, which parameterization is due to Euler. (Note that similarly, the two variables describing a curve in the plane, y and x, could be parameterized on one independent variable.) The square of the line element in $u - v$ space, I, is called the first fundamental form and is

$$\begin{aligned} I &= ds^2 = d\vec{s} \cdot d\vec{s} = (\vec{s}_u \, du + \vec{s}_v \, dv) \cdot (\vec{s}_u \, du + \vec{s}_v \, dv) \\ &= (\vec{s}_u \cdot \vec{s}_u) \, du^2 + 2 (\vec{s}_u \cdot \vec{s}_v) \, du \, dv + (\vec{s}_v \cdot \vec{s}_v) \, dv^2 \end{aligned} \tag{2.18a}$$

or,

$$I = E \, du^2 + 2F \, du \, dv + G \, dv^2 \tag{2.18b}$$

where the notation $\vec{s}_u$ denotes the partial derivative of $\vec{s}$ with respect to u and similarly for $\vec{s}_v$. The total differential of $\vec{s}$, $d\vec{s}$, is given by, since $\vec{s}$ depends on u and v only,

$$d\vec{s} = \frac{\partial \vec{s}}{\partial u} du + \frac{\partial \vec{s}}{\partial v} dv = \vec{s}_u \, du + \vec{s}_v \, dv \tag{2.19}$$

Also,

$$E = \vec{s}_u \cdot \vec{s}_u \tag{2.20a}$$

$$F = \vec{s}_u \cdot \vec{s}_v \tag{2.20b}$$

and

$$G = \vec{s}_v \cdot \vec{s}_v \tag{2.20c}$$

are called the first fundamental coefficients. The u and v parameter lines are perpendicular to each other in the surface of interest if $F = 0$. We note as an aside that the coefficients of the differentials in the expression for the square of the line element are the elements of the metric tensor characterizing the surface–in this case in $u - v$ space. If $F = 0$, then the $u - v$ space metric tensor has no off-diagonal elements and thus represents an orthogonal coordinate system. (See Appendix A for a review of selected elements of tensor analysis.) A unit vector normal (perpendicular) to the surface of interest is

$$\hat{n} = \frac{\vec{s}_u \times \vec{s}_v}{|\vec{s}_u \times \vec{s}_v|} \qquad (2.21)$$

where $\vec{s}_u \times \vec{s}_v$ is the usual vector cross product, and $|\vec{s}_u \times \vec{s}_v|$ is the absolute value of $\vec{s}_u \times \vec{s}_v$. The quantity $-d\vec{s} \cdot d\hat{n}$ is called the second fundamental form and is given by

$$
\begin{aligned}
\mathrm{II} &= -d\vec{s} \cdot d\hat{n} = -(\vec{s}_u\,du + \vec{s}_v\,dv) \cdot (\hat{n}_u\,du + \hat{n}_v\,dv) \\
&= (-\vec{s}_u \cdot \hat{n}_u)\,du^2 - (\vec{s}_u \cdot \hat{n}_v + \vec{s}_v \cdot \hat{n}_u)\,du\,dv - (\vec{s}_v \cdot \hat{n}_v)\,dv^2 \\
&= L\,du^2 + 2M\,du\,dv + N\,dv^2
\end{aligned}
\qquad (2.22)
$$

where

$$L = -\vec{s}_u \cdot \hat{n}_u \qquad (2.23a)$$

$$M = -\frac{1}{2}(\vec{s}_u \cdot \hat{n}_v + \vec{s}_v \cdot \hat{n}_u) \qquad (2.23b)$$

and

$$N = -\vec{s}_v \cdot \hat{n}_v \qquad (2.23c)$$

are called the second fundamental coefficients. It is shown in differential geometry that

$$(EG - F^2)\,\kappa^2 - (EN + GL - 2FM)\,\kappa + (LN - M^2) = 0 \qquad (2.24)$$

which is a quadratic equation having two non-negative roots, the principal curvatures at a point P, κ_1 and κ_2, or a single root of multiplicity two, if $\kappa_1 = \kappa_2$. When this latter is the case, the point P at which the curvature is being evaluated is called an *umbilical point* and in that special case,

$$\kappa = \kappa_1 = \kappa_2 = \frac{L}{E} = \frac{M}{F} = \frac{N}{G} \qquad (2.25)$$

Equation (2.24) can be written as

$$\kappa^2 - 2H\kappa + \kappa_g = 0 \qquad (2.26a)$$

where

$$H = \frac{1}{2}\left(\kappa_1 + \kappa_2\right) = \frac{EN + GL - 2FM}{2(EG - F^2)} \tag{2.26b}$$

is the average (mean) curvature at the point P and

$$\kappa_g = \kappa_1 \kappa_2 = \frac{LN - M^2}{EG - F^2} \tag{2.26c}$$

is the Gaussian curvature at the point P. (The Gaussian curvature plays an important role in the study of the intrinsic (coordinate-independent) geometry of surfaces.) We now identify the quantity $2H$, or,

$$\frac{1}{r_1} + \frac{1}{r_2} = \kappa_1 + \kappa_2 = \frac{EN + GL - 2FM}{EG - F^2} \tag{2.27}$$

with the same quantity appearing in the right-hand side of Eq. (2.5). Thus if $\vec{s}$, the equation for the surface separating two media is given, along with the interfacial tension, γ, Eq. (2.27) can be used to compute the pressure difference, $p_1 - p_2$, across the interface.

As an example of this procedure, we consider a particular form of $\vec{s}$, namely

$$\vec{s} = u\,\hat{i} + v\,\hat{j} + f(u, v)\,\hat{k} \tag{2.28}$$

where the $\hat{i}$, $\hat{j}$, and $\hat{k}$ are the constant unit vectors along the x, y, and z, directions in Cartesian coordinates, $f(u, v)$ describes a surface above the $u - v$ plane, and in this special case, $x = u$, $y = v$, and $z = f(u, v)$. For this form of $\vec{s}$ we have

$$\vec{s}_u = \hat{i} + \frac{\partial f(u, v)}{\partial u}\hat{k} \tag{2.29a}$$

and

$$\vec{s}_v = \hat{j} + \frac{\partial f(u, v)}{\partial v}\hat{k} \tag{2.29b}$$

so that, from Eqs. (2.20a)–(2.20c)

$$E = 1 + \left(\frac{\partial f(u, v)}{\partial u}\right)^2 \tag{2.30a}$$

$$F = \left(\frac{\partial f(u, v)}{\partial u}\right)\left(\frac{\partial f(u, v)}{\partial v}\right) \tag{2.30b}$$

$$G = 1 + \left(\frac{\partial f(u, v)}{\partial v}\right)^2 \tag{2.30c}$$

Also,

$$\vec{s}_u \times \vec{s}_v = \det \begin{bmatrix} \hat{i} & \hat{j} & \hat{k} \\ 1 & 0 & \frac{\partial f}{\partial u} \\ 0 & 1 & \frac{\partial f}{\partial v} \end{bmatrix} = -\hat{i}\left(\frac{\partial f}{\partial u}\right) - \hat{j}\left(\frac{\partial f}{\partial v}\right) + \hat{k} \qquad (2.31\text{a})$$

and

$$|\vec{s}_u \times \vec{s}_v| = [(\vec{s}_u \times \vec{s}_v)\cdot(\vec{s}_u \times \vec{s}_v)]^{1/2} = D \qquad (2.31\text{b})$$

where

$$D = \left[\left(\frac{\partial f}{\partial u}\right)^2 + \left(\frac{\partial f}{\partial v}\right)^2 + 1\right]^{\frac{1}{2}} \qquad (2.31\text{c})$$

so that

$$\hat{n} = \left(-\frac{\partial f}{\partial u}\hat{i} - \frac{\partial f}{\partial v}\hat{j} + \hat{k}\right) D^{-1} \qquad (2.31\text{d})$$

Thus

$$\begin{aligned} \hat{n}_u &= \left(\frac{\partial f}{\partial u}\hat{i} + \frac{\partial f}{\partial v}\hat{j} - \hat{k}\right)\left[\frac{\partial f}{\partial u}\left(\frac{\partial^2 f}{\partial u^2}\right) + \frac{\partial f}{\partial v}\left(\frac{\partial^2 f}{\partial u\,\partial v}\right)\right] D^{-3} \\ &\quad - \left(\frac{\partial^2 f}{\partial u^2}\hat{i} + \frac{\partial^2 f}{\partial u\,\partial v}\hat{j}\right) D^{-1} \end{aligned} \qquad (2.32\text{a})$$

and

$$\begin{aligned} \hat{n}_v &= \left(\frac{\partial f}{\partial u}\hat{i} + \frac{\partial f}{\partial v}\hat{j} - \hat{k}\right)\left[\frac{\partial f}{\partial u}\left(\frac{\partial^2 f}{\partial v\,\partial u}\right) + \frac{\partial f}{\partial v}\left(\frac{\partial^2 f}{\partial v^2}\right)\right] D^{-3} \\ &\quad - \left(\frac{\partial^2 f}{\partial u\,\partial v}\hat{i} + \frac{\partial^2 f}{\partial v^2}\hat{j}\right) D^{-1} \end{aligned} \qquad (2.32\text{b})$$

Thus, using Eqs. (2.23a), (2.29a), and (2.32a),

$$L = \left[\frac{\partial^2 f}{\partial u^2}\right] D^{-1} \qquad (2.33\text{a})$$

$$M = \left[\frac{\partial^2 f}{\partial u \partial v}\right] D^{-1} \qquad (2.33\text{b})$$

$$N = \left[\frac{\partial^2 f}{\partial v^2}\right] D^{-1} \qquad (2.33\text{c})$$

Now, from Eqs. (2.30a)–(2.30c)

$$EG - F^2 = 1 + \left(\frac{\partial f}{\partial u}\right)^2 + \left(\frac{\partial f}{\partial v}\right)^2 = D^2 \qquad (2.34)$$

so that Eq. (2.27) becomes in this case,

$$
\frac{1}{r_1} + \frac{1}{r_2} = \frac{\left[1 + \left(\frac{\partial f}{\partial u}\right)^2\right] \frac{\partial^2 f}{\partial v^2} + \left[1 + \left(\frac{\partial f}{\partial v}\right)^2\right] \frac{\partial^2 f}{\partial u^2} - 2\frac{\partial^2 f}{\partial u \partial v}\frac{\partial f}{\partial u}\frac{\partial f}{\partial v}}{D^3} \tag{2.35}
$$

from which the mean curvature of the surface at a point in the surface and hence the difference in pressure across the interface between two media at each point in the interface can be computed directly if $\tilde{s}(u, v)$ is of the form given in Eq. (2.28).

2.5 SHAPE OF AN INTERFACE

The development given in the previous section is for computing the mean curvature at each location of a curved interface of known shape. One way of finding the shape of the interface in the first place is to solve a particular partial differential equation subject to appropriate boundary conditions, as we shall shortly describe. The differential equation to be solved for the equilibrium shape of the interface between two media in a gravitational field can be derived as follows.

The equations describing a mechanical system can be derived by considering an arbitrary variation of the coordinates (and in some problems the momenta and velocities as well) describing the system. This variation, since it does not actually occur in the motion of the system (and certainly not in the motion of a system in equilibrium since in that case there is none) is called a *virtual displacement* of the system. The particular variation

$$
\delta W = F \, \delta q \tag{2.36}
$$

where δW is an arbitrary variation in the work done in response to a force F acting through a corresponding arbitrary (virtual) displacement δq is called the *principle of virtual work*. From Eq. (2.6) we see that the force across a curved interface is

$$
F = \int (p_1 - p_2) \, dA = \gamma \int \left(\frac{1}{r_1} + \frac{1}{r_2}\right) dA \tag{2.37a}
$$

where $dA = ds_1 \, ds_2$ (from Figure 2.1) is an element of interfacial area and the corresponding element of virtual work, δW, is

$$
\delta W = \int \gamma \left(\frac{1}{r_1} + \frac{1}{r_2}\right) dA \, \delta q \tag{2.37b}
$$

If the interface is in a gravitational field, the body forces per unit area acting on each of the two media making up the system are $\rho_1 gz$ and $\rho_2 gz$, respectively so that the corresponding total body forces are

$$F_1 = \int_{S_1} \rho_1 gz \, dA \tag{2.38a}$$

$$F_2 = \int_{S_1} \rho_2 gz \, dA \tag{2.38b}$$

where the densities ρ_1 and ρ_2 are those appropriate to the materials in regions S_1 and S_2 on the two sides of the interface of elementary area dA, z is the elevation of the elementary interfacial area measured above an arbitrary datum, g is the acceleration due to gravity, and the virtual change (variation) in energy, δU_1, associated with a virtual displacement of the interface a distance δq is

$$\begin{aligned} \delta U_1 &= \int_{S_1} \rho_1 gz \, dA \, \delta q_1 + \int_{S_2} \rho_2 gz \, dA \, \delta q_2 \\ &= \int_{S_1} \rho_1 gz \, dA \, \delta q_1 - \int_{S_1} \rho_2 gz \, dA \, \delta q_1 \end{aligned} \tag{2.39}$$

in which δq_1 is the negative of δq_2, a reflection of the fact that in a displacement of the interface between two media, a positive displacement of one medium in a given direction is a negative displacement in that direction for the other medium. The virtual variation in volume of medium 1 is

$$\delta V = dA \, \delta q \tag{2.40}$$

This condition enters as an equation of constraint so that we have for the total variation, in which W and U_1 are to be minimized subject to the constraint $\delta V = 0$,

$$\delta W + \delta U_1 + \lambda \delta V = 0 \tag{2.41a}$$

or

$$\begin{aligned} \int \gamma \left(\frac{1}{r_1} + \frac{1}{r_2} \right) dA \, \delta q_1 + \int \rho_1 gz \, dA \, \delta q_1 \\ - \int \rho_2 gz \, dA \, \delta q_1 + \int \lambda \, dA \, \delta q_1 \\ = 0 \end{aligned} \tag{2.41b}$$

where λ is a Lagrange undetermined multiplier (see Appendix C for a brief discussion of Lagrange multiplier formalism) and, since δq is an arbitrary variation, the integrand of

$$\int \left[\gamma \left(\frac{1}{r_1} + \frac{1}{r_2} \right) + (\rho_1 - \rho_2) gz + \lambda \right] dA \, \delta q_1 = 0 \tag{2.42}$$

must equal zero. This yields the differential equation for the shape of the interface, which is

$$\gamma \left(\frac{1}{r_1} + \frac{1}{r_2} \right) + (\rho_1 - \rho_2)\, gz + \lambda = 0 \tag{2.43}$$

where r_1 and r_2 are measured from the interior of medium 1 and γ is the interfacial energy per unit area. In the case of a surface in which z of (2.43) is $z = f(x, y)$, the expression for $1/r_1 + 1/r_2$ is given by (2.35) with $u = x$ and $v = y$, which is then combined with (2.43) to yield the full nonlinear partial differential equation to be solved for the function describing the shape of the interface, $f(x, y)$.

2.6 INTERFACE A SURFACE OF REVOLUTION

We now consider, from a non-differential-geometry point of view, since this approach is encountered in the literature as well, an example of solving the partial differential equation for the shape of an interface that is a surface of revolution. In Figure 2.4 medium 1 is air, medium 2 is water in a capillary tube, and γ is the usual air–water interfacial tension. The surface of revolution depicted in Figure 2.4 is formed by revolving the curve $z(\xi)$ about the z axis with $z(0) = z_0$ so that z_0 is taken to be the height of the lowest part of the surface above the origin of the $z - \xi$ coordinate system. One of the two principal radii of curvature through the point $P(\xi, z)$, shown in Figure 2.4, is drawn perpendicular to the surface at $P(\xi, z)$, terminates on the axis of revolution, and is given by

$$r_1 = \frac{\xi}{\sin(\theta)} = \frac{\xi\sqrt{1 + z'(\xi)^2}}{z'(\xi)} \tag{2.44}$$

where Eq. (2.16b) with $y'(x)$ replaced by $z'(\xi)$ has been used for $\sin(\theta)$. The principal radius of curvature r_1 sweeps out the surface of revolution in a direction perpendicular to the plane of the page. The circles swept out by $P(\xi, z)$ are called the *parallels of the surface of revolution*. The other principal radius of curvature, r_2, is r_a, as given in (2.14) with $y(x)$ of (2.13) replaced by $z(\xi)$. The various images of the original curve $z = z(\xi)$ as it rotates to form the surface of revolution are called the *meridians of the surface*. Substituting r_2, the reciprocal of Eq. (2.13), with $y(x)$ replaced by $z(\xi)$, and r_1 from (2.44) into (2.43), yields for the differential equation to be solved for the shape of the surface of revolution, $z(\xi)$,

$$\gamma \left(\frac{z''(\xi)}{[1 + z'(\xi)^2]^{3/2}} + \frac{z'(\xi)}{\xi\,[1 + z'(\xi)^2]^{1/2}} \right) + (\rho_1 - \rho_2)g\, z(\xi) + \lambda = 0 \tag{2.45}$$

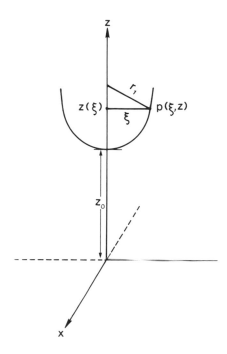

Figure 2.4: Cross-section of a surface of revolution of height $z = z(\xi)$ above the $z = 0$ plane.

or

$$\frac{1}{\xi}\frac{d}{d\xi}\left(\frac{\xi z'(\xi)}{[1 + z'(\xi)^2]^{1/2}}\right) + (\rho_1 - \rho_2)\frac{g\,z(\xi)}{\gamma} + \frac{\lambda}{\gamma} = 0 \qquad (2.46)$$

Equation (2.46) has never been solved exactly, even for the case of water rising in a capillary of circular cross section, and the shape of the meniscus at the air–water interface is only approximately known. The assumption that this surface is a section of a sphere, in which case $\sin(\theta)$ of (2.44) equals $\cos(\varphi)$, where φ is the usual angle of capillarity, is the lowest-level approximation to its shape.

We now consider, in an approximate fashion, the case of a capillary of circular cross section and radius a having a diameter large compared to its height (compare Rayleigh, 1915). At a distance from its walls such that $z'(\xi) << 1$, where ξ is the radial coordinate measured from the center of the capillary and z is the height of the part of the capillary meniscus at a distance ξ from the capillary axis above the flat water surface containing the bottom, of the capillary. Eq. (2.46) then becomes approximately

$$\frac{1}{\xi}\frac{d}{d\xi}\left[\frac{\xi\, z'(\xi)}{1}\right] = \frac{1}{\xi}\left[\xi\, z''(\xi) + z'(\xi)\right] \simeq \beta^2 z(\xi) - \frac{\lambda}{\gamma} \qquad (2.47)$$

where

$$\beta^2 = \frac{(\rho_l - \rho_v)g}{\gamma} \qquad (2.48)$$

and $\rho_2 = \rho_l$, the mass density of the liquid in the capillary, and $\rho_1 = \rho_v$ the mass density of the liquid vapor above the capillary meniscus. Thus,

$$z''(\xi) + \frac{1}{\xi}z'(\xi) \simeq \beta^2 z(\xi) - \frac{\lambda}{\gamma} \qquad (2.49)$$

Equation (2.49) is an ordinary differential equation to be solved for the shape of the meniscus surface. As a next step toward this solution we now evaluate λ as follows. From Eq. (2.43) we have

$$\lambda = -\frac{2\gamma}{r_c} + (\rho_l - \rho_v)gz_0 \qquad (2.50)$$

where $r_c = r_1 = r_2$ is the radius of curvature of the meniscus at $\xi = 0$ and the origin of coordinates is such that $z(0) = z_0$. Note further that the difference in pressure across the interface at $\xi = 0$ is $2\gamma/r_c$ and is given by the height of rise of the capillary column at the lowest point of the meniscus, z_0, via

$$(\rho_l - \rho_v)gz_0 = \frac{2\gamma}{r_c} \qquad (2.51)$$

On comparing (2.50) with (2.51), we see that $\lambda = 0$. Equation (2.49) with λ set equal to zero and rearranged to be a homogeneous differential equation can be solved to within two arbitrary constants using the generalized Bessel equation, which has as its solution the modified Bessel functions of the first and second kinds of order n as follows. (See Appendix B for a brief discussion of Bessel functions.) The generalized Bessel equation

$$Y''(x) - \frac{2\alpha - 1}{x}Y'(x) - \left(\beta^2\varrho^2 x^{2\varrho-2} - \frac{n^2 - \alpha^2\varrho^2}{x^2}\right)Y(x) = 0 \qquad (2.52)$$

has the solution

$$Y(x) = Ax^\alpha I_n(\beta x^\varrho) + Bx^\alpha K_n(\beta x^\varrho) \qquad (2.53)$$

where the arbitrary constants A and B are fitted from the boundary conditions appropriate to the problem being solved, $I_n(x)$ is the modified Bessel function of the first kind of order n, and $K_n(x)$ is the modified Bessel function of the second kind of order n. Equation (2.52) is used by choosing α, β, ϱ, and n such that it assumes the form of the equation to be solved and using the values thus found in Eq. (2.53). Thus, in our case, with $y(x) = z(\xi)$, $\alpha = 0$, $\beta = \beta$, $\varrho = 1$, $n = 0$, and $x = \xi$, we have for the solution of the homogeneous form of Eq. (2.49)

$$z(\xi) = AI_0(\beta\xi) + BK_0(\beta\xi) \tag{2.54}$$

Since $K_0(\beta\xi) \to \infty$ as $\xi \to 0$, we set $B = 0$, and since $I_0(\beta\xi) \to 1$ as $\xi \to 0$, while $z(0) = z_0$, we set $A = z_0$ and arrive at

$$z(\xi) = z_0 I_0(\beta\xi) \tag{2.55}$$

for the solution of (2.49). In order to compute the pressure difference across an interface having the shape given in (2.55), we now develop an expression for twice the mean curvature of a surface of revolution in the context of differential geometry. This expression will turn out to be the same as was found by adding (2.14) and the reciprocal of (2.44). We recall (2.27) for twice the mean curvature, which is

$$\frac{1}{r_1} + \frac{1}{r_2} = \frac{EN + GL - 2FM}{EG - F^2} \tag{2.27}$$

where E, F, and G are the usual first fundamental coefficients given by Eqs. (2.20a)–(2.20c), and L, M, and N are the second fundamental coefficients given by Eqs. (2.23a)–(2.23c). It is shown in differential geometry that the general equation for a surface of revolution about the ξ axis is

$$\vec{S}(\xi, \theta) = f(\xi)\cos(\theta)\hat{i} + f(\xi)\sin(\theta)\hat{j} + g(\xi)\hat{k} \tag{2.56}$$

where ξ is the usual radial coordinate in the $x - y$ plane, and θ is the angle between the x axis and ξ. In the equations for computing the first and second fundamental coefficients, u is replaced by ξ and v is replaced by θ, so that, for example,

$$E = \vec{S}_\xi \cdot \vec{S}_\xi = \frac{\partial\vec{S}}{\partial\xi} \cdot \frac{\partial\vec{S}}{\partial\xi} \tag{2.57a}$$

$$L = -\vec{S}_\xi \cdot \hat{n}_\theta = -\frac{\partial\vec{S}}{\partial\xi} \cdot \frac{\partial\hat{n}}{\partial\theta} \tag{2.57b}$$

etc., and the normal to the surface is given by

$$\hat{n} = \frac{\vec{S}_\xi \times \vec{S}_\theta}{\left|\vec{S}_\xi \times \vec{S}_\theta\right|} \tag{2.58}$$

We thus compute $\vec{S}_\xi$, S_θ, $\hat{n}_\xi$, and $\hat{n}_\theta$ and use them to compute the first and second fundamental coefficients for a surface of revolution about the ξ axis. The results are

$$E = f'(\xi)^2 + g'(\xi)^2 \tag{2.59a}$$

$$F = 0 \tag{2.59b}$$

$$G = f(\xi)^2 \tag{2.59c}$$

$$L = \frac{f'(\xi)g'(\xi) - f''(\xi)g'(\xi)}{[f'(\xi)^2 + g'(\xi)^2]^{1/2}} \tag{2.60a}$$

$$M = 0 \tag{2.60b}$$

$$N = \frac{f(\xi)g'(\xi)}{[f'(\xi)^2 + g'(\xi)^2]^{1/2}} \tag{2.60c}$$

With F and M equal to zero, (2.27) becomes

$$\frac{1}{r_1} + \frac{1}{r_2} = \frac{N}{G} + \frac{L}{E} \tag{2.61}$$

It can be shown that $F = M = 0$ whenever the $u = $ const. and $v = $ const. (or $r = $ const. and $\theta = $ const.) families of curves are orthogonal to each other and further that in that case

$$\kappa_1 = \frac{1}{r_1} = \frac{N}{G} \tag{2.62}$$

$$\kappa_2 = \frac{1}{r_2} = \frac{L}{E} \tag{2.63}$$

where κ_1 and κ_2 are the curvatures corresponding to the radii of curvature r_1 and r_2, respectively, and the subscripts 1 and 2 have been chosen to correspond to those of (2.44) and (2.14), respectively. Setting $f(\xi) = \xi$ and $g(\xi) = z(\xi)$ from (2.55) and evaluating N, G, L, and E, we have

$$\frac{1}{r_1} + \frac{1}{r_1} = \frac{z'(\xi)}{\xi[1 + z'(\xi)^2]^{1/2}} + \frac{z''(\xi)}{[1 + z'(\xi)^2]^{3/2}} \tag{2.64}$$

which is the same as the expression appearing in (2.45). To proceed, we now evaluate $z'(\xi)$ and $z''(\xi)$ from (2.55) as follows. We recall the general relation

$$I_0'(x) = I_1(x) \tag{2.65}$$

Setting $\beta\xi = x$ and applying the chain rule for differentiating a function of a function (I_0 is the outer function and $\beta\xi$ is the inner function) yields

$$\frac{dI_0(\beta\xi)}{d\xi} = \frac{dI_0(\beta\xi)}{d(\beta\xi)}\frac{d(\beta\xi)}{d\xi} = I_0'(\beta\xi)\beta = I_1(\beta\xi)\beta \tag{2.66}$$

where the prime denotes differentiation with respect to the entire argument of the (outer) function. Thus,

$$z'(\xi) = z_0 \frac{dI_0(\beta\xi)}{d\xi} = z_0\beta I_1(\beta\xi) \tag{2.67}$$

We now recall one of the recursion relations for Bessel functions

$$Z'_\nu(x) = Z_{\nu-1}(x) - \frac{\nu}{x}Z_\nu(x) \tag{2.68}$$

where $Z_\nu(x)$ can be any of $J_\nu(x)$, $Y_\nu(x)$, $I_\nu(x)$, or $K_\nu(x)$. Setting $x = \beta\xi$, $Z_\nu(x) = I_\nu(x)$, $\nu = 1$, and (again) employing the chain rule yields

$$\frac{dI_1(\beta\xi)}{d\xi} = \frac{dI_1(\beta\xi)}{d(\beta\xi)}\frac{d(\beta\xi)}{d\xi} = I'_1(\beta\xi)\,\beta = \left(I_0(\beta\xi) - \frac{1}{\beta\xi}I_1(\beta\xi)\right)\beta \tag{2.69}$$

Thus, using (2.69), we have for $z''(\xi)$

$$z''(\xi) = z_0\beta\frac{dI_1(\beta\xi)}{d\xi} = z_0\,\beta^2\left(I_0(\beta\xi) - \frac{1}{\beta\xi}I_1(\beta\xi)\right) \tag{2.70}$$

Substituting (2.67) and (2.70) into (2.64) yields

$$\frac{1}{r_1} + \frac{1}{r_2} = \frac{z_0\beta I_1(\beta\xi)}{\xi\left[1 + z_0^2\beta^2 I_1(\beta\xi)^2\right]^{1/2}} + \frac{z_0\,\beta^2\left[I_0(\beta\xi) - \frac{1}{\beta\xi}I_1(\beta\xi)\right]}{\left[1 + z_0^2\beta^2 I_1(\beta\xi)^2\right]^{3/2}} \tag{2.71}$$

Using particular choices for ξ and z_0, (2.71) multiplied by γ would yield the pressure difference across a (shallow) curved interface due to surface tension. We now develop an expression for z_0 as follows. From (2.46), with $\lambda = 0$, using also (2.48) we have

$$\frac{1}{\xi}\frac{d}{d\xi}\left(\frac{\xi z'(\xi)}{[1 + z'(\xi)^2]^{1/2}}\right) = \beta^2 z(\xi) = \frac{1}{\xi}\frac{d}{d\xi}[\xi\sin(\theta)] \tag{2.72}$$

where θ is the angle between the z axis and a radius of curvature drawn from the z axis perpendicular to the meniscus, as shown in Figure 2.4. Indicating the integration of (2.72) from $\xi' = 0$ to ξ, the radius of the "large"-diameter capillary, we have

$$\int_0^\xi \frac{d}{d\xi'}[\xi'\sin(\theta)]\,d\xi' = \beta^2\int_0^\xi \xi' z(\xi')\,d\xi' \tag{2.73}$$

where ξ' is a dummy variable of integration. Substituting (2.55) for $z(\xi)$, and carrying out the integration indicated on the left-hand side of (2.73) yields

$$[\xi'\sin(\theta)]_0^\xi = \xi\cos(\varphi) = \beta^2 z_0\int_0^\xi \xi' I_0(\beta\xi')\,d\xi' \tag{2.74}$$

where φ is the usual angle of capillarity as shown in Figure 2.4. Changing the variable of integration from ξ' to $\tau = \beta\xi'$ and recalling the integral

$$\int_0^x \tau^\nu I_{\nu-1}(\tau)\, d\tau = x^\nu I_\nu(x) \tag{2.75}$$

yields, with $\nu = 1$, and $x = \beta\xi$,

$$\xi\cos(\varphi) = z_0 \beta\xi I_1(\beta\xi) \tag{2.76}$$

or

$$z_0 = \frac{\cos(\varphi)}{\beta I_1(\beta\xi)} \tag{2.77}$$

Equation (2.77) is substituted into Eq. (2.71), which is then multiplied by γ to yield the expression for the pressure difference across the curved interface.

We now consider the case of a capillary of circular cross section and radius small enough for the shape of the meniscus to be approximated by a section of a hemisphere as considered by Rayleigh (1915). Beginning with Eq. (2.73), we have

$$\xi\cos(\varphi) = \beta^2 \int_0^\xi \xi' z(\xi')\, d\xi' \tag{2.78}$$

From Figure 2.4 we see that the center of curvature of the capillary meniscus is located at the $z - \xi$ coordinates $(z_0 + r_c, 0)$. The equation of a circle of radius r_c centered at (z_1, ξ_1) in a $z(\xi) - \xi$ coordinate system is given by

$$r_c^2 = [z(\xi) - z_1]^2 \pm (\xi - \xi_1)^2 \tag{2.79}$$

where $z_1 = z_0 + r_c$ and $\xi_1 = 0$. Choosing the upper sign on the right-hand side of Eq. (2.79) (the lower sign corresponds to the semicircle above the one describing the cross section of the meniscus) and rearranging, we have for $z(\xi)$ in the spherical approximation,

$$z(\xi) = z_0 + r_c - (r_c^2 - \xi^2)^{1/2} \tag{2.80}$$

We note further, for later use, that also in this approximation, for $\varphi < \pi/2$,

$$\xi = r_c \cos(\varphi) \tag{2.81}$$

Substituting (2.80) into (2.78), carrying out the indicated integration, and rearranging terms yields for the height, z_0, measured from $z = 0$ to the base of the meniscus, to which fluid wetting the wall of the capillary ($\varphi < \pi/2$) will rise

$$z_0 = \frac{2\gamma\cos(\varphi)}{(\rho_l - \rho_v)g\xi} - \frac{2\xi}{3\cos^3(\varphi)}\left[1 - \sin^3(\varphi)\right] - \frac{\xi}{\cos(\varphi)} \tag{2.82}$$

where (2.81) has been used to eliminate r_c as well.

2.7   VAPOR PRESSURE OVER CURVED SURFACES

We now consider the change in the vapor pressure of a substance over its surface when the surface is curved. This vapor pressure is in general increased over that of a substance over a "flat" surface if the surface is convex (negative radii of curvature) and is reduced below that of a "flat" surface of the same material if the surface is concave (positive radii of curvature). We consider the experimental arrangement shown in Figure 2.5 in which a vacuum bell jar contains a pan of water into which is inserted a capillary tube of circular cross section and radius ξ. The pressure of the liquid just under the meniscus is p_l, the pressure of the water vapor just above the meniscus is p_v, while at the flat liquid surface the pressure of the vapor just above the surface, p_{ov}, equals that of the liquid just under the surface, p_{ol}.

The bell jar contains water and water vapor only, so that γ is the true surface tension; p_v is the vapor pressure in the bell jar and is a function of height above the flat water surface. We write

$$p_v - p_l = \int \rho_v g \, dh - \int \rho_l g \, dh \qquad (2.83)$$

where dh is an element of height above the "flat" water surface, ρ_v is the mass density of the vapor in the bell jar and is a function of height above the flat water surface, ρ_l is the mass density of the liquid in the capillary tube, and the rest of the symbols are as previously defined. Writing

$$dh = \frac{dp_v}{\rho_v g} \qquad (2.84a)$$

and using the ideal gas law to give

$$mp_v = \frac{mN_v kT}{V_g} = \rho_v kT \qquad (2.84b)$$

where N_v vapor molecules, each of mass m, are contained in a volume V_g, yields

$$dh = \frac{kT \, dp_v}{mp_v g} \qquad (2.84c)$$

so that (2.83) becomes

$$p_v - p_l = \int_{p_{ov}}^{p_v} (\rho_v - \rho_l) \frac{dp_v'}{\rho_v} = \int_{p_{ov}}^{p_v} dp_v' - \frac{\rho_l kT}{m} \int_{p_{ov}}^{p_v} \frac{dp_v'}{p_v'} \qquad (2.85)$$

or

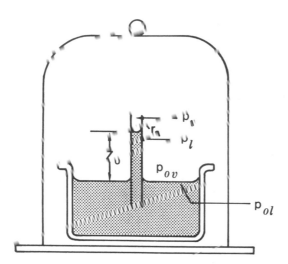

Figure 2.5: A vacuum bell jar containing water vapor, water in a b...
containing a capillary tube in which the water at the base of the meni...
has risen to an equilibrium height v above the "flat" surface of water...
the beaker.

$$\eta \quad p_l = p_v - p_{ov} - \frac{\rho_l kT}{m} \ln\left(\frac{p_v}{p_{ov}}\right) \tag{2.86}$$

For the case of a small-diameter capillary of circular cross section, if the meniscus is assumed to be part of a hemisphere, we have approximately

$$p_v - p_l \approx \frac{2\gamma \cos(\varphi)}{\zeta} \tag{2.87}$$

so that

$$\frac{2\gamma \cos(\varphi)}{\xi} = p_v - p_{ov} - m_l \ln\left(\frac{p_v}{p_{ov}}\right) \tag{2.88}$$

Equation (2.88) cannot be solved explicitly for p_v using algebra alone, as a little experimentation will show. Lagrange's expansion, however, affords a convenient means of effecting this solution as follows. If

$$\zeta = \alpha + \beta \Phi(\zeta) \tag{2.89}$$

where α and β are parameters that are not functions of ζ, then

$$\zeta = \alpha + \sum_{n=1}^{\infty} \frac{\beta^n d^{n-1}}{n! d\alpha^{n-1}} [\Phi(\alpha)^n] \tag{2.90}$$

Writing (2.88) as

$$p_v = p_{0v} D e^{p_v/\Psi_0} \tag{2.91}$$

where

$$D = e^{-A/\xi - p_{0v}/\Psi_0} \tag{2.92}$$

and

$$A = \frac{2\gamma m \cos(\varphi)}{\rho_l kT} \tag{2.93}$$

where $\varphi < \pi/2$, yields, on comparing (2.91) with (2.85), $\zeta = p_v, \alpha = 0, \beta = p_{0v}D, \Phi(\zeta) = e^{\zeta/\Psi_0}$, so that $\Phi(\alpha)^n = (e^{\alpha/\Psi_0})^n = e^{n\alpha/\Psi_0}$. Substituting the results into (2.91) yields

$$p_v = \sum_{n=1}^{\infty} \frac{(p_{0v}D)^n}{n!} \left(\frac{d^{n-1}}{d\alpha^{n-1}} \left(e^{\alpha n/\Psi_0} \right) \right) \Bigg|_{\alpha=0}$$

$$= \sum_{n=1}^{\infty} \frac{(p_{0v}D)^n}{n!} \left(\frac{n}{\Psi_0} \right)^{n-1} \tag{2.94}$$

or

$$\frac{p_v}{p_{0v}} = \sum_{n=1}^{\infty} \frac{(n)^{n-1}}{n!} \left(\frac{p_{0v}}{\Psi_0} \right)^{n-1} e^{-n(A/\xi + p_{0v}/\Psi_0)} \tag{2.95}$$

which is the vapor pressure above the meniscus divided by the vapor pressure of the vapor over a flat surface of water containing the same concentrations of impurities and at the same temperature as that in the capillary. Approximating the sum on the right-hand side of (2.95) by its first term yields

$$\frac{p_v}{p_{0v}} \simeq e^{-A/\xi} C(T) \tag{2.96a}$$

where

$$C(T) = e^{-p_{0v}/\Psi_0} \tag{2.96b}$$

and is a temperature-dependent correction factor to the usual form of the Kelvin relation, which is generally given as

$$\frac{p_v}{p_{0v}} \simeq e^{-A/\xi} \tag{2.97}$$

and is seen to represent a very special case.

The utility of the Kelvin relation is that it can be used with a suitable frequency distribution function to compute theoretically the average vapor pressure in an unsaturated porous medium, which in turn is a reflection of the matric potential.

In Chapter 1 it was noted that at constant temperature and chemistry the matric potential of the ith species of liquid in a porous medium, Ψ, is related to the average vapor pressure, $\bar{p}_v$, of the liquid in the porous medium by

$$^{(i)}\Psi = {}^{(i)}\Psi_0 \ln\left(\frac{{}^{(i)}\bar{p}_v(r)}{{}^{(i)}p_{0v}}\right) \tag{2.98}$$

where $^{(i)}\bar{p}_v(r)$ is made up of contributions from capillary meniscus surfaces of all shapes and sizes containing liquid, as shown in Eq. (1.3). Since p_{0v} depends on soil water chemistry as well as temperature, as will be discussed in Chapter 3, Ψ depends on these parameters as well and has additional temperature dependence through Ψ_0 and γ, which last depends strongly on soil water chemistry as well.

2.8   POISEUILLE'S LAW AND CAPILLARITY

We now derive Poiseuille's law describing the flow of liquid through a capillary tube of circular cross section. Consider two plates each of area $A/2$ oriented parallel to the y axis. Let the viscous force resisting the motion of the fluid be f_v, where

$$f_v = -\eta A \frac{dv_x}{dy} \tag{2.99}$$

in which η is the measured dynamic viscosity, v_x is the velocity of the fluid in the x direction, and, dv_x/dy is the change in v_x with y. We can fix the units of η by writing

$$\eta = \frac{f_v}{A\frac{dv_x}{dy}} \Rightarrow \frac{\text{dyne sec cm}}{\text{cm}^2} \frac{\text{cm}}{\text{cm}} = \frac{\text{g}}{\text{cm sec}} = \text{poise}$$

where ρ is the mass density of the fluid, and the quantity η/ρ is called the *kinematic viscosity*. The metric unit of kinematic viscosity is called the stoke and is equal to 1 poise divided by 1 g/cm^2 or 1 cm^2/sec. We now consider a circular tube of radius ξ oriented along the z axis of a polar coordinate system, as shown in Figure 2.7. We consider an imaginary cylinder of fluid of radius ζ and length ℓ, coaxial with the z axis and subjected to a pressure difference $P_2 - P_1$ across its ends, as shown in Figure 2.6, and contained within the tube of radius ξ. If the velocity of the fluid in the z direction is constant, the pressure forces are exactly balanced by the viscous drag force f_v (accelerations result from unbalanced forces, and the acceleration required to start the fluid moving is neglected in this treatment) and the net force on the fluid is zero. Thus,

$$(P_2 - P_1)\pi\zeta^2 = f_v \tag{2.100}$$

or

$$(P_2 - P_1)\pi\zeta^2 + \eta(2\pi\zeta\ell)\frac{dv_z}{d\zeta} = 0 \tag{2.101}$$

where

$$A = 2\pi\zeta\ell \tag{2.102}$$

is the lateral area of the imaginary cylinder of fluid. Thus,

$$\frac{dv_z}{d\zeta} = -\frac{(P_2 - P_1)\zeta}{2\eta\ell} \tag{2.103}$$

which becomes, on separating variables and indicating integration,

$$-\int_{v_0}^{v_z} dv_z' = \frac{P_2 - P_1}{2\eta\ell}\int_0^\zeta \zeta'\, d\zeta' \tag{2.104}$$

where v' and ζ' are dummy variables of integration, v_0 is the velocity of the fluid along the axis of the tube where $\zeta' = 0$, and v_z is the velocity of the fluid at $\zeta' = \zeta$, so that

$$v_0 - v_z = \frac{\zeta^2(P_2 - P_1)}{4\eta\ell} \tag{2.105}$$

If it is now assumed that $v_z = 0$ at the wall of the tube, where $\zeta = \xi$, we have

$$v_0 = \frac{\xi^2(P_2 - P_1)}{4\eta\ell} \tag{2.106}$$

so that combining (2.105) and (2.106) yields

$$v_z(\zeta) = \frac{(P_2 - P_1)}{4\eta\ell}(\xi^2 - \zeta^2) \tag{2.107}$$

from which it is seen that the fluid velocity is a maximum along the axis of the tube ($\zeta = 0$) and is (assumed) to be zero on the walls ($\zeta = \xi$). We now compute the volume of fluid per unit time, q, flowing through the pipe as follows. By inspection,

$$q = \int v_z(\zeta)\, dA_p \tag{2.108}$$

where A_p is the cross-sectional area of the pipe and

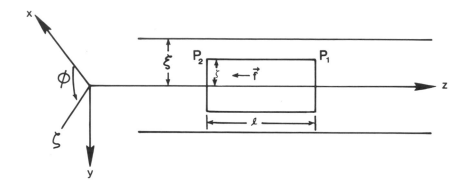

Figure 2.6: A tube of circular cross section of radius ξ oriented along the z axis of a polar coordinate system having coordinates ζ and θ.

$$dA_p = 2\pi\zeta \, d\zeta \tag{2.109}$$

so that

$$q = \int_0^\xi \frac{(P_2 - P_1)}{4\eta\ell}(\xi^2 - \zeta^2)2\pi\zeta \, d\zeta = \frac{(P_2 - P_1)\pi\xi^4}{8\eta\ell} \tag{2.110}$$

or

$$q(\xi) = \frac{(P_2 - P_1)\pi\xi^4}{8\eta\ell} \tag{2.111}$$

which is Poiseuille's law giving the volume of fluid per unit time flowing in a tube of circular cross section of radius ξ under the influence of a pressure difference $P_2 - P_1$ acting across a length ℓ of fluid. The volume of fluid per unit area per unit time flowing in a tube of circular cross section is the Darcy velocity for a single capillary and is given by

$$v_a(\xi) = \frac{\int v_z \, dA_p}{\int dA_p} = \frac{q(\xi)}{\int_0^\xi 2\pi\zeta \, d\zeta} = \frac{(P_2 - P_1)\xi^2}{8\eta\ell} \tag{2.112}$$

which we now identify as the velocity of fluid in the tube averaged over the cross-sectional area of the tube, or the average velocity of fluid flowing in the tube. Making particular choices for P_2 and P_1, appropriate to capillarity and gravity, respectively, yields

$$P_2 \simeq -\frac{2\gamma \cos(\varphi)}{\xi} \tag{2.113}$$

for $\varphi < \pi/2$ and

$$P_1 = \rho g h \, \cos(\theta) \tag{2.114}$$

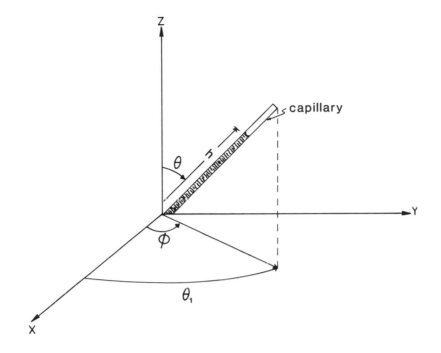

Figure 2.7: A capillary of radius ξ making an angle θ with the vertical.

where θ is the angle a capillary tube of radius ξ and height h (the height to which the water has risen in it in a time t) makes with the vertical, as shown in Figure 2.7. Thus, to compute $h(t)$ we now write, following Peiris and Tennakone (1980)

$$\frac{dh(t)}{dt} = v_a(\xi) \qquad (2.115)$$

Identifying ℓ of (2.191) with $h(t)$ yields

$$\frac{dh(t)}{dt} = \frac{2\gamma \cos(\varphi)\xi}{8\eta h(t)} - \frac{\rho g \, h(t) \cos(\theta)\xi^2}{8\eta h(t)} \qquad (2.116)$$

Setting

$$A_1 = \frac{\xi\gamma \cos(\varphi)}{4\eta} \qquad (2.117a)$$

and

$$B_1 = \frac{\rho g \xi^2 \cos(\theta)}{8\eta} \qquad (2.117b)$$

in (2.116) yields

$$\frac{dh(t)}{dt} = \frac{A_1 - B_1 h(t)}{h(t)} \tag{2.118}$$

Separating variables and indicating integration yields

$$\int_0^h \frac{h' \, dh'}{A_1 - B_2 h'} = \int_0^t dt' \tag{2.119}$$

Integrating and rearranging terms yields

$$-\frac{B_1^2 t + B_1 h}{A_1} = \ln\left(\frac{A_1 - B_1 h}{A_1}\right) \tag{2.120}$$

Setting

$$\frac{A_1}{B_1} = h_0 \tag{2.121a}$$

and

$$\frac{B_1^2}{A_1} = \frac{1}{t_0} \tag{2.121b}$$

and rearranging yields for the height of rise of liquid in a circular capillary as a function of time

$$h(t) = h_0 \left(1 - e^{-h(t)/h_0 - t/t_0}\right) \tag{2.122}$$

We note that (2.122) is an implicit relation for $h(t)$. As in (2.88), ordinary algebraic techniques are unavailing in solving (2.122) for $h(t)$, but Lagrange's expansion can be applied in this case (Case, 1990). We set

$$c = e^{-t/t_0} \tag{2.123}$$

so that (2.122) becomes

$$\frac{h}{h_0} = 1 - ce^{-h/h_0} \tag{2.124}$$

We now compare (2.89) and (2.124) to make the identifications $\zeta = h/h_0$, $\beta = -c$, $\Phi(\zeta) = e^{-\zeta}$, $\alpha = 1$, $f(\zeta) = \zeta$. Thus, $f(\alpha) = \alpha$, $f'(\alpha) = 1$, $\Phi(\alpha) = e^{-\alpha}$, $\Phi(\alpha)^n = e^{-\alpha n}$ while

$$\frac{d^{n-1}}{d\alpha^{n-1}} e^{-n\alpha} = (-n)^{n-1} e^{-\alpha n} \tag{2.125}$$

Using the identifications made results in

$$\frac{h}{h_0} = 1 + \sum_{n=1}^{\infty} \frac{(-c)^n (-1)^{n-1} e^{-n}}{n!} \tag{2.126}$$

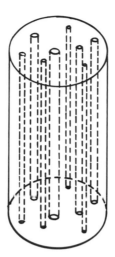

Figure 2.8: A cross section of porous medium made up of a bundle of capillaries having a pore radius frequency distribution $f(\xi)$. The conducting capillaries may be scattered across the face of area A' or concentrated in one part of it.

Thus

$$\frac{h}{h_0} = 1 - \sum_{n=1}^{\infty} \frac{c^n n^{n-1} e^{-n}}{n!} \qquad (2.127a)$$

or

$$h(t, \xi) = h_0 - h_0 \sum_{n=1}^{\infty} \frac{e^{-nt/t_0} n^{n-1} e^{-n}}{n!} \qquad (2.127b)$$

which is the required expression. Differentiating (2.127) with respect to time (t) yields for the velocity of fluid (averaged over the cross-sectional area of the tube) rising in a tube of radius ξ due to capillary action, the Darcy velocity in a single capillary,

$$v_a(\xi) = \frac{h_0}{t_0} \sum_{n=1}^{\infty} \frac{e^{-nt/t_0} n^n e^{-n}}{n!} \qquad (2.128)$$

The volume per unit time, $q(\xi)$, flowing through a single capillary of constant circular cross section is thus given by

$$q(\xi) = \pi\xi^2 v_a(\xi) \tag{2.129}$$

We now develop an expression for the Darcy velocity, using the bundle of capillaries model of a porous medium, as follows. Let A' be the cross-sectional area of a sample core of material, as shown schematically in Figure 2.8. The conducting area of the face, which is assumed to be perpendicular to the direction of flow, is

$$^{(o)}\epsilon A' = N' \int_{r_m}^{R} \pi\xi^2 f_c(\xi)\, d\xi \tag{2.130}$$

where $^{(o)}\epsilon$ is the connected porosity of the sample, in the notation of Chapter 1, N' is the number of conducting capillaries cut by the sample face, and the rest of the symbols are as previously defined. The total volume of liquid per unit time crossing the area A' is

$$Q = N' \int_{r_m}^{r} q(\xi) f_c(\xi)\, d\xi = N' \int_{r_m}^{r} \pi\xi^2 v_a(\xi) f_c(\xi)\, d\xi \tag{2.131}$$

and the Darcy velocity, $^{(1)}J$, the volume per unit area per unit time, is

$$^{(1)}J = \frac{Q}{A'} = \frac{^{(o)}\epsilon \int_{r_m}^{r} \xi^2 f_c(\xi)\, ^{(1)}v_a(\xi)d\xi}{\int_{r_m}^{R} \xi^2 f_c(\xi)d\xi}$$

$$= {}^{(o)}\epsilon <{}^{(1)}v_a> = {}^{(o)}\epsilon\, {}^{(1)}v_p\, {}^{(1)}\theta \tag{2.132}$$

where $i = 1$ represents liquid water, (2.130) and (2.131) have been used, the symbol $<{}^{(1)}v_a>$ stands for the average of $^{(1)}v_a(\xi)$ over the distribution $\xi^2 f_c(\xi)$, and is equal to the average velocity of fluid relative to the soil sample, or pore velocity, $^{(1)}v_p$ multiplied by the degree of saturation corresponding to the upper limit of the integral in the numerator of (2.132) in the sense of Eq. (1.6). Substituting (2.128) into (2.132) yields

$$^{(1)}J(r) = \frac{\frac{^{(o)}\epsilon\rho_l g \cos(\theta)}{8\eta} \sum_{n=1}^{\infty} \frac{n^n e^{-n}}{n!} \int_{r_m}^{r} \xi^4 f_c(\xi) e^{-n d_0 t \xi^3}\, d\xi}{\int_{r_m}^{R} \xi^2 f_c(\xi)\, d\xi} \tag{2.133}$$

Note that this result generalizes the original Green–Ampt result (Green and Ampt, 1911) from that obtained using as a physical model of a porous medium a bundle of capillaries of circular cross sections and equal radii to that of a bundle of capillaries of circular cross sections having a radius frequency distribution $f_c(\xi)$. It is assumed that all the capillaries in the bundle make the same angle with the vertical.

If the capillaries are assumed to be oriented randomly, then

$$^{(1)}J(r) = \frac{\frac{^{(o)}\epsilon \rho_l g}{8\eta} \sum_{n=1}^{\infty} \frac{n^n e^{-n}}{n!} \int_{r_m}^{r} \int_{0}^{\pi/2} \xi^4 f_c(\xi,\theta) e^{-nd_0 t \xi^2} \cos(\theta) \, d\theta \, d\xi}{\int_{r_m}^{R} \xi^2 f_c(\xi) d\xi}$$

(2.134)

where now $f_c(\xi,\theta)$ is a joint probability distribution giving the fraction of capillaries of radius ξ making an angle θ with the vertical. If geometric information is available, namely $r(\vec{r})$, where $\vec{r}$ is a position vector, then in the special case that $K(\theta_s)$ is a scalar function, we may write

$$\hat{n} \cdot {}^{(i)}\vec{J} = - {}^{(i)}K({}^{(i)}\theta)(\vec{\nabla} {}^{(i)}H) \cdot \hat{n}$$

(2.135)

or

$$^{(i)}K({}^{(i)}\theta) = \frac{-\hat{n} \cdot {}^{(i)}\vec{J}}{\hat{n} \cdot \vec{\nabla} {}^{(i)}H}$$

(2.136)

where

$$^{(i)}H = \frac{^{(i)}\Psi}{^{(i)}\rho_l g} + z$$

(2.137)

and $\hat{n}$ is a unit vector along the direction of $\vec{\nabla} {}^{(i)}H$, which is also the direction of $\cos(\theta)$ in (2.134). In a vertical soil column imbibition experiment, the average height of wetting as a function of time can be calculated in terms of the bundle of capillaries model as

$$\bar{h}(t) = \int_{r_m}^{r} h(t,\xi) f_c(\xi) \, d\xi$$

(2.138)

where $h(t,\xi)$ is given by (2.127). The degree of saturation of the column at any height above its base and time is given by substituting (2.127) into (2.138), where r is the radius of the largest capillary containing water at the location of interest in the column. Such a soil column imbibition experiment can lead to a measurement of $f_c(\xi)$ as follows. From (2.127), the time at which the matric potential at a given height is measured determines the largest value of ξ, namely r, that can transmit water to that level since the largest vapor pressure contribution to $^{(i)}\Psi$ arises from the capillaries having the largest radii. Thus, if matric potential is measured at a given height as a function of time, a set of $^{(i)}\Psi - r$ data is generated so that Eq. (2.98) can in principle be solved as an integral equation for $f_c(\xi)$. If the degree of saturation at a given height is measured as a function of time instead of matric potential, then the resulting $^{(i)}\theta - r$ data set can be used to solve (1.6) as an integral equation for $f_c(\xi)$, where now $v(\xi)$ is interpreted as

$$v_c(\xi,t) = \pi\xi^2 h(t,\xi)$$

(2.139)

and $r = r(t)$, as discussed. Simultaneous measurements of $^{(i)}\Psi$ and $^{(i)}\theta$ at a given height during imbibition of water, denoted by $i = 1$, in a soil column followed by their simultaneous measurements during drainage would yield a hysteresis loop for the soil of interest as well.

SELECTED REFERENCES

Buff, F. P., "The Theory of Capillarity", *Handbuch der Physik*, Vol. X, Structure of Liquids, pp. 281–304, Springer-Verlag, Göttingen, 1960.

Case, C. M., "Rate of rise of liquid in a capillary tube–revisited," *Am. J. Physics* 58 (9), 888–89, September, 1990.

Green, W. H., and G. A. Ampt, "Studies of soil physics: I. Flow of air and water through soils," *J. Agricultural Science* 4, 1-24, 1911.

Kreyszig, E., *Differential Geometry*, University of Toronto Press, Toronto, Canada, 1959, Chapter VI.

Lamb, H., *Hydrodynamics*, 6th ed., Dover Publications, New York, 1945.

Lipschutz, M., *Theory and Problems of Differential Geometry*, McGraw-Hill Book Co., New York, 1969.

Newman, F., and V. Searle, *The General Properties of Matter*, The MacMillian Co., New York, 1939.

Peiris, M. G. C., and K. Tennakone, "Rate of Rise of Liquid in a Capillary Tube," *American J. Physics* 48, 415, 1980.

Rayleigh, "On the Theory of the Capillary Tube," *Proc. Royal Soc. of London*, Series A 92, 184–195, 1915.

PROBLEMS

2.1 Compute the mean and Gaussian curvatures for the surface for which $\vec{s} = u\hat{i} + v\hat{j} + (u^2 + v^2)\hat{k}$. Interpret your results in terms of the shape of the surface. Is the point $u = v = 0$ an umbilical point?

2.2 Solve Eq. (2.24) of the text for κ_1 and κ_2 for the case in which $\vec{s} = u\hat{i} + v\hat{j} + \hat{k}(u^2 + v^2)$. Do the results of the computation agree with those of Problem 2.1?

2.3 In Eq. (2.37a) of the text, the element of interfacial area is given by

$$dA = |\vec{s}_u \times \vec{s}_v| \, du \, dv$$

Using Eqs. (2.20a)–(2.20c) as required, show that

$$dA = \sqrt{EG - F^2} \, du \, dv$$

where E, F, and G are the coefficients of the first fundamental form.

2.4 The unit vector

$$\hat{n} = \frac{\vec{s}_u \times \vec{s}_v}{|\vec{s}_u \times \vec{s}_v|}$$

represents the direction of the directed element of area, $d\vec{A}$, of Problem 2.3, relative to the coordinate system having the unit vectors $\hat{i}$, $\hat{j}$, and $\hat{k}$. Compute this unit vector for the surface $\vec{S}$ of Problem 2.1. Make a sketch showing $\hat{n}$ for various locations on $\vec{S}$, that is, for various values of u and v.

2.5 Show that for

$$\vec{s} = u\hat{i} + v\hat{j} + (u^2 + v^2)\hat{k}, \quad \hat{n} \cdot \delta\hat{n} = \hat{n} \cdot \left(\frac{\partial \hat{n}}{\partial u}\delta u + \frac{\partial \hat{n}}{\partial v}\delta v \right) = 0$$

2.6 The function $C\xi e^{-\xi/r_0}$ is taken to be a Poisson-like average effective pore/capillary radius frequency distribution normalized to one over the interval r_m to R where $r_m < \xi < R$. Find an expression for C, the normalization constant, by carrying out the integration in

$$1 = C \int_{r_m}^{R} \xi e^{-\xi/r_0} \, d\xi$$

What is the physical significance of r_0?

2.7 Show that the integral

$$\int_0^\beta \xi e^{-\xi/r_0 - A/\xi} \, d\xi$$

can be expressed as

$$\sum_{l=0}^{\infty} \frac{(-1)^l \beta^{l+2}}{(r_0^l)l!} E_{l+3}(A/\beta)$$

where

$$E_n(z) = \int_1^\infty \frac{e^{-zt}}{t^n} \, dt$$

is the exponential integral of order n. Note in passing that for $n = 1$, a change of variable of integration ($u = zt$) yields

$$E_1(z) = \int_z^\infty \frac{e^{-u}}{u} \, du$$

which is the "well function," so-called, appearing in the Theis equation of saturated groundwater flow. The integral that is the subject of this problem arises in the expression for the matric potential derived from Eq. (1.3) with $f(\xi)$ given by a Poisson-like capillary radius frequency distribution and the vapor pressure over a single capillary of radius r given by the Kelvin relation of Eq. (2.97), namely

$$\frac{^{(1)}\bar{p}_v(r)}{^{(1)}p_{0v}} = \frac{1}{r_0^2} \int_{r_m}^{r} \xi e^{-\xi/r - A/\xi} \, d\xi$$

Substituting this result into Eq. (2.98) with $i = 1$ yields an expression for the matric potential in terms of the average pressure of vapor evolved from the pore/capillary water surfaces in the bundle of capillaries porous medium.

2.8 Show, using integration by parts, that

$$E_{n+1}(z) = \frac{1}{n} \left[e^{-z} - z E_n(z) \right]$$

where $E_n(z)$ is the exponential integral of Problem 2.7.

2.9 Make a qualitative argument, assuming that a single species, water vapor in the soil, for example, behaves as an ideal gas, and so obeys the ideal gas law, leading to the conclusion that

$$\left(\frac{\partial \, ^{(i)}\theta}{\partial T} \right)_\Psi < 0$$

2.10 Show, if

$$\theta = \tan^{-1} \left(\frac{dy}{dx} \right)$$

that

$$\frac{d\theta}{dx} = \frac{\frac{d^2 y}{dx^2}}{1 + \left(\frac{dy}{dx} \right)^2}$$

2.11 Show, if

$$\theta = \tan^{-1} \left(\frac{dy}{dx} \right)$$

that

$$\sin(\theta) = \frac{y'(x)}{\sqrt{1 + [y'(x)]^2}}$$

and

$$\cos(\theta) = \frac{1}{\sqrt{1 + [y'(x)]^2}}$$

2.12 If the parametric equation for a surface of revolution is $\vec{s}(r, \theta) = f(r)\cos(\theta)\hat{i} + f(r)\sin(\theta)\hat{j} + g(r)\hat{k}$, show that the corresponding first fundamental coefficients are given by $E = f'^2 + g'^2$, $F = 0$, and $G = f^2$.

2.13 Given $\vec{s}(r, \theta) = f(r)\cos(\theta)\hat{i} + f(r)\sin(\theta)\hat{j} + g(r)\hat{k}$ as the parametric equation for a surface of revolution, show that

(a) $$\hat{n} = \frac{\vec{s}_r \times \vec{s}_\theta}{|\vec{s}_r \times \vec{s}_\theta|} = \frac{-\hat{i}g'\cos(\theta) - \hat{j}g'\sin(\theta) + \hat{k}f}{(f'^2 + g'^2)^{\frac{1}{2}}}$$

(b) $$\frac{\partial \hat{n}}{\partial r} = \hat{n}_r = \frac{[\hat{i}g'\cos(\theta) + \hat{j}g'\sin(\theta) - \hat{k}f'](f'f'' + g'g'')}{(f'^2 + g'^2)^{\frac{3}{2}}}$$

$$+ \frac{-\hat{i}g''\cos(\theta) - \hat{j}g''\sin(\theta) + \hat{k}f''}{(f'^2 + g'^2)^{\frac{1}{2}}}$$

(c) $$\frac{\partial \hat{n}}{\partial \theta} = \hat{n}_\theta = \frac{\hat{i}g'\sin(\theta) - \hat{j}g'\cos(\theta)}{(f'^2 + g'^2)^{\frac{1}{2}}}$$

2.14 Show that for the case of a surface of revolution about the z axis, the second fundamental coefficients are given by

$$L = \frac{f'g' - f''g'}{(f'^2 + g'^2)^{\frac{1}{2}}}, \quad M = 0, \quad N = \frac{fg'}{(f'^2 + g'^2)^{\frac{1}{2}}}$$

2.15 Show that by solving the quadratic equation in κ, namely $(EG - F^2)\kappa^2 - (EN + GL - 2FM)\kappa + (LN - M^2) = 0$ with $F = M = 0$, that the principal curvatures for a surface of revolution are N/G and L/E.

2.16 Compute the amount by which mercury is depressed below its flat surface inside a capillary of circular cross section when the capillary is inserted in mercury if (a) the capillary radius is 0.5 cm, and (b) the capillary radius is 0.1 cm, if $\gamma = 480$ dyne/cm, $\rho_l = 13.6$ g/cm^3, and $\rho_v \ll \rho_l$.

2.17 Recalling that as $x \to 0$,

$$I_\nu(x) \sim \frac{\left(\frac{x}{2}\right)^\nu}{\Gamma(\nu+1)}$$

where $\sim$ means "asymptotically equal to" and $\Gamma(\nu+1)$ is the usual gamma function, show that for $\xi << 1$, Eq. (2.71) becomes

$$\frac{1}{r_1} + \frac{1}{r_2} = z_o\beta^2 - \frac{z_o^3\beta^6\xi^2}{4} + \cdots$$

by expanding the denominators of the right-hand side of Eq. (2.71) in a binomial series and retaining only the first two terms in each series.

3

EQUILIBRIUM
THERMODYNAMICS AND
STATISTICAL MECHANICS

3.1 EQUILIBRIUM THERMODYNAMICS

3.1.1 Combined First and Second Laws of Thermodynamics

There exist two conventions for writing the first law of thermodynamics, which itself is a statement of conservation of energy. The first embodies the convention that the work done on a system by an external agency, for example, compressing a gas from an initial volume $_iV$ to a smaller final volume $_fV$, is negative and the work done by the system in expanding gas in a cylinder from a smaller initial volume $_iV$ to a larger final volume $_fV$ is positive. The second convention, which will be adopted in this work, is the opposite of the first in that the work done on the system is positive and the work done by the system is negative. Using this convention, the differential form of the first law of thermodynamics may be written as

$$\delta Q + \delta W = dU \qquad (3.1)$$

where δQ is the small amount of heat (energy) entering the system during a process that takes the system from an initial to a final state, δW is the corresponding small amount of work (positive) performed on the system during this process, and dU is the differential increase in the internal energy of the system during the process. The symbol δ appearing in the quantities δQ and δW is used here to indicate that these quantities, called inexact differentials, are not the total differentials of any mathematical function

53

and that their effects, once the process generating them has occurred, cannot be separated from each other in the change in internal energy of the system, dU, which is an exact differential, and so can be thought of as the differential of a mathematical function. This convention is such that for the following systems δW, which may be written as an exact differential, dW, if the dependence of the generalized force on the generalized displacement is known is

$$\delta W = -P \, dV \tag{3.2}$$

where P is the pressure exerted on a system by an external agency, and dV is the resulting change in the volume of a chemical system, which by definition is one for which the work term in the combined first and second laws of thermodynamics is given by Eq. (3.2), when it is compressed (work done on the system)

$$\delta W = \gamma \, dA \tag{3.3}$$

where γ is interfacial tension and dA is the increase in area that occurs when work is done on a surface,

$$\delta W = F \, dl \tag{3.4}$$

where F is the force exerted on a wire and l is the change in length it undergoes when work is done on it. Note that in both conventions for the first law, the work done on the system is the negative of the work done by the system. This means, for example, that when a chemical system is compressed, the differential of the work done on the system is positive and is given by (3.2) and simultaneously the differential of the work done by the system is negative and is given by $P \, dV$. Both conventions for the first law lead to the same expressions when a given system is being considered, since the sign convention of (3.1) is consistent with those adopted in (3.2)–(3.4). The inexact differentials indicated in (3.2)–(3.4) can be made exact if the paths in the $P - V$, $\gamma - A$, $F - l$, etc. generalized force–displacement planes, respectively, are specified. This specification of path is accomplished by specifying the equation of state for the working substance. For example, for an ideal gas, the ideal gas law, its equation of state, is used to write the pressure of the gas as a function of its volume. Equation of state information is not part of thermodynamics as such but is required as an addition to thermodynamic formalism in any specific application. For the cases in which the matric potential–degree of saturation (the degree of saturation of a porous medium containing liquid of species i is the fraction of the pore space filled with the liquid of species i) hysteresis is ignored, examples of which, when the liquid is water, are, field work, numerical modeling of unsaturated flow, and certain soils in which the saturation–desaturation cycle has been repeated a sufficient number of times so that the area of the hysteresis loop is small, we write for the unsaturated porous–medium–liquid system

$$\delta W = - \, {}^{(i)}\Psi \, d \, {}^{(i)}\theta \tag{3.5}$$

where ${}^{(i)}\Psi$ is the matric potential of liquid in the unsaturated porous medium, and ${}^{(i)}\theta$ is the degree of saturation, as noted previously. The increment of work per unit elementary volume of porous medium is written as an inexact differential to indicate that the path in the ${}^{(i)}\Psi - {}^{(i)}\theta$ plane must be specified (specifying ${}^{(i)}\Psi$ as a given function of ${}^{(i)}\theta$), in order to make ${}^{(i)}\Psi \, d \, {}^{(i)}\theta$ an exact differential before the work involved in going from an initial to a final state, characterized by ${}^{(i)}\Psi_i$, ${}^{(i)}\theta_i$, and ${}^{(i)}\Psi_f$, ${}^{(i)}\theta_f$, respectively, can be calculated by integrating (3.5). Combining (3.1) and (3.5) yields for the first law of thermodynamics as applied to unsaturated flow when hysteresis is not taken into account but dW is considered exact,

$$\delta Q = dU + {}^{(i)}\Psi \, d \, {}^{(i)}\theta \tag{3.6}$$

It is usual, when writing the first law, to recognize the generalized force in the work term, ${}^{(i)}\Psi$ in this case, as an intensive quantity (a quantity not depending on the size of the system) and the generalized displacement as an extensive quantity. Matric potential is an intensive quantity, as is the degree of saturation. However, ${}^{(i)}\theta$ may be multiplied by the bulk volume of the porous medium being considered, V_B, and the porosity of the porous medium, ${}^{(o)}\epsilon$, to convert it to the required extensive quantity, namely the volume of the liquid of species i in the porous medium sample. We will not perform this multiplication here, but instead use ${}^{(i)}\theta$ as the generalized displacement and adopt the convention that the thermodynamic quantities that appear, internal energy, U, in (3.6), for example, and those that follow are understood to be normalized to unit elementary volume of porous medium sample containing fluid. A number of thermodynamic relations may now be derived from (3.6). For example, the heat capacity per unit elementary volume of porous medium at a constant degree of saturation of species i is given by

$$C_{{}^{(i)}\theta} = \left(\frac{\delta Q}{dT} \right)_{{}^{(i)}\theta} = \left(\frac{\partial U}{\partial T} \right)_{{}^{(i)}\theta} \tag{3.7}$$

Similarly, the heat capacity per unit elementary volume of porous medium at constant matric potential of species i is given by

$$C_{{}^{(i)}\Psi} = \left(\frac{\delta Q}{\delta T} \right)_{{}^{(i)}\Psi} = \left(\frac{\partial U}{\partial T} \right)_{{}^{(i)}\Psi} + {}^{(i)}\Psi \left(\frac{\partial \, {}^{(i)}\theta}{\partial T} \right)_{{}^{(i)}\Psi} \tag{3.8}$$

The change in ${}^{(i)}\theta$ with temperature may be written in terms of the isobaric coefficient of thermal expansion of the liquid of species i as follows. The isobaric coefficient of thermal expansion, $^{(i)}\beta_l$, is defined in (3.9), where $^{(i)}V_l$ is the volume of liquid of species i being considered, T is temperature, and P is the pressure external to the liquid with which it is in equilibrium.

$$^{(i)}\beta_l = \frac{1}{^{(i)}V_l} \left(\frac{\partial\, ^{(i)}V_l}{\partial T} \right)_P \tag{3.9}$$

Setting

$$^{(i)}V_l = {}^{(o)}\epsilon\, V_B\, {}^{(i)}\theta \tag{3.10}$$

as discussed, noting that ϵ and V_B are approximately constant relative to θ, so far as temperature variations are concerned, and identifying P, to within an additive constant, with $^{(i)}\Psi$, we have on substituting (3.10) into (3.9),

$$^{(i)}\beta_l - {}^{(o)}\beta_s = \frac{1}{^{(o)}\epsilon\, V_B\, {}^{(i)}\theta} \left(\frac{\partial({}^{(o)}\epsilon\, V_B\, {}^{(i)}\theta)}{\partial T} \right)_{^{(i)}\Psi}$$

$$\simeq -\frac{1}{^{(i)}\theta} \left(\frac{\partial\, ^{(i)}\theta}{\partial T} \right)_{^{(i)}\Psi}$$

$$\simeq -\, {}^{(i)}\beta_l \tag{3.11}$$

where $^{(o)}\beta_s$ is the bulk modulus of the matrix material and for soil is about a tenth of the value for liquid water, $^{(1)}\beta_l$. Thus, on substituting (3.12) into (3.8), we have for $C_{(i)\Psi}$,

$$C_{(i)\Psi} \simeq \left(\frac{\partial U}{\partial T} \right)_{(i)\Psi} + {}^{(i)}\Psi\, {}^{(i)}\beta_l\, {}^{(i)}\theta \tag{3.12}$$

Since hysteresis is not being considered, we can write

$$\delta Q = T\, dS \tag{3.13}$$

where S is the entropy of a soil–water system per unit elementary volume of porous medium and dS is an exact differential. We can then write

$$C_{(i)\theta} = \left(\frac{\delta Q}{\delta T} \right)_{(i)\theta} = T \left(\frac{\partial S}{\partial T} \right)_{(i)\theta} \tag{3.14a}$$

and

$$C_{(i)\Psi} = \left(\frac{\delta Q}{\delta T} \right)_{(i)\Psi} = T \left(\frac{\partial S}{\partial T} \right)_{(i)\Psi} \tag{3.14b}$$

An expression for the difference between $C_{(i)\Psi}$ and $C_{(i)\theta}$ in terms of experimentally available quantities can now be derived as follows. From (3.14a) we have

$$C_{(i)\theta} = T\left(\frac{\partial S}{\partial T}\right)_{(i)\theta} = T\frac{\partial(S,\ ^{(i)}\theta)}{\partial(T,\ ^{(i)}\theta)} = T\frac{\partial(S,\ ^{(i)}\theta)}{\partial(T,\ ^{(i)}\Psi)}\frac{\partial(T,\ ^{(i)}\Psi)}{\partial(T,\ ^{(i)}\theta)}$$

$$= T\frac{\partial(S,\ ^{(i)}\theta)}{\partial(T,\ ^{(i)}\Psi)}\left(\frac{\partial(T,\ ^{(i)}\theta)}{\partial(T,\ ^{(i)}\Psi)}\right)^{-1} \tag{3.15}$$

where the notation for the Jacobian determinant in two dimensions is

$$\det\begin{bmatrix} \frac{\partial(x,y)}{\partial x} & \frac{\partial u(x,y)}{\partial y} \\ \frac{\partial v(x,y)}{\partial x} & \frac{\partial v(x,y)}{\partial y} \end{bmatrix} = \det\begin{bmatrix} \left(\frac{\partial u}{\partial x}\right)_y & \left(\frac{\partial u}{\partial y}\right)_x \\ \left(\frac{\partial v}{\partial x}\right)_y & \left(\frac{\partial v}{\partial y}\right)_x \end{bmatrix} \tag{3.16}$$

in which the notation $u = u(x,y)$ and $v = v(x,y)$ along with the identities

$$\frac{\partial(u,y)}{\partial(x,y)} = \frac{\partial u(x,y)}{\partial x} = \left(\frac{\partial u}{\partial x}\right)_y \tag{3.17a}$$

and

$$\frac{\partial(u,v)}{\partial(x,y)} = \left(\frac{\partial(x,y)}{\partial(u,v)}\right)^{-1} \tag{3.17b}$$

have been used. Applying the identity

$$\frac{\partial(u,v)}{\partial(x,y)} = -\frac{\partial(v,u)}{\partial(x,y)} \tag{3.18}$$

to both the numerator and denominator of the term $[\partial(T,\ ^{(i)}\theta)/\partial(T,\ ^{(i)}\Psi)]^{-1}$ in conjunction with (3.17a) and (3.17b) to yield

$$\left(\frac{\partial(T,\ ^{(i)}\theta)}{\partial(T,\ ^{(i)}\Psi)}\right)^{-1} = \left(\frac{\partial(\ ^{(i)}\theta,T)}{\partial(\ ^{(i)}\Psi,T)}\right)^{-1} = \left(\frac{\partial\ ^{(i)}\theta}{\partial\ ^{(i)}\Psi}\right)_T^{-1} \tag{3.19}$$

gives for (3.15)

$$C_{(i)\theta} = T\left(\frac{\partial\ ^{(i)}\Psi}{\partial\ ^{(i)}\theta}\right)_T\frac{\partial(S,\ ^{(i)}\theta)}{\partial(T,\ ^{(i)}\Psi)}$$

$$= T\left(\frac{\partial\ ^{(i)}\Psi}{\partial\ ^{(i)}\theta}\right)_T \det\begin{bmatrix} \left(\frac{\partial S}{\partial T}\right)_{(i)\Psi} & \left(\frac{\partial S}{\partial\ ^{(i)}\Psi}\right)_T \\ \left(\frac{\partial\ ^{(i)}\theta}{\partial T}\right)_{(i)\Psi} & \left(\frac{\partial\ ^{(i)}\theta}{\partial\ ^{(i)}\Psi}\right)_T \end{bmatrix}$$

$$= T\left(\frac{\partial\ ^{(i)}\Psi}{\partial\ ^{(i)}\theta}\right)_T\left[\left(\frac{\partial S}{\partial T}\right)_{(i)\Psi}\left(\frac{\partial\ ^{(i)}\theta}{\partial\ ^{(i)}\Psi}\right)_T - \left(\frac{\partial S}{\partial\ ^{(i)}\Psi}\right)_T\left(\frac{\partial\ ^{(i)}\theta}{\partial T}\right)_{(i)\Psi}\right] \tag{3.20}$$

where $(\partial\ ^{(i)}\Psi/\partial\ ^{(i)}\theta)_T$ is the slope of the moisture characteristic at constant temperature. Substituting (3.14b) and the Maxwell relation

$$-\left(\frac{\partial S}{\partial\ ^{(i)}\Psi}\right)_T = \left(\frac{\partial\ ^{(i)}\theta}{\partial T}\right)_{(i)\Psi} \tag{3.21}$$

obtained by considering the Gibbs free energy to be a function of T and $^{(i)}\Psi$ (this will be discussed in more detail shortly) into (3.20) yields

$$C_{(i)\theta} = C_{(i)\Psi} + T^{(i)}\beta_l^2 {}^{(i)}\theta^2 \left(\frac{\partial {}^{(i)}\Psi}{\partial {}^{(i)}\theta} \right)_T \qquad (3.22a)$$

or

$$C_{(i)\theta} - C_{(i)\Psi} = T^{(i)}\beta_l^2 {}^{(i)}\theta^2 \left(\frac{\partial {}^{(i)}\Psi}{\partial {}^{(i)}\theta} \right)_T \qquad (3.22b)$$

The slope of the moisture characteristic is in general greater than zero, which is obscured by the way in which it is customarily graphed, so that in (3.22b) $C_{(i)\theta} > C_{(i)\Psi}$, since the other quantities appearing on the right–hand side of that equation are positive as well.

3.1.2 Legendre Transforms and Thermodynamic Potentials

The usual thermodynamic potentials, namely the Gibbs free energy, G, the Helmholtz free energy, F, and the enthalpy, H, can be obtained from the internal energy function U by appropriate Legendre transformations, which we now discuss.

We consider a function of three independent variables (the generalization to more independent variables is obvious) $f(x, y, z)$, the total differential of which is

$$df(x, y, z) = \frac{\partial f}{\partial x} dx + \frac{\partial f}{\partial y} dy + \frac{\partial f}{\partial z} dz \qquad (3.23a)$$

If we set

$$u = \frac{\partial f}{\partial x}, v = \frac{\partial f}{\partial y}, w = \frac{\partial f}{\partial z} \qquad (3.23b)$$

the total differential of f becomes

$$df = u \, dx + v \, dy + w \, dz \qquad (3.23c)$$

Three factorial (3!) new functions can now be derived from f, each of which has as its independent variables one or more of the u, v, w and any remaining x, y, z. For example, considering the total differential of a new function g defined as $f - ux$ yields

$$\begin{aligned} dg &= d(f - ux) = u \, dx + v \, dy + w \, dz - x \, du - u \, dx \\ &= -x \, du + v \, dy + w \, dz \end{aligned} \qquad (3.24a)$$

Since, from (3.24a) the partial derivatives of g are

$$\frac{\partial g}{\partial u} = -x \qquad (3.24b)$$

$$\frac{\partial g}{\partial y} = v \qquad (3.24c)$$

$$\frac{\partial g}{\partial z} = w \qquad (3.24d)$$

so that

$$g = g(u, y, z) \qquad (3.24e)$$

We see that one of the partial derivatives of f has become one of the independent variables on which g depends and that the negative of the conjugate independent variable on which f depends, x in this case, is now the partial derivative of g with respect to the partial derivative of f with respect to the conjugate variable. This process can be carried out for each original partial-derivative–independent-variable pair, and each choice yields a new function with its own set of independent variables. In a given application, such as equilibrium thermodynamics, the initial partial-derivative–independent-variable pairs to be operated on depend on the requirements of the problem being considered. We now consider the application of this formalism to equilibrium thermodynamics.

The fundamental relation fully characterizing the unsaturated porous medium thermodynamic system of interest is the internal energy function $U(S, {}^{(1)}\theta, {}^{(0)}X, {}^{(2)}X, \ldots, {}^{(N_s)}X)$, normalized to unit elementary volume of porous medium in this case, where the ${}^{(i)}X$ with $i \neq 1$ due to the explicit labeling of degree of saturation of a single liquid component of interest, water, for example, as ${}^{(1)}\theta$, are the numbers of mols (mol numbers or gram-molecular weights) of the ith species of the N_s constituents of the system, and S is the entropy of the system per unit elementary volume of porous medium as previously defined. The internal energy is an extensive quantity, and partial derivatives of it are intensive quantities. Thus, from the combined first and second laws and the functional dependence of U on the mol numbers we have

$$T = \left(\frac{\partial U}{\partial S} \right)_{{}^{(1)}\theta,\, {}^{(i)}X} , \qquad i \neq 1 \qquad (3.25a)$$

$$- {}^{(1)}\Psi = \left(\frac{\partial U}{\partial {}^{(1)}\theta} \right)_{S, {}^{(i)}X} \quad , \qquad i \neq 1 \qquad (3.25\text{b})$$

$$ {}^{(i)}\mu = \left(\frac{\partial U}{\partial {}^{(i)}X} \right)_{ {}^{(1)}\theta, S} \quad , \qquad i \neq 1 \qquad (3.25\text{c})$$

where the ${}^{(i)}\mu$ are the chemical potentials (here normalized to unit elementary volume of porous medium) of the various components of the system. Using (3.25a)–(3.25c), we write the total differential of U as

$$dU = T\, dS - {}^{(1)}\Psi\, d\, {}^{(1)}\theta + \sum_{i=0}^{N_s} {}^{(i)}\mu\, d\, {}^{(i)}X \quad , \qquad i \neq 1 \qquad (3.26)$$

The Legendre transform operation can now be carried out to produce additional thermodynamic potential functions, the utility of which depends on the problem being considered. Note that the intensive parameters appearing in (3.26), namely T, ${}^{(1)}\Psi$, and the ${}^{(i)}\mu$, which are the partial derivatives of U with respect to the mol numbers on which they depend, will become the independent variables(s) of the thermodynamic potential functions obtained from U by the application of the Legendre transformation formalism. By way of example, we consider a Legendre transform of (3.26) such that both T and ${}^{(1)}\Psi$ become the independent variables in the new formalism. Thus we write

$$
\begin{aligned}
dG &= d(U - TS + {}^{(1)}\Psi\, {}^{(1)}\theta) \\
&= T\, dS - {}^{(1)}\Psi\, d\, {}^{(1)}\theta - T\, dS - S\, dT \\
&\quad + {}^{(1)}\Psi\, d\, {}^{(1)}\theta + {}^{(1)}\theta\, d\, {}^{(1)}\Psi + \sum_{i=0}^{N_s} {}^{(i)}\mu\, d\, {}^{(i)}X \\
&= -S\, dT + {}^{(1)}\theta\, d\, {}^{(1)}\Psi + \sum_{i=0}^{N_s} {}^{(i)}\mu\, d\, {}^{(i)}X \quad , \qquad i \neq 1 \qquad (3.27)
\end{aligned}
$$

where

$$G = U - TS + {}^{(1)}\Psi\, {}^{(1)}\theta \qquad (3.28)$$

is called the Gibbs free energy of the system, here normalized to unit elementary volume of porous medium, and represents the work available in a reversible process from a system at constant temperature and matric po-

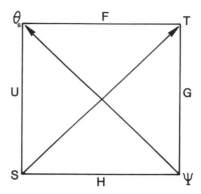

Figure 3.1: Mnemonic diagram in which the thermodynamic potentials F (Helmholtz free energy), G (Gibbs free energy), H (enthalpy), and U (internal energy) lie between their "natural" independent variables, namely those that maximize their information content.

tential. Note that for a chemical system, which, by definition, is one in which (3.2) represents the work term in the first law,

$$G = U - TS + PV \qquad (3.29)$$

which is the more familiar form. Other Legendre transforms of U are possible, and their computation will be explored in the exercises. We note finally that the Legendre transform preserves the information content of the original state functions in the form(s) of the transformed functions.

A device for displaying the set of total differentials that arises when the various possible Legendre transformations on $U(S, {}^{(1)}\theta)$, with the mol numbers of the other constituents of the system held constant, are carried out can be adapted for unsaturated flow from one given by Max Born in 1929 and is given in Figure 3.1. The total differentials of the thermodynamic potentials at constant mol numbers (for $i \neq 1$) are read from the diagram as follows. The total differential of a thermodynamic potential function depends linearly on the sum of the differentials of its "natural" (information maximizing) dependent variables, which occupy the corners of the diagram adjacent to it. The quantities multiplying these total differentials are connected to them by the arrows. If the arrow points away from a variable the total differential of which is being considered, a plus sign is used. If the arrow points toward a total differential variable, a minus sign

is used. Applying this sign convention yields for the total differentials of
the thermodynamic potentials

$$dU = TdS - {}^{(1)}\Psi \, d \, {}^{(1)}\theta + \sum_{i=0}^{N_s} {}^{(i)}\mu \, d \, {}^{(i)}X \quad , \quad i \neq 1 \tag{3.30a}$$

$$dF = -S \, dT - {}^{(1)}\Psi \, d \, {}^{(1)}\theta + \sum_{i=0}^{N_s} {}^{(i)}\mu \, d \, {}^{(i)}X \quad , \quad i \neq 1 \tag{3.30b}$$

$$dG = -S \, dT + {}^{(i)}\theta \, d \, {}^{(1)}\Psi + \sum_{i=0}^{N_s} {}^{(i)}\mu \, d \, {}^{(i)}X \quad , \quad i \neq 1 \tag{3.30c}$$

$$dH = T \, dS + {}^{(1)}\theta \, d \, {}^{(1)}\Psi + \sum_{i=0}^{N_s} {}^{(i)}\mu \, d \, {}^{(i)}X \quad , \quad i \neq 1 \tag{3.30d}$$

in which at constant mol numbers the $d \, {}^{(i)}X$ would equal zero so that the
third term in each of the expressions (3.30a)–(3.30d) would vanish.

3.1.3   Maxwell Relations

As mentioned, the Maxwell relations are a set of second derivative relations
among the thermodynamic variables that can be derived from the first
derivative relations expressed in the total differentials in (3.30a)–(3.30d).
Consider the function (3.23a) and note that in general, since the order in
which the partial derivatives of a function are taken to yield a given result
can be arbitrary

$$\frac{\partial}{\partial y} \left(\frac{\partial f}{\partial x} \right) = \frac{\partial}{\partial x} \left(\frac{\partial f}{\partial y} \right) \tag{3.31a}$$

so that, in the notation of (3.23b)

$$\frac{\partial u}{\partial y} = \frac{\partial v}{\partial x} \tag{3.31b}$$

with similar relations possible for the other independent variables on which
f depends. Applying this relation to (3.30a), for example, yields, at constant mol numbers,

$$\frac{\partial}{\partial \, {}^{(1)}\theta} \left(\frac{\partial U}{\partial S} \right) = \frac{\partial}{\partial S} \left(\frac{\partial U}{\partial \, {}^{(1)}\theta} \right) \tag{3.32a}$$

or

$$\frac{\partial T}{\partial \, {}^{(1)}\theta} = - \frac{\partial \, {}^{(1)}\Psi}{\partial S} \tag{3.32b}$$

which is customarily written in thermodynamics as

$$\left(\frac{\partial T}{\partial \,^{(1)}\theta}\right)_S = -\left(\frac{\partial \,^{(1)}\Psi}{\partial S}\right)_{^{(1)}\theta} \tag{3.32c}$$

where the subscripts outside the parentheses indicate which of the original independent variables are held constant when the indicated differentiations are carried out. Similar relations can be derived from (3.30b)–(3.30d), and are summarized in Figure 3.1 as follows. We consider the corners of the square only and move counterclockwise to write, in order, the quantity being differentiated, the quantity with respect to which differentiation is carried out, and the quantity being held constant. This result is set equal to the partial derivative that results when the variables are written in the same way except that one starts with the variable immediately clockwise of the variable differentiated in the counterclockwise traverse around the square, and goes around clockwise. The sign in front of each derivative term is determined as follows. If the quantity being held constant is at the tail of a diagonal arrow, a plus sign is used. If the quantity being held constant in the differential expression is at the head of one of the diagonal arrows, a negative sign is used. We illustrate this process by starting with $^{(1)}\theta$ in the upper left-hand corner of the square of Figure 3.1. Going counterclockwise, we have $(\partial \,^{(1)}\theta/\partial S)_{^{(1)}\Psi}$ with a plus sign since $^{(1)}\Psi$ is at the tail of a diagonal arrow. For the clockwise traverse we begin with T, the first quantity clockwise to $^{(1)}\theta$, and write $(\partial T/\partial \,^{(1)}\Psi)_S$, which again is positive since S is at the tail of an arrow. Thus we have, as could be derived directly from (3.30d) in the usual way,

$$\left(\frac{\partial \,^{(1)}\theta}{\partial S}\right)_{^{(1)}\Psi} = \left(\frac{\partial T}{\partial \,^{(1)}\Psi}\right)_S \tag{3.33a}$$

Similarly, beginning with S in Figure 3.1, we have

$$-\left(\frac{\partial S}{\partial \,^{(1)}\Psi}\right)_T = \left(\frac{\partial \,^{(1)}\theta}{\partial T}\right)_{^{(1)}\Psi} \tag{3.33b}$$

which can be derived directly from (3.30c). Next we have

$$-\left(\frac{\partial \,^{(1)}\Psi}{\partial T}\right)_{^{(1)}\theta} = -\left(\frac{\partial S}{\partial \,^{(1)}\theta}\right)_T \tag{3.33c}$$

or

$$\left(\frac{\partial \,^{(1)}\Psi}{\partial T}\right)_{^{(1)}\theta} = \left(\frac{\partial S}{\partial \,^{(1)}\theta}\right)_T \tag{3.33d}$$

which can be derived from (3.30b). Finally we have

$$\left(\frac{\partial T}{\partial \,^{(1)}\theta}\right)_S = -\left(\frac{\partial \,^{(1)}\Psi}{\partial S}\right)_{^{(1)}\theta} \tag{3.33e}$$

which was derived in the usual way from (3.30a).

An interesting application of the Maxwell relation formalism is to derive the energy relation that in the context of the current work provides an expression for the change in the internal energy of the soil–water system per unit volume of porous medium, U, with degree of saturation at constant temperature. We begin by writing

$$\left(\frac{\partial U}{\partial\,^{(1)}\theta}\right)_T = \frac{\partial(U,T)}{\partial(\,^{(1)}\theta, T)} = \frac{\partial(U,T)}{\partial(S,\,^{(1)}\theta)}\,\frac{\partial(S,\,^{(1)}\theta)}{\partial(\,^{(1)}\theta, T)}$$

$$= -\frac{\partial(S,\,^{(1)}\theta)}{\partial(T,\,^{(1)}\theta)}\,\frac{\partial(U,T)}{\partial(S,\,^{(1)}\theta)}$$

$$= -\left(\frac{\partial S}{\partial T}\right)_{(1)\theta}\det\left[\begin{array}{cc}\left(\frac{\partial U}{\partial S}\right)_{(1)\theta} & \left(\frac{\partial U}{\partial\,^{(1)}\theta}\right)_S \\ \left(\frac{\partial T}{\partial S}\right)_{(1)\theta} & \left(\frac{\partial T}{\partial\,^{(1)}\theta}\right)_S\end{array}\right] \quad (3.34)$$

Recalling that $U = U(S,\,^{(1)}\theta)$ and writing

$$dU = \left(\frac{\partial U}{\partial S}\right)_{(1)\theta} dS + \left(\frac{dU}{\partial\,^{(1)}\theta}\right)_S d\,^{(1)}\theta \quad (3.35a)$$

we note on comparison with (3.30a) that

$$\left(\frac{\partial U}{\partial S}\right)_{(1)\theta} = T \quad (3.35b)$$

and that

$$\left(\frac{\partial U}{\partial\,^{(1)}\theta}\right)_S = -\,^{(1)}\Psi \quad (3.35c)$$

Expanding the determinant in (3.34), substituting (3.35b) and (3.35c), recalling

$$\left(\frac{\partial x}{\partial y}\right)_z = \left(\frac{\partial y}{\partial x}\right)_z^{-1} \quad (3.36a)$$

so that

$$\left(\frac{\partial S}{\partial T}\right)_{(1)\theta} = \left(\frac{\partial T}{\partial S}\right)_{(1)\theta}^{-1} \quad (3.36b)$$

and rearranging terms yields for (3.34)

$$\left(\frac{\partial U}{\partial\,^{(1)}\theta}\right)_T = -T\left(\frac{\partial S}{\partial T}\right)_{(1)\theta}\left(\frac{\partial T}{\partial\,^{(1)}\theta}\right)_S - \,^{(1)}\Psi \quad (3.37)$$

Recalling the general result that

$$\left(\frac{\partial x}{\partial y}\right)_z \left(\frac{\partial y}{\partial z}\right)_x \left(\frac{\partial z}{\partial x}\right)_y = -1 \tag{3.38a}$$

so that, on applying (3.36a)

$$\left(\frac{\partial S}{\partial T}\right)_{(1)\theta} \left(\frac{\partial T}{\partial \,{}^{(1)}\theta}\right)_S = -\left(\frac{\partial S}{\partial \,{}^{(1)}\theta}\right)_T \tag{3.38b}$$

and recalling from (3.33d) that

$$\left(\frac{\partial S}{\partial \,{}^{(1)}\theta}\right)_T = \left(\frac{\partial \,{}^{(1)}\Psi}{\partial T}\right)_{(1)\theta}$$

we have on substituting (3.38b) and (3.33d) into (3.37)

$$\left(\frac{\partial U}{\partial \,{}^{(1)}\theta}\right)_T = T\left(\frac{\partial \,{}^{(1)}\Psi}{\partial T}\right)_{(1)\theta} - \,{}^{(1)}\Psi \tag{3.39}$$

which is the desired result.

3.1.4 Gibbs–Duhem Relation

We now consider the integrated form of the various thermodynamic potentials as follows, beginning with the internal energy U. Since U is an extensive quantity, we have for a chemical system

$$U(\lambda S, \lambda V, \lambda^{(1)}X, \lambda^{(2)}X, \dots, \lambda^{(N_s)}X) = \lambda U(S, V, {}^{(1)}X, \dots, {}^{(N_s)}X) \tag{3.40a}$$

where λ is an arbitrary multiplier, and differentiating both sides with respect to λ yields,

$$\frac{\partial U}{\partial(\lambda S)}\frac{\partial(\lambda S)}{\partial \lambda} + \frac{\partial U}{\partial(\lambda V)}\frac{\partial(\lambda V)}{\partial \lambda} + \frac{\partial U}{\partial(\lambda\,{}^{(1)}X)}\frac{\partial(\lambda\,{}^{(1)}X)}{\partial \lambda} + \cdots$$

$$= \lambda U(S, V, {}^{(1)}X, \cdots {}^{(N_s)}X) \tag{3.40b}$$

Setting $\lambda = 1$ yields

$$\frac{\partial U}{\partial S}S + \frac{\partial U}{\partial V}V + \frac{\partial U}{\partial \,{}^{(1)}X}\,{}^{(1)}X + \cdots + \frac{\partial U}{\partial \,{}^{(N_s)}X}\,{}^{(N_s)}X$$

$$= U(S, V, {}^{(1)}X, \cdots {}^{(N_s)}X) \tag{3.41}$$

We recall that

$$\frac{\partial U}{\partial S} = T \tag{3.42a}$$

$$\frac{\partial U}{\partial V} = -P \tag{3.42b}$$

$$\frac{\partial U}{\partial \, ^{(i)}X} = \, ^{(i)}\mu \tag{3.42c}$$

so that on substituting (3.40a) and (3.40b) and (3.42a) and (3.42b) into (3.41) we have

$$ST - PV + \sum_{i=1}^{N_s} \, ^{(i)}\mu \, ^{(i)}X = U \tag{3.43}$$

which is the integrated form of the fundamental relation for a chemical system. The usual form of the Gibbs–Duhem relation, which relates the Legendre-transformed combined first and second laws to the differentials of the chemical potentials can be derived from (3.26), with $P \, dV$ substituted for $\, ^{(1)}\Psi \, d \, ^{(1)}\theta$, and (3.43) as follows. Computing the total differential of (3.43) yields

$$T \, dS + S \, dT - P \, dV - V \, dP + \sum_{i=1}^{N_s} \, ^{(i)}\mu \, d \, ^{(i)}X + \sum_{i=1}^{N_s} \, ^{(i)}X \, d \, ^{(i)}\mu = dU \tag{3.44a}$$

Combining (3.50a) and (3.26) yields

$$S \, dT - V \, dP + \sum_{i=1}^{N_s} \, ^{(i)}X \, d \, ^{(i)}\mu = 0 \tag{3.44b}$$

or

$$-S \, dT + V \, dP = \sum_{i=1}^{N_s} \, ^{(i)}X \, d \, ^{(i)}\mu \tag{3.45}$$

which is the desired (Gibbs–Duhem) relation.

3.1.5   Equilibrium of a Thermodynamic System

We now derive the condition for thermodynamic equilibrium of parts of a system with each other, namely the equality of the chemical potentials of the parts of the system (phases), using the formalism of Lagrange multipliers. We consider a system made up of N_s constituents and M_p phases, where the term *phase* is not taken to mean state of matter (solid, liquid, gas, or plasma) only, but can also refer to a salt–ice mixture, surface phase, etc. in which the constituents do not react chemically with each other. We further assume that all N_s constituents are present in all M_p phases. The Gibbs free energy for such a system is given by

$$G = \sum_{i=1}^{N_s} \sum_{j=1}^{M_p} {}^{(i)}X_j \, {}^{(i)}\mu_j \tag{3.46}$$

where ${}^{(i)}X_j$ is the number of mols of constituent i in phase j and ${}^{(i)}\mu_j$ is the chemical potential–Gibbs free energy per mol–of the ith constituent in the jth phase.

The Gibbs free energy of a chemical system is a function of the temperature and pressure of the system, and for a system in equilibrium at constant temperature and pressure, the total differential of G, dG, equals zero. The N_s equations of constraint, one for each species, are

$$_1g = \sum_{j=1}^{M_p} {}^{(1)}X_j = {}_1c \tag{3.47a}$$

$$_2g = \sum_{j=1}^{M_p} {}^{(2)}X_j = {}_2c \tag{3.47b}$$

$$\vdots$$

$$_{N_s}g = \sum_{j=1}^{M_p} {}^{(N_s)}X_j = {}_{N_s}c \tag{3.47c}$$

where the $_ic$ are constants. Since we have assumed that no chemical reactions occur, the only possible changes in the ${}^{(i)}X_j$ are transfers of one or more constituents from one phase to another. (The pressure-induced melting leading to the release of minerals and consequent increases in the concentration of chemical constituents in the water at the base of a glacier is an example of this.) Thus, using the Lagrange multiplier formalism of Appendix B, we have

$$dG + d\sum_{i=1}^{N_s} {}^{(i)}\lambda \, {}_ig = 0 \tag{3.48}$$

or on substituting (3.46) and (3.47) into (3.48)

$$d\left(\sum_{i=1}^{N_s} \sum_{j=1}^{M_p} {}^{(i)}X_j \, {}^{(i)}\mu_j + {}^{(i)}\lambda \, {}^{(i)}X_j \right) = 0 \tag{3.49a}$$

or

$$\sum_{i=1}^{N_s} \sum_{j=1}^{M_p} d^{(i)}X_j({}^{(i)}\mu_j + {}^{(i)}\lambda) = 0 \qquad (3.49b)$$

where the $^{(i)}\lambda$ are the Lagrange undetermined multipliers for the system. Since changes in the $^{(i)}X_j$ can occur, $d\ ^{(i)}X_j$ is not necessarily equal to zero. Thus, the only way in which (3.49b) is guaranteed to hold is if

$$^{(i)}\mu_j = -\ {}^{(i)}\lambda \qquad (3.50a)$$

for all i and j, that is

$$-\ {}^{(1)}\lambda = {}^{(1)}\mu_1 = {}^{(1)}\mu_2 = {}^{(1)}\mu_3 \cdots = {}^{(1)}\mu_{M_p} \qquad (3.50b)$$

$$-\ {}^{(2)}\lambda = {}^{(2)}\mu_1 = {}^{(2)}\mu_2 = {}^{(2)}\mu_3 \cdots = {}^{(2)}\mu_{M_p} \qquad (3.50c)$$

$$\vdots$$

$$-\ {}^{(N_s)}\lambda = {}^{(N_s)}\mu_1 = {}^{(N_s)}\mu_2 = {}^{(N_s)}\mu_3 \cdots = {}^{(N_s)}\mu_{M_p} \qquad (3.50d)$$

Equations (3.50a)–(3.50d) express the important physical fact that for a system in equilibrium at constant temperature and pressure, the chemical potential of a given constituent has the same value in every phase. For example, for water in an unsaturated porous medium, the chemical potential of the water vapor is the same as the chemical potential of the water held in the pores by capillary forces. This fact will be the starting point in a later subsection for the derivation of the equation relating the vapor pressure of the water vapor in a porous medium to the matric potential of the water in the medium–the basic equation of the thermocouple psychrometer given as Eq. (1.1).

3.2   APPLICATIONS OF THERMODYNAMICS

3.2.1   Thermodynamic Properties of Liquid Water

The water molecule, H_2O, is shown in the form of a "ball" and stick" model in Figure 3.2. The pressure–volume diagram for water near its triple point is shown in Figure 3.3. The pressure-volume-temperature surface for water is shown in Figure 3.4. The critical point of water occurs at a pressure of about 218.3 atm $\sim 215.5 \times 10^5 \mathrm{nt/m^2}$ and a temperature of about 374.15 °C. These values are well above those encountered in unsaturated flow field situations, and so the behavior of water in the vadose zone near the critical point will not be considered further.

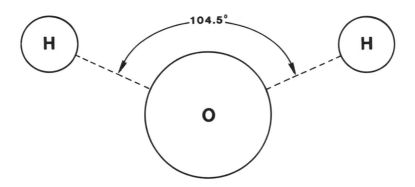

Figure 3.2: A "ball and stick" depiction of a water molecule. The hydrogen–oxygen bond length is 0.95718×10^{-8} cm.

A possible exception to the above would be the pressure–temperature conditions that could exist in the vicinity of "waste" nuclear fuel rods emplaced in fractured tuff near the water table. Even there, however, it is not anticipated that the unsaturated hydraulic conductivities would be so low as to contain the water vapor generated from water in the soil by heat from the fuel rods and thus raise the pressure of the vapor to the vicinity of the critical point.

These properties of water are all for situations in which the water surfaces being considered are sufficiently large so that surface shape effects related to interfacial tension (the term *surface tension* refers to the interfacial tension of a pure substance in contact with its own vapor) are not involved. The air–water interfacial tension of water as a function of temperature is shown in Figure 3.5. (The triple point of water is the combination of pressure and temperature at which it exists simultaneously as a liquid, a vapor, and a solid.) An empirical representation of the surface tension of water is (Zemansky, 1957, p. 291)

$$\gamma = \gamma_o \left(1 - \frac{T}{T_o} \right)^{m_e} \tag{3.51}$$

where γ_o is the surface tension of water at 0 °C, 75.5 dynes/cm, $T_o = 368$ °C, $0\,°C \le T \le 368\,°C$, and $m_e = 1.2$.

The energy "stored" in a liquid water surface, the surface energy, is not the same as its surface tension, but can be computed as follows. For

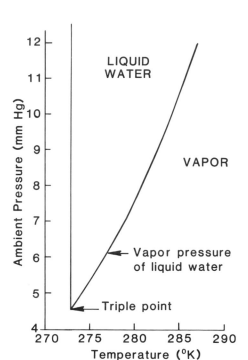

Figure 3.3: Pressure–temperature diagram in the vicinity of the triple point. The triple point temperature is 0.0098 °C and the triple point pressure is 4.58 mm Hg.

a chemical system the work term is given, by definition, by (3.2) so that at constant mol numbers (constant composition) the combined first and second laws become

$$dU = T \, dS - P \, dV \tag{3.52}$$

If S is now viewed as a function of T and V, we have

$$dS = \left(\frac{\partial S}{\partial T} \right)_V dT + \left(\frac{\partial S}{\partial V} \right)_T dV \tag{3.53a}$$

so that

$$T \, dS = T \left(\frac{\partial S}{\partial T} \right)_V dT + T \left(\frac{\partial S}{\partial V} \right)_T dV \tag{3.53b}$$

Recalling that

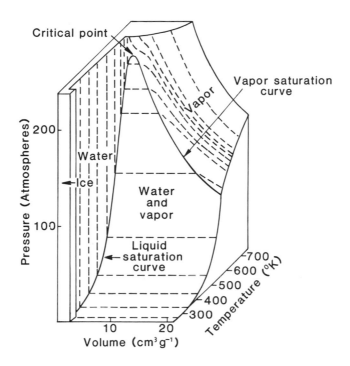

Figure 3.4: Pressure–volume–temperature surface for water (after Eisenberg and Kauzmann, 1969, p. 59).

$$\delta Q = T \, dS \tag{3.13}$$

we have

$$\left(\frac{\delta Q}{dT}\right)_V = C_V = T\left(\frac{dS}{dT}\right)_V \tag{3.54}$$

where C_V is the heat capacity of the system at constant volume. Legendre transforming (3.52) on S yields for the total differential of the Helmholtz free energyfor a chemical system

$$dF = -S \, dT - P \, dV \tag{3.55}$$

From (3.53) we find the Maxwell relation

$$\left(\frac{\partial P}{\partial T}\right)_V = \left(\frac{\partial S}{\partial V}\right)_T \tag{3.56}$$

Using (3.54) and (3.56) in (3.53b) yields

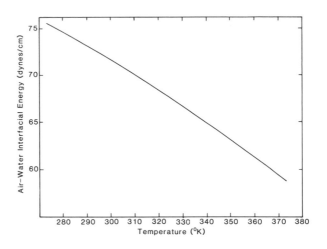

Figure 3.5: Air-water interfacial tension as a function of temperature.

$$T \, dS = C_V \, dT + T \left(\frac{\partial P}{\partial T} \right)_V dV \qquad (3.57)$$

This is called the first $T \, dS$ equation and is written for a chemical system. Carrying out a similar development when the work term is

$$dW = \gamma \, dA \qquad (3.3)$$

yields for the $T \, dS$ equation for a system in which surface tension is being considered

$$T \, dS = C_A \, dT - T \left(\frac{\partial \gamma}{\partial T} \right)_A dA \qquad (3.58)$$

where C_A is the heat capacity at constant surface area. Writing $\delta Q = T \, dS$ in (3.58) and assuming an isothermal process yields

$$\delta Q = -T \left(\frac{\partial \gamma}{\partial T} \right)_A dA \qquad (3.59)$$

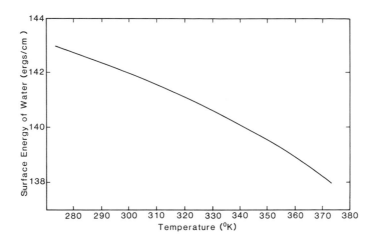

Figure 3.6: Surface energy for a water surface of infinite radius of curvature ("flat") as a function of temperature.

The first law of thermodynamics written for this system is

$$dU = \delta Q + \gamma \, dA \tag{3.60}$$

so that using (3.59a) with (3.60) and rearranging yields for the surface energy per unit area as a function of temperature

$$\frac{dU}{dA} = \gamma - T \left(\frac{\partial \gamma}{\partial T} \right)_A \tag{3.61}$$

which is plotted for water in Figure 3.6 and is the total energy per unit area stored in the surface when it is formed. Both the boiling point (the temperature at which the vapor pressure at a water surface is equal to the ambient atmospheric pressure) of water and its freezing point change, irrespective of chemical solution effects, from the measured flat surface values when water is placed in a porous medium. The vapor pressure of water in a porous medium is in general lower at a given temperature than when the water surface is "flat" as quantified for each meniscus surface by the generalized Kelvin relation given in Eq. (2.95).

From the negative slope of the fusion curve of Figure 3.3, we see qualitatively that since the pressure of water under a concave meniscus is at a lower pressure than water under a surface of infinite radius of curvature at a given temperature, if the temperature of the water in the soil is somewhat above the triple point temperature, lowering the pressure sufficiently will put the water into the ice I vapor region without freezing it. If the temperature of the water is above 0 °C but below the triple point temperature, then lowering the pressure will move the water in the direction of the ice I region, effectively increasing the freezing temperature but in no case increasing it beyond the triple point temperature of 0.0098 °C.

3.2.2   Impurity Concentration at a Liquid Surface

As an example of the use of (3.45), we now derive the expression for the difference between the concentration of an impurity in the surface of a solvent from that in the bulk and the concomitant change in the surface tension of the solvent with bulk concentration of impurities. The relevant work term in the first law is, from (3.3), $dW = \gamma\,dA$, so that

$$dU = T\,dS + \gamma\,dA + \sum_{i=1}^{2} {}^{(i)}\mu\,d\,{}^{(i)}X \tag{3.62a}$$

which becomes in this case

$$dU = T\,dS + \gamma\,dA + {}^{(2)}\mu\,d(A\,{}^{(2)}C) \tag{3.62b}$$

with

$${}^{(1)}X = A\,{}^{(1)}C \tag{3.63a}$$

$${}^{(2)}X = A\,{}^{(2)}C \tag{3.63b}$$

where ${}^{(1)}C$ is the concentration of the solvent, water in this case, in the surface phase, in mols per unit area, ${}^{(2)}C$ is the concentration of solute in the surface phase, above or below the concentration of the solute in the bulk, ${}^{(1)}\mu$ is the chemical potential of the solvent, ${}^{(1)}X \simeq 0$, and the rest of the quantities are as previously defined. Writing the Gibbs–Duhem relation for this case yields

$$-S\,dT - A\gamma = \sum_{j=1}^{2} (A\,{}^{(j)}C)\,d\,{}^{(j)}\mu \tag{3.64}$$

where ${}^{(2)}\mu$ is the chemical potential of the solute, which becomes, for constant temperature and interfacial area A

$$-d\gamma = {}^{(1)}C\,d\,{}^{(1)}\mu + {}^{(2)}C\,d\,{}^{(2)}\mu \tag{3.65}$$

If we choose the location of the imaginary surface separating the surface phase from the bulk phase to be such that the bulk phase contains a uniform concentration of solvent, we have $d\,^{(1)}\mu = 0$ (the change in $^{(1)}\mu$ is small compared to that of $^{(2)}\mu$), and (3.65) becomes

$$-d\gamma = \,^{(2)}C\ d\,^{(2)}\mu \qquad (3.66)$$

but

$$d\,^{(2)}\mu = \left(\frac{\partial\,^{(2)}\mu}{\partial\,^{(2)}n}\right) d\,^{(2)}n = \frac{R_g T}{^{(2)}n}\,d\,^{(2)}n \qquad (3.67)$$

where $^{(2)}n$ is the concentration of the solute in the bulk of the solution, R_g is the ideal gas constant written in terms of the same concentration units in which $^{(2)}C$ is measured, and T is (absolute) temperature in degrees Kelvin so that on combining (3.66) and (3.67) and rearranging, we have

$$^{(2)}C = \frac{-\,^{(2)}n}{R_g T}\frac{\partial\gamma}{\partial\,^{(2)}n} \qquad (3.68)$$

The minus sign in (3.68) shows that if the change of γ with $^{(2)}n$ is such that γ decreases as $^{(2)}n$ increases, so that if $\partial\gamma/\partial\,^{(2)}n$ is negative, then $^{(2)}C > 0$ and the impurity tends to concentrate in the surface layer. Conversely, if γ increases as $^{(2)}n$ increases, then $^{(2)}C < 0$ and the concentration of the solute in the surface layer is below that of the bulk.

Various empirical and semiempirical equations for the change in interfacial energy with changes in solution composition have been proposed to fit the results of specific experiments or sets of experiments but will not be discussed here.

3.2.3 Matric Potential at Constant Temperature

We now consider the derivation of the relation between total matric potential and the vapor pressure of a single fluid, water, for example, in the pores of a porous medium, the psychrometer relation. The Gibbs free energy for a chemical system is given by $G = U - TS + PV$, as mentioned. The total differential of G is

$$dG = V\,dP - S\,dT = \left(\frac{\partial G}{\partial P}\right)_T dP + \left(\frac{\partial G}{\partial T}\right)_P dT \qquad (3.69)$$

and we write g_l as the Gibbs free energy per unit mass of liquid in the pore space of the medium, v_l as the volume per unit mass of the liquid in the pore space of the medium, s_l as the entropy per unit mass of liquid in the pore space of the medium, and similarly define g_v, v_v, and s_v as the analogous quantities per unit mass of vapor in the porous medium. (We recall that by convention vapor above the critical temperature is called a gas.) We then have by application of (3.69) to the liquid and vapor in the porous medium respectively,

$$dg_l = v_l \, dp_l - s_l \, dT_l = \left(\frac{\partial g_l}{\partial p_l}\right)_{T_l} dp_l + \left(\frac{\partial g_l}{\partial T_l}\right)_{p_l} dT_l \qquad (3.70a)$$

and

$$dg_v = v_v \, dp_v - s_v \, dT_v = \left(\frac{\partial g_v}{\partial p_v}\right)_{T_v} dp_v + \left(\frac{\partial g_v}{\partial T_v}\right)_{p_v} dT_v \qquad (3.70b)$$

at constant temperature $dT_l = dT_v = 0$ and, in general, the condition for equilibrium, whether the temperature is constant or not is

$$dg_l = dg_v \qquad (3.71)$$

so that from these conditions and (3.70a) and (3.70b) we have

$$v_l \, dp_l = v_v \, dp_v \qquad (3.72)$$

Noting that ρ_l, the mass density of the liquid, is

$$\rho_l = \frac{1}{v_l} \qquad (3.73)$$

and that the mass density of the vapor, assuming it obeys the ideal gas law, is given by

$$\frac{1}{v_v} = \rho_v = \frac{mp_v}{kT} \qquad (3.74)$$

where m is the mass of a vapor molecule, p_v is the pressure of the vapor, k is Boltzmann's constant, and T is the absolute temperature in degrees Kelvin. Substituting (3.73) and (3.74) into (3.72) and indicating integration results in

$$\int_{p_{0l}}^{\bar{p}_l} \frac{dp_l}{\rho_l} = \frac{kT}{m} \int_{p_{0v}}^{\bar{p}_v} \frac{dp_v}{p_v} \qquad (3.75)$$

where $\bar{p}_l$ is the average pressure of the liquid in the medium under the curved water menisci, p_{0l} is the pressure of the liquid at the same temperature and containing the same chemical impurities under a flat surface (infinite radius of curvature), $\bar{p}_v$ is the average vapor pressure of the vapor in the porous medium, and p_0 is the pressure of the vapor over a flat water surface, where the water of the reference solution contains the same concentrations of impurities and is at the same temperature as the water in the porous medium. Noting that ρ_l is essentially constant under the range of pressures being considered, carrying out the integrations indicated in (3.75), identifying $\bar{p}_l - p_{0l}$ with Ψ, the total matric potential, and adding the superscript 1 for water, as done previously, yields

$$^{(1)}\Psi = {}^{(1)}\bar{p}_l - {}^{(1)}p_{0l} = \frac{\rho_l kT}{m} \ln \left(\frac{{}^{(1)}\bar{p}_v}{{}^{(1)}p_{0v}} \right) = {}^{(1)}\Psi_0 \ln \left(\frac{{}^{(1)}\bar{p}_v}{{}^{(1)}p_{0v}} \right) \quad (3.76)$$

where

$$^{(1)}\Psi_0 = \frac{\rho_l kT}{m} \quad (1.1a)$$

as noted previously.

3.3 DILUTE SOLUTIONS IN POROUS MEDIA

There are four effects on the flow of liquid through unsaturated porous media that occur when impurities dissolve in the water in the pores of the porous medium to form a dilute, nonelectrolytic solution. They are (1) the change in impurity concentration in meniscus surface layers discussed and quantified in Eq. (3.68), (2) the lowering of the vapor pressure of a dilute solution below that of the pure solvent which vapor pressure replaces $^{(1)}p_{0v}$ of Eq. (1.1a) in the computation of matric potential, (3) the equilibrium concentration of dissolved impurities in soil water changes with applied pressure, specifically under the capillary menisci formed at air–soil solution interfaces, and (4) the change in solution concentration with height in a gravitational field, a consequence of the thermodynamic requirement that solution equilibrium is equilibrium with respect to all external fields. This last modifies the matric potential via effects based on items (1)–(3). Aspects of the equilibrium chemistry of dilute solutions will be discussed first followed by discussion of items (2)–(4) in turn.

3.3.1 Equilibrium Chemistry of Dilute Solutions

Up to now, except for the calculation leading to Eq. (3.68), we have considered the fluid in a porous medium to be a pure substance containing no impurities. However, for example, water in soil does contain impurities–at the very least those dissolved from the pore/capillary walls. The nature and concentrations of these dissolved impurities depend on the soil type, itself a reflection of the relative amounts of minerals present, which, along with temperature, affect the observed vapor pressure of the soil solution vapor in a given geometrical porous medium situation. The approximation of a dilute solution, to be defined thermodynamically, holds, generally speaking, for water in soils, and even the water in a soil paste or a lake sediment contains higher concentrations of impurities than does soil water.

We consider first a few aspects of dilute solution theory without regard to the change in equilibrium solution concentration that occurs due to the lowering of solution pressure under a concave meniscus. We begin by considering the lowering of the vapor pressure of a dilute solution below that of its solute. To the extent that non–electrolytic ideal solutions containing a number of solutes are considered, the net lowering of the vapor pressure of a given solute in a given geometric situation is to a first approximation the sum of the reductions in vapor pressure produced by each solute. The thermodynamic definition of a dilute solution is that the chemical potential of each constituent takes the form

$$^{(i)}\mu = {}^{(i)}g + R_g T \ln \left(\frac{^{(i)}X}{X} \right) = {}^{(i)}g + R_g T \ln({}^{(i)}n) \qquad (3.77)$$

where

$$X = \sum_{i=1}^{N_s} {}^{(i)}X \qquad (3.78)$$

so that $^{(i)}X/X = {}^{(i)}n$ is the mol fraction of the ith constituent, and $^{(i)}g$ is the Gibbs free energy per mol of the ith constituent in its pure state. Note that the $^{(i)}\mu$ depend on all of the $^{(i)}X$, one of which is the solvent, as well as on T and P for a chemical system, whereas $^{(i)}g$ depends only on T and P. We recall that the condition for equilibrium of the ith constituent in the liquid and vapor phases simultaneously is

$$^{(i)}\mu_v = {}^{(i)}\mu_l \qquad (3.79)$$

where $^{(i)}\mu_v$ is the chemical potential of the ith constituent in the vapor phase and $^{(i)}\mu_l$ is the chemical potential of the ith constituent in the liquid phase. Thus, if we identify $i = 1$ with the solvent, water in this case, and $i = 2$ with a single soil mineral solute, we have, assuming that the solute is nonvolatile, that is, that water vapor is the only vapor present,

$$^{(1)}\mu_v = {}^{(1)}\mu_l = {}^{(1)}g_l + R_g T \ln \left(1 - \frac{^{(2)}X}{X} \right) \qquad (3.80)$$

or, since water vapor only is (assumed) present

$$^{(1)}g_v + R_g T \ln \left(\frac{^{(1)}X}{X} \right) = {}^{(1)}g_l + R_g T \ln \left(1 - \frac{^{(2)}X}{X} \right)$$

$$= {}^{(1)}g_l + R_g T \ln(1 - {}^{(2)}n) \qquad (3.81)$$

where $^{(1)}g_v$ and $^{(1)}g_l$ are Gibbs free energies per mol of pure water vapor and liquid, respectively, $^{(2)}n$ is the mol fraction of component 2, and for a single constituent, $^{(1)}X$ in this case, $X = {}^{(1)}X$ and $\ln\left({}^{(1)}X/{}^{(1)}X\right) = 0$. Taking the total differential of (3.81) at constant temperature yields

$$d({}^{(1)}g_v - {}^{(1)}g_l) = R_g T \, d\left[\ln(1 - {}^{(2)}n)\right] \tag{3.82}$$

or, using (3.70a) and (3.70b)

$$({}^{(1)}v_v - {}^{(1)}v_l)\, dP = R_g T \, d[\ln(1 - {}^{(2)}n)] \tag{3.83}$$

where dP is the change in pressure of either the vapor or the liquid, $^{(1)}v_v$ is the volume per mol of the water vapor, and $^{(1)}v_l$ is the volume per mol of the liquid. Assuming that the water vapor behaves as an ideal gas and $^{(1)}v_l$ is constant, we have on substituting in (3.83) and indicating integration,

$$\int_{P_{0v}}^{P_{sv}} \left(\frac{R_g T}{P_v'} - {}^{(1)}v_l\right) dP_v' = -R_g T \int_0^{{}^{(2)}n} \frac{d\,{}^{(2)}n'}{1 - {}^{(2)}n'} \tag{3.84}$$

where P_{0v} is the vapor pressure over a flat water surface when the water contains no solute, and P_{sv} is the vapor pressure over the surface when the water contains a mol fraction of solute $^{(2)}n$. Carrying out the indicated integration and rearranging terms yields

$$R_g T \, \ln\left(\frac{P_{sv}}{P_{0v}}\right) = R_g T \, \ln(1 - {}^{(2)}n) + {}^{(1)}v_l(P_{sv} - P_{0v}) \tag{3.85}$$

If we neglect the second term on the right as being proportional to a small difference, then

$$\frac{P_{sv}}{P_{0v}} \simeq 1 - {}^{(2)}n \tag{3.86a}$$

or

$$\frac{P_{0v} - P_{sv}}{P_{0v}} = {}^{(2)}n \tag{3.86b}$$

which is Raoult's law for the change in the vapor pressure over a flat water surface when impurities sufficient to form a dilute solution are added to the water. Rearranging (3.85) to yield

$$\frac{P_{sv}}{P_{0v}} = 1 - \beta \, \ln(1 - {}^{(2)}n) + \beta \, \ln\left(\frac{P_{sv}}{P_{0v}}\right) \tag{3.87a}$$

where

$$\beta = \frac{R_g T}{{}^{(1)}v_l P_{0v}} \tag{3.87b}$$

yields a form suitable for Lagrange's expansion. This is seen by recalling that if

$$\zeta = \alpha + \beta\Phi(\zeta) \tag{3.88a}$$

then

$$f(\zeta) = f(\alpha) + \sum_{n=1}^{\infty} \frac{\beta^n}{n!} \frac{d^{n-1}}{d\alpha^{n-1}} \{f'(\alpha)[\Phi(\alpha)]^n\} \tag{3.88b}$$

and making the identifications $\zeta = P_{sv}/P_{ov}$, $\Phi(\zeta) = \ln(\zeta)$, $\alpha = 1 - \beta\ln(1 - {}^{(2)}n)$, $\beta = \beta$ in (3.88a), and $f(\alpha) = \alpha$ in (3.88b) so that on substituting into (3.88b) we have

$$\zeta = \alpha + \sum_{n=1}^{\infty} \frac{\beta^n}{n!} \frac{d^{n-1}}{d\alpha^{n-1}} [\ln(\alpha)^n] \tag{3.89}$$

Thus, the first-order correction for P_{sv}/P_{ov} is

$$\frac{P_{sv}}{P_{ov}} = 1 - \frac{R_g T}{{}^{(1)}v_l P_{ov}} \ln(1 - {}^{(2)}n)$$

$$+ \left(\frac{R_g T}{{}^{(1)}v_l P_{ov}}\right) \left(\ln\left(1 - \frac{R_g T}{{}^{(1)}v_l P_{ov}} \ln(1 - {}^{(2)}n)\right)\right) \tag{3.90}$$

The important physical consideration is that since the expression for the matric potential depends on the ratio $\bar{p}_v/p_{ov}$, a change in solution chemistry changes the value of p_{ov} to P_{sv}, as given in (3.86a) or (3.90). If impurity concentrations vary from place to place in the area of interest, then gradients of Ψ can result.

We now derive the relation expressing the increase in the boiling point of a dilute nonelectrolytic solution over that of the pure solvent. Computing the total differential $d({}^{(1)}g_v - {}^{(1)}g_l)$ of (3.82) at constant pressure rather than at constant temperature yields

$$(- {}^{(1)}s_v + {}^{(1)}s_l) \, dT = R_g \ln(1 - {}^{(2)}n) \, dT + R_g T \, d[\ln(1 - {}^{(2)}n)] \tag{3.91}$$

where ${}^{(1)}s_v$ and ${}^{(1)}s_l$ are the entropies per mol of the water vapor and liquid water, respectively. Substituting $R_g \ln(1 - {}^{(2)}n)$ from (3.81) into (3.91) yields

$$(- {}^{(1)}s_v - {}^{(1)}s_l) \, dT = \frac{{}^{(1)}g_v - {}^{(1)}g_l}{T} \, dT + R_g T d \left[\ln(1 - {}^{(2)}n)\right] \tag{3.92}$$

We now recall that

$$g = u - Ts + Pv = h - Ts \tag{3.93}$$

where u is the internal energy per mol, v is the volume per mol, and h is the enthalpy per mol. Writing (3.93) for the liquid and vapor phases of water, respectively, yields

$$^{(1)}g_l = {}^{(1)}h_l - T\,{}^{(1)}s_l \tag{3.94a}$$

and

$$^{(1)}g_v = {}^{(1)}h_v - T\,{}^{(1)}s_v \tag{3.94b}$$

Combining (3.94a) and (3.93) with (3.92) and rearranging yields

$$0 = \frac{{}^{(1)}h_v - {}^{(1)}h_l}{T_g} \, dT + R_g T \, d[\ln(1 - {}^{(2)}n)] \tag{3.95}$$

Recalling that

$$\ln(1 - x) = -x - \frac{x^2}{2} - \frac{x^3}{3} - \cdots \tag{3.96}$$

where $0 < x < 1$ we have

$$d[\ln(1 - {}^{(2)}n)] = -d\,{}^{(2)}n - {}^{(2)}n \, d\,{}^{(2)}n - \cdots \tag{3.97}$$

so that (3.95) becomes, on identifying ${}^{(1)}h_v - {}^{(1)}h_l$ with the latent heat of vaporization of water, $l_v(T)$, when the water surface is flat, and noting that ${}^{(2)}n \ll 1$,

$$\frac{-l_v(T)}{R_g T^2} \frac{dT}{d\,{}^{(2)}n_2} \simeq -1 \tag{3.98}$$

or

$$\frac{\Delta T_v}{\Delta\,{}^{(2)}n} \simeq \frac{R_g T^2}{l_v(T)} \tag{3.99}$$

which is the increase in the vaporization temperature of the solution over that of the pure solute as the concentration of the solute is increased by an amount $\Delta\,{}^{(2)}n$ so that the temperature at which the solution vaporizes is given by

$$T_s = T + \Delta T_v \tag{3.100}$$

If a dilute solution is in equilibrium with pure ice, we have, similarly to (3.80), that

$$^{(1)}\mu_l = {}^{(1)}g_l + R_g T \, \ln(1 - {}^{(2)}n) \tag{3.101a}$$

so that

$$^{(1)}\mu_s = {}^{(1)}\mu_l = {}^{(1)}g_l + R_g T \, \ln(1 - {}^{(2)}n) \tag{3.101b}$$

where ${}^{(1)}\mu_s$ and ${}^{(1)}\mu_l$ are the chemical potentials of component 1, water, in the solid and liquid phases, respectively. Since we have assumed that the ice is pure,

$$^{(1)}g_s = {}^{(1)}\mu_s \tag{3.102}$$

so that

$$^{(1)}g_s = {}^{(1)}g_l + R_g T \, \ln(1 - {}^{(2)}n) \qquad (3.103)$$

It can be shown by carrying out a series of steps similar to those resulting in (3.95) that

$$-\frac{{}^{(1)}h_s - {}^{(1)}h_l}{T} \, dT = R_g T \, d[\ln(1 - {}^{(2)}n)] \qquad (3.104)$$

and by identifying $^{(1)}h_l - {}^{(1)}h_s$ as the latent heat of fusion, $l_f(T)$, of pure water, that

$$\frac{\Delta T_f}{\Delta \, {}^{(2)}n} \simeq -\frac{R_g T^2}{l_f(T)} \qquad (3.105)$$

where ΔT_f is the amount by which the freezing point of the dilute solution is below that of pure water, so that the new freezing point is

$$T_f = T - \Delta T_f \qquad (3.106)$$

3.3.2   Pressure Dependence of Solution Concentration

We now develop the relation between the chemical concentration of a dilute solution, viewed as an open system, and the pressure applied to the solution by an external agency. We begin with the total differential of the internal energy for a chemical system, namely

$$dU = T \, dS - P \, dV - m_s h \, dg + \sum_{i=1}^{N_s} \mu_i \, d\nu_i \qquad (3.107)$$

where μ_i is here the chemical potential per particle instead of per mol of species i, ν_i, in the notation of Gibbs, is the number of particles of type i being considered, N_s is the number of different types of particles, m_s is the fictitious "average" mass of a soil solution "particle" that is a distance h above an arbitrary datum, g is the acceleration due to gravity, and dg is the change of g with height above the arbitrary datum. We consider the case in which $N_s = 2$, a solvent, water, for example, and a single solute. In this case the number of molecules of water in the system, ν_1, is approximately constant, and ν_2 is the number of molecules of solute, not necessarily conserved; that is, the system is open and can exchange particles with its surroundings, which in this case means that solute can adsorb onto pore/capillary walls and thus be removed from solution, or be dissolved into solution from the walls and thus be added to the solution so that the pore/capillary walls act as the particle reservoir for the solute. Equation (3.107) thus becomes

$$dU \simeq T\,dS - P\,dV + \mu_2\,d\nu_2 \qquad (3.108)$$

Legendre transforming U of (3.108) on P and μ_2 yields

$$d\left(U + PV - \mu_2\nu_2\right) = dY = T\,dS + V\,dP - \nu_2\,d\mu_2 \qquad (3.109)$$

where Y has been set equal to the derived thermodynamic potential function $U + PV - \mu_2\nu_2$. The Maxwell relation connecting P and ν_2 is, from (3.109),

$$\left(\frac{\partial \nu_2}{\partial P}\right)_{S,\,^{(2)}\mu} = -\left(\frac{\partial \nu_2}{\partial V}\right)_{S,P} \qquad (3.110)$$

where P is the pressure in the water in a capillary just below its meniscus, ν_2 is the number of molecules of solute there, V is the volume of soil water just below the meniscus, and $^{(2)}\mu$ is the chemical potential of the solute. Note that the Legendre transform leading to (3.110) has cast $^{(2)}\mu$ into the role of an independent variable and ν_2 into the role of a dependent variable given by

$$\nu_2 = -\left(\frac{\partial Y}{\partial\,^{(2)}\mu}\right)_{S,P} \qquad (3.111)$$

Additional familiarity with the idea of an open system and the pressure dependence of solution concentration can be gained by examining this situation from the viewpoint of statistical mechanics, to which we now turn.

3.4 EQUILIBRIUM STATISTICAL MECHANICS

Statistical mechanics is the discipline concerned with forming equation sets describing the interactions at the microscopic–atomic and molecular–level of macroscopic systems–crystals, gases, liquids, etc., and carrying out suitable averages–ensemble averages so-called–which are averages over the possible values of the coordinates and momenta that the individual atoms and molecules of which the macroscopic system is composed can have by averaging over a mentally pictured collection of systems-the ensemble–infinite in number–which are macroscopic replicas of the given system, to yield macroscopic equilibrium thermodynamic properties of the system. The assumptions made in constructing the microscopic model of the system and the choice of ensemble averaging function are determined by the nature of the system being considered. The function giving the relative probability of a given macroscopic system having particular values of the coordinates and momenta of the particles comprising it can be written as $f(\vec{q}, \vec{p}, t)$. It can be shown that if $f(\vec{q}, \vec{p}, t)$ is to be independent of time $f(\vec{q}, \vec{p}) = f_e(E)$, where E is the total energy of the macroscopic system and its ensemble average value, $\bar{E}$, is the internal energy, U, of the system in the sense of equilibrium thermodynamics. Ensemble averages of quantities depending on coordinates and momenta, $F(\vec{q}, \vec{p})$, for example, can be written as

$$\overline{F} = \frac{\int\int F(\vec{q}, \vec{p})\, f(\vec{q}, \vec{p})\, d\vec{q}\, d\vec{p}}{\int\int f(\vec{q}, \vec{p})\, d\vec{q}\, d\vec{p}} \tag{3.112}$$

where the denominator of (3.112) is called the *partition function* and the specific choices of the function $f_e(E)$ are determined by the assumptions made as to the macroscopic properties of the systems making up ensembles of particular kinds. These in turn determine the specific relations of the partition function of each type of ensemble to thermodynamics.

We first discuss the so-called microcanonical ensemble, which is a mental collection of systems macroscopically identical to the one being considered, as is the case for all ensembles. Each of the systems comprising this ensemble is assumed to have constant energy E, constant volume V, and the set of particles of each species present, denoted by $[\nu]$, is taken to be constant as well. Thus, $f_e(E)$ is written for this case as $f_{me}(E) = $ constant for E in the range $E_0 \leq E \leq E_0 + \delta E$ (here δ denotes a small change rather than an inexact differential) and $f_{me}(E) = 0$ otherwise, where f_{me} stands for the distribution function appropriate to the microcanonical ensemble. The thermodynamic functions directly calculable from the partition function $\Omega(\nu, V, E)$, the integral over f_{mc} as noted above, in the microcanonical ensemble, are

$$S = k \ln\{\Omega([\nu], V, E)\} \tag{3.113a}$$

$$\frac{1}{T} = \left(\frac{\partial(k \ln\{\Omega([\nu], V, E)\})}{\partial E}\right)_{V,[\nu]} = \left(\frac{\partial S}{\partial E}\right)_{V,[\nu]} \tag{3.113b}$$

$$\frac{P}{T} = \left(\frac{\partial\{k \ln[\Omega(\nu, V, E)]\}}{\partial V}\right)_{E,[\nu]} = \left(\frac{\partial S}{\partial V}\right)_{E,[\nu]} \tag{3.113c}$$

$$-\frac{\mu_i}{T} = \left(\frac{\partial(k \ln\{\Omega([\nu], V, E)\})}{\partial\nu_i}\right)_{E,V,[\nu_j]} \tag{3.113d}$$

where k is Boltzmann's constant. In Eq. (3.113d) $^{(i)}\mu$ is the chemical potential of the particles of species i, and the numbers of particles of all other species, the set $[\nu_j]$, where $i \neq j$, are held constant in the differentiation. Due to the requirement that the systems being considered each have constant energy, to within a small amount, δE, the partition function in the microcanonical ensemble proves difficult to calculate in practice.

The most frequently used partition function is the partition function for the canonical ensemble. It is assumed that of the systems composing this ensemble, each has constant volume, constant temperature, and constant numbers each species of particles of which the system is composed. In this case $f_e(E)$ takes the form

$$f_{ce}(E) = A_c e^{-E/kT} \tag{3.114}$$

where f_{ce} denotes the distribution function for the the canonical ensemble, A_c is a suitable normalizing constant for the distribution and can be shown to equal $e^{-F/kT}$, where F is the Helmholtz free energy when f_{ce} is normalized to one, and the rest of the symbols are as previously defined. The partition function itself is the integral over $e^{-E/kT}$ so that, on denoting this partition function by $Q(T, V, [\nu])$ we have

$$F = -kT \ln [Q(T, V, [\nu])] \tag{3.115}$$

Since the systems considered in both the microcanonical and canonical ensembles have constant numbers of particles, neither is suitable for the open system that we wish to consider. What is required for the problem at hand is a partition function for an ensemble of open systems–ones that can exchange particles with an external "reservoir," so that the number of particles the system contains is a dependent variable, and, for the case at hand, one for which the pressure of the system is an independent variable as well. In statistical mechanics, the usual partition function used to describe an open system is the partition function of the grand canonical ensemble. The thermodynamic potential function, PV, calculated directly from the partition function in the grand canonical ensemble, Ξ, is derived by carrying out a double Legendre transform of the internal energy on S and ν. This partition function, therefore, is for systems having as independent variables volume and temperature as well as chemical potential, the independent variable corresponding to the (variable) number of particles, but not the pressure as required in this case. In order to construct the appropriate partition function, we sum (or integrate as appropriate) over the partition functions in the microcanonical ensemble, each constructed for a different number of particles and volume as follows (see Hill, 1958).

$$e^{\chi/kT} = W = \sum_{\nu_i} \sum_V \Omega(\nu_i, V, E) e^{^{(i)}\mu\nu_i/kT} e^{-pV/kT} \tag{3.116}$$

where χ/T is the double Legendre transform of the entropy S, instead of the internal energy U, on ν_i and V as was given in (3.109) above and W is the required partition function. Thus,

$$d\left(\frac{\chi}{T}\right) = \frac{1}{T}\,dE - \frac{V}{T}\,dP + \frac{1}{T}\sum_{i=1}^{N_s} \nu_i\,d^{(i)}\mu \qquad (3.117a)$$

and, since the temperature is assumed constant

$$d\chi = dE - V\,dP + \sum_{i=1}^{N_s} \nu_i\,d^{(i)}\mu \qquad (3.117b)$$

Thus,

$$\left(\frac{\partial \chi}{\partial E}\right)_{P,\,{}^{(i)}\mu} = 1 \qquad (3.118a)$$

$$\left(\frac{\partial \chi}{\partial P}\right)_{E,\,{}^{(i)}\mu} = -V \qquad (3.118b)$$

$$\left(\frac{\partial \chi}{\partial\,{}^{(i)}\mu}\right)_{E,P} = \nu_i \qquad (3.118c)$$

From (3.116)

$$\chi = kT\ln(W) \qquad (3.119)$$

and thus, from (3.118b), (3.118c) and (3.119) we have

$$V = -kT\left(\frac{\partial \ln(W)}{\partial P}\right)_{E,\,{}^{(i)}\mu} \qquad (3.120a)$$

$$\bar{\nu}_i = kT\left(\frac{\partial \ln(W)}{\partial\,{}^{(i)}\mu}\right)_{E,P} \qquad (3.120b)$$

where $\bar{\nu}_i$ is the average number of particles of species i in the system. Note that while χ may be set equal to E, its explicit dependence on ${}^{(i)}\mu$ and P in place of ν_i and V means that it is actually a different function than the E [or U of (3.109)] of a chemical system and so should be given a distinct symbol. Thus, in the unsaturated porous medium systems of interest, N_s chemical species are dissolved in the water in the soil and their origin is material either adsorbed on the walls of the pores/capillaries containing the water, or material comprising the walls. Thus, the walls of the pores/capillaries containing the soil water act as the "particle reservoir" for the "open system" in the sense of statistical mechanics, that in this case is the soil water in the pore space of the medium. Equilibrium in this case means that the chemical potential of the species i in solution, ${}^{(i)}\mu$, is the same as the chemical potential for that species on/in the wall. The change in the number of particles in equilibrium in the solution as a function of pressure has been given in (3.110) when a Legendre transform of the internal energy on the particle numbers of the species being exchanged is considered. From (3.117b) we have

$$\left(\frac{\partial \nu_i}{\partial P}\right)_{E,\,^{(i)}\mu} = -\left(\frac{\partial V}{\partial\,^{(i)}\mu}\right)_{E,P} \tag{3.121}$$

which expresses the change in the number of particles of species i in solution being exchanged with the material of the pore/capillary walls containing the soil solution as a function of pressure in what is called the entropy representation rather than the energy representation of (3.110). The pressure lowering under a concave meniscus is thus seen to affect the equilibrium chemistry of that part of the soil solution of which it is composed. Whether this represents an increase or decrease in the concentration of a given species depends on the species and can depend on which other species are present as well. In order to apply these considerations to elucidate the chemical equilibrium of soil solution in a capillary, a mathematical model of the binding energies of the molecules in the minerals making up the capillary walls and the interactions with the water molecules in the soil solution is required. This form of E would then be inserted into (3.116), the partition function computed, and (3.121) applied. This procedure amounts to predicting equilibrium solution chemistry of water in the presence of soil minerals, and, because it can in principle include geometric effects of the porous medium on the equilibrium chemistry of the soil solution, it represents an extension of the usual numerical chemical equilibrium models customarily applied to aqueous soil solutions that do not take the geometrical effects of the porous medium on the equilibrium solution chemistry into account. This application of statistical mechanics would also represent its extension to a class of problems not usually considered.

In order to provide further insight into statistical mechanics as such and pave the way for certain of the vapor transport considerations of Chapter IV, we now consider an ideal gas of identical particles with no internal degrees of freedom using successively the formalism of the microcanonical ensemble, the canonical ensemble, and the ensemble of Eq. (3.116).

Since an ideal gas is assumed to be have no potential energy of interaction between its molecules, the solute molecules of a dilute solution can be treated in a similar way, and it is this circumstance that leads to the ideal gas form of equation for the osmotic pressure that exists after the osmosis of a solute through a semipermeable membrane has occurred. This lends additional relevance to the ideal gas considerations to be given using statistical mechanics.

For a microcanonical ensemble of systems of such particles, we have, where the constant value of f_{me} has been removed from the integrand and its form taken into account by the requirement that the integrals over coordinates and momenta be performed subject to the energy range condition on E, and the quantity $h^{3\nu}$ appearing in the denominator on the right hand side of (3.122) normalizes the integral to a unit volume in phase space,

$$\Omega(\nu, V, E) = \frac{1}{\nu! h^{3\nu}} \int_{E-\delta E}^{E} \cdots \int d\vec{p}\, d\vec{q} \qquad (3.122)$$

where ν is the number of particles in the system (solute particles for example), h is Planck's constant, equal to 6.625×10^{-34} J sec, the range of integration is over all values of generalized coordinates, $\vec{q}$, and generalized momenta, $\vec{p}$, consistent with the systems making up the ensemble having a constant energy to within a range of energy δE, and

$$d\vec{p} = dp_{1x}\, dp_{1y}\, dp_{1z}\, dp_{2x}\, dp_{2y}\, dp_{2z} \cdots dp_{\nu x}\, dp_{\nu y}\, dp_{\nu z} \qquad (3.123a)$$

$$d\vec{q} = dq_{1x}\, dq_{1y}\, dq_{1z}\, dq_{2x}\, dq_{2y}\, dq_{2z} \cdots dq_{\nu x}\, dq_{\nu y}\, dq_{\nu z} \qquad (3.123b)$$

where the subscript ix, for example, refers to the x component of coordinates or momenta of particle i. Since the system is enclosed in volume V, we have

$$V = \int dq_{ix}\, dq_{iy}\, dq_{iz} \qquad (3.124)$$

where $1 \leq i \leq \nu$, so that the integrals over the coordinates of the ν particles become V^ν and the partition function becomes

$$\Omega(\nu, V, E) = \frac{V^\nu}{\nu! h^{3\nu}} \int_{E-\delta E}^{E} \cdots d\vec{p} \qquad (3.125)$$

Since the energy of the particles is kinetic (energy of motion) only–no potential energy of interaction (which is normally coordinate dependent) by assumption, then the energy, E_i of the ith particle is

$$E_i = \frac{p_i^2}{2m} \qquad (3.126a)$$

where m is the mass of the particle and p_i is its momentum, its mass multiplied by its velocity, v_i, and hence a vector quantity, and

$$p_i^2 = p_{ix}^2 + p_{iy}^2 + p_{iz}^2 \qquad (3.126b)$$

where p_{ix} is the x component of p_i, and similarly for p_{iy} and p_{iz}. Thus, the constant value of energy of the system, E, is given by

$$E = \sum_{i=1}^{\nu} E_i = \sum_{i=1}^{\nu} \frac{p_i^2}{2m} \tag{3.127}$$

For any given energy, the "length" of the corresponding momentum coordinates is

$$p_i = (2mE_i)^{1/2} \tag{3.128}$$

and in the 3ν-dimensional momentum hyperspace (a hyperspace is a space of more than three dimensions) the "distance" of the constant energy surface of the microcanonical ensemble from the origin of momentum coordinates is $(2mE)^{1/2}$. Thus, the region of integration in (3.122) is the hypervolume in momentum space between the energy surfaces of constant E and $E - \delta E$. The volume, V_n, of the n-dimensional sphere of radius R_n is (Courant, 1947, pp. 302–4)

$$V_n = \frac{2\pi^{n/2}(R_n)^n}{n \, \Gamma\left(\frac{n}{2}\right)} = \frac{\pi^{n/2}(R_n)^n}{\Gamma\left(\frac{n+2}{2}\right)} \tag{3.129}$$

Identifying p_i with R_n, n with 3ν, and the value of $\int_{E-\delta E}^{E}$ with the value of the integrand evaluated at the limits, since δE is taken to be small, yields

$$\Omega(\nu, V, E) = \frac{V^{\nu}\pi^{3\nu/2}}{\nu! h^{3\nu} \Gamma\left(\frac{3\nu+2}{2}\right)} \left\{ (2mE)^{3\nu/2} - [2m(E - \delta E)]^{3\nu/2} \right\}$$

$$= \frac{V^{\nu}(2\pi m E)^{3\nu/2}}{\nu! h^{3\nu} \Gamma\left(\frac{3\nu}{2} + 1\right)} \left[1 - \left(1 - \frac{\delta E}{E}\right)^{3\nu/2} \right] \tag{3.130}$$

We now recall the first term of Stirling's asymptotic formula for $\Gamma(ax + b)$ which is

$$\Gamma(ax + b) \sim \sqrt{2\pi} c^{-ax}(ax)^{ax+b-\frac{1}{2}} \tag{3.131a}$$

which, for $a = 1$ and $b = 1$, becomes

$$\Gamma(x + 1) \sim x^x e^{-x} \sqrt{2\pi x} \tag{3.131b}$$

Thus, since for x an integer, ν, for example,

$$\Gamma(\nu + 1) = \nu! \tag{3.131c}$$

so that from (3.131b) and (3.131c) we have,

$$\ln(\nu!) \simeq \nu \ln(\nu) - \nu + \frac{1}{2}\ln(2\pi\nu) \simeq \nu \ln(\nu) - \nu \tag{3.132a}$$

and

$$\ln\left[\Gamma\left(\frac{3\nu}{2}+1\right)\right] \sim \frac{3\nu}{2}\ln\left(\frac{3\nu}{2}\right) - \frac{3\nu}{2} \tag{3.132b}$$

Thus, with an eye toward computing the various thermodynamic quantities of (3.113a)–(3.113d), we compute $\ln[\Omega(\nu, V, E)]$ to the accuracy of Stirling's approximation to yield

$$\ln[\Omega(\nu, V, E)] = \nu \ln\left(\frac{V(2\pi mE)^{3/2}}{\nu^{5/2}h^3(\frac{3}{2})^{3/2}}\right) + \frac{5\nu}{2} \tag{3.133}$$

From (3.113a), with rearrangement of (3.133) we have for the entropy of an ideal gas

$$S = \nu k \ln\left(\frac{Ve(2\pi mE)^{3/2}}{\nu^{5/2}h^3(\frac{3}{2})^{3/2}}\right) + \frac{3\nu k}{2} \tag{3.134}$$

where e is the base of natural logarithms. Applying (3.113b) to (3.134) yields upon rearrangement

$$E = \frac{3\nu kT}{2} \tag{3.135}$$

which expresses the equipartition of energy of an ideal gas of particles among its degrees of freedom. Substituting (3.135) into (3.134) yields

$$S = \nu k \ln\left(\frac{Ve(2\pi mkT)^{3/2}}{\nu h^3}\right) + \frac{3\nu k}{2} = \nu k \ln\left(\frac{Ve}{\nu\lambda^3}\right) + \frac{3\nu k}{2} \tag{3.136}$$

where

$$\lambda = \frac{h}{(2\pi mkT)^{1/2}} \tag{3.137}$$

is the thermal de Broglie wavelength of a particle of mass m at absolute temperature T and (3.136) is the Sackur–Tetrode equation for the entropy of an ideal gas. Applying (3.113c) to (3.134) yields

$$\frac{P}{T} = \frac{\partial}{\partial V}\left[\nu k \ln\left(\frac{e}{\nu\lambda^3}\right) + \nu k \ln(\nu)\right] = \frac{\nu k}{V} \tag{3.138a}$$

or

$$PV = \nu kT \tag{3.138b}$$

which is the ideal gas law as well as being the expression for the osmotic pressure, P, of ν particles in a solution of volume V after osmosis has taken place.

We now consider the ideal gas in the canonical ensemble. The partition function in the canonical ensemble may be thought of as a sum or integral over partition functions in the microcanonical ensemble, each having a different constant energy multiplied by an energy-dependent exponential factor. Thus, we write for the partition function in the canonical ensemble of systems each containing ν particles at constant volume and temperature,

$$Q(T, V, \nu) = \frac{1}{\nu! h^{3\nu}} \int \int \cdots \int e^{-E(\vec{p}, \vec{q})/kT} \, d\vec{p} \, d\vec{q} \qquad (3.139)$$

where all the symbols have been previously defined and the integral over microcanonical ensembles is carried out by extending the requirement of constant energy to be one of constant energy for each ensemble with different energies for different ensembles and integrating over all energies consistent with the regions of phase space $(\vec{p}, \vec{q})$ accessible to the system. For an ideal gas of particles, the $\vec{q}$ integrations again yield V^{ν}, and the 3ν-fold p integrations over all energies become,

$$\int \int \cdots \int e^{-E(\vec{p}, \vec{q})/kT} \, d\vec{p} = \left(\int_{-\infty}^{\infty} e^{-p^2/2mkT} \, dp \right)^{3\nu} = (2\pi mkT)^{3\nu/2}$$

$$(3.140)$$

Thus, the partition function for an ideal gas of identical particles having no internal degree of freedom is, in the canonical ensemble,

$$Q(T, V, \nu) = \frac{V^{\nu} (2\pi mkT)^{3\nu/2}}{\nu! h^{3\nu}} \qquad (3.141)$$

Treating $\ln(\nu!)$ as in (3.132a) above we have for the Helmholtz free energy of an ideal gas

$$F = -\nu kT \ln \left(\frac{(2\pi mkT)^{3/2} V e}{\nu h^3} \right) = -\nu kT \ln \left(\frac{V e}{\nu \lambda^3} \right) \qquad (3.142)$$

where the symbols are as previously defined. Recalling that for a chemical system

$$P = -\left(\frac{\partial F}{\partial V} \right)_{T, \nu} \qquad (3.143a)$$

we have

$$P = -(-\nu kT) \frac{\partial}{\partial V} \left[\ln \left(\frac{e}{\nu \lambda^3} \right) + \ln(V) \right] = \frac{\nu kT}{V} \qquad (3.143b)$$

once again the equation of state of an ideal gas. Further recalling that

$$S = -\left(\frac{\partial F}{\partial T} \right)_{V, \nu} \qquad (3.144)$$

we have upon differentiating (3.136) and rearrangement

$$S = \frac{3}{2}\nu k - \frac{F}{T} \tag{3.145}$$

Recalling that for a chemical system at constant numbers of particles

$$F = E - TS \tag{3.146}$$

we combine (3.145) and (3.146) and recover (3.135), the energy per particle for ν particles of an ideal gas.

We now consider the ideal gas using the ensemble of Eq. (3.116). We begin with a double sum (or integral as appropriate) over the partition function in the microcanonical ensemble using weighting factors depending exponentially on the volume and the number of particles of the systems making up the microcanonical ensembles as given in (3.114) and thus write

$$W = e^{\chi/kT}$$

$$= \sum_{\nu_1=0}^{m_1} \sum_{\nu_2=0}^{m_2} \cdots \sum_{\nu_m=0}^{m_m} \frac{e^{[\,^{(1)}\mu\nu_1 + \,^{(2)}\mu\nu_2 + \cdots + \,^{(m)}\mu\nu_m]/kT}}{\nu_1!\nu_2!\cdots\nu_m!}$$

$$\times \int e^{-pV/kT}\Omega(\nu_i, V, E)\,dV \tag{3.147}$$

where m_i is the total number of particles of species ν_i in each of the systems making up that ensemble, where $1 \leq i \leq n$, p is the pressure of each system in the ensemble, the same for all systems, V is the volume of each of the systems making up a given microcanonical ensemble, and the rest of the symbols are as previously defined. For a set of ensembles composed of systems of one species of particles with no internal degrees of freedom treated as an ideal gas, $n = 1$, and from (3.133) we have

$$\Omega(\nu_i, V, E) = \left(\frac{Ve(2\pi mE)^{3/2}}{\nu_i^{5/2}h^3\left(\frac{3}{2}\right)^{3/2}}\right)^{\nu_i} e^{3\nu_i/2} \tag{3.148}$$

Since a single species of particle is being considered, we can write $E = 3\nu kT/2$ and find

$$\Omega(\nu, V, E) = \left(\frac{Ve^{5/2}(2\pi mkT)^{3/2}}{\nu h^3}\right)^{\nu} \tag{3.149}$$

where the subscript on ν_i has been dropped since only one species of particle is being considered and the partition function, W, for an open system at constant temperature (energy) and pressure becomes

$$W = \sum_{\nu=0}^{N_s} \frac{e^{\mu\nu/kT} A^{\nu}}{\nu! h^{3\nu} \nu^{\nu}} \int_{0}^{\infty} e^{-pV/kT} V^{\nu} \, dV \tag{3.150}$$

where

$$A = e^{5/2} (2\pi mkT)^{3/2} \tag{3.151}$$

Setting $y = at$, where a is positive and real, in

$$\Gamma(z) = \int_{0}^{\infty} e^{-t} t^{z-1} \, dt \tag{3.152}$$

where $\Gamma(z)$ is the gamma function, yields

$$\frac{\Gamma(z)}{a^z} = \int_{0}^{\infty} e^{-ay} y^{z-1} \, dy \tag{3.153}$$

Identifying a with p/kT, $z - 1$ with ν, and y with V in (3.150) yields

$$W = \frac{kT}{P} \sum_{\nu=0}^{N_s} \frac{B^{\nu}}{\nu^{\nu}} \tag{3.154}$$

where

$$B = \frac{e^{(5/2)+\nu/kT} (2\pi mkT)^{3/2} kT}{h^3 P} \tag{3.155}$$

Replacing ν^{ν} by $\nu! e^{\nu}$ from (3.132a) with $n = \nu$, and letting $N_s \to \infty$ yields

$$W \simeq \frac{kT}{P} \sum_{\nu=0}^{\infty} \frac{(B/e)^{\nu}}{\nu!} = \frac{kT}{P} e^{B/e} \tag{3.156}$$

Thus, from (3.120a) and (3.159) we have

$$V = -kT \frac{\partial \ln(W)}{\partial P} = \frac{kT}{P} [\ln(W) + c] \simeq \frac{kT}{P} \ln(W) \tag{3.157a}$$

and from (3.123b) and (3.156)

$$\bar{\nu} = kT \frac{\partial \ln(W)}{\partial \mu} = \ln(W) \tag{3.157b}$$

so that combining (3.157a) and (3.157b) yields

$$PV = \bar{\nu} kT \tag{3.158}$$

as law as previously given. Note that in the derivation of nstruction of (3.147) it has been assumed by Legendre all the $^{(i)}\mu$ that the systems being considered are open to all of the species of particles being considered. If instead it required in a given problem that only some species of particles exchange with "reservoirs" external to the system (osmosis is an example), then S is Legendre transformed only on the set of $^{(i)}\mu$ corresponding to those species and only those $^{(i)}\mu$ (and ν_i) appear in (3.147).

3.5 CHEMICAL EQUILIBRIUM IN A SOIL COLUMN

We now consider the effect of gravity on the concentration of an ideal solution in a column of soil, in which capillary effects are initially neglected, as a function of height above its base. The work terms appearing in the differential form of the combined first and second laws are in this case $-P\,dV$ and $-m_s h\,dg$, where h is the height above the base of the column of solution, g is the acceleration due to gravity, m_s is the mass of an element of solution a height h above the base of the column, P is the pressure in the solution a height h above the base of the column, and V is the volume of an element of soil water solution at that same location. We thus have

$$dU = T\,dS - P\,dV - m_s h\,dg + \sum_{i=1}^{N_s} {}^{(i)}\mu\,d^{(i)}X \qquad (3.159)$$

Carrying out a Legendre transform on S, V, and g to compute the differential form of the Gibbs free energy for this case yields

$$dG = d(U - TS + PV + m_s hg)$$

$$= -SdT + VdP + m_s g\,dh + \sum_{i=1}^{N_s} {}^{(i)}\mu d^{(i)}X \qquad (3.160)$$

from which we see that the independent variables on which G depends in this case are T, P, h, and the $^{(i)}X$'s (mol numbers). The chemical potentials of the ith constituents, which we recall can be expressed as the change in the Gibbs free energy with a change in the number of mols of the ith constituent of the system, $\partial G/\partial\,^{(i)}X$, thus depend on T, P, h, and the mol fractions so that the total differential of the chemical potential of the ith species of the soil water solution at constant mol numbers is

$$d\,^{(i)}\mu = \frac{\partial\,^{(i)}\mu}{\partial h}dh + \frac{\partial\,^{(i)}\mu}{\partial P}dP + \frac{\partial\,^{(i)}\mu}{\partial T}dT + \frac{\partial\,^{(i)}\mu}{\partial\,^{(i)}n}d^{(i)}n \qquad (3.161)$$

When the system is in equilibrium at constant temperature, $d^{(i)}\mu$ and dT are zero. Noting that

$$dP = \frac{-w}{v} g \, dh \qquad (3.162)$$

where w is the mass per mol of the solution at a height h above the base of the column and v is the volume per mol of the solution at the same location. From the definition of an ideal solution [Eq. (3.77)]

$$\frac{\partial^{(i)}\mu}{\partial^{(i)}n} = \frac{R_g T}{{}^{(i)}n} \qquad (3.163)$$

since ${}^{(i)}g$ as the Gibbs free energy per mol of the ith pure component does not depend on ${}^{(i)}n$. From Eq. (3.30c) with ${}^{(1)}\theta$ replaced by V, Ψ replaced by P, and the resulting Maxwell relation,

$$\left(\frac{\partial G}{\partial P}\right)_T = V \qquad (3.164a)$$

written as Gibbs free energy and volume per mol, we have

$$\frac{\partial^{(i)}\mu}{\partial P} = {}^{(i)}v \qquad (3.164b)$$

where ${}^{(i)}v$ is the volume per mol of the ith species in solution. Finally,

$$\frac{\partial^{(i)}\mu}{\partial h} = {}^{(i)}w \, g \qquad (3.165)$$

where ${}^{(i)}w$ is the mass per mol of the ith component of the solution. Using these relations (3.161) becomes

$$0 = {}^{(i)}w \, g \, dh - {}^{(i)}v \left(\frac{w}{v}\right) g \, dh + \frac{R_g T}{{}^{(i)}n} d^{(i)}n \qquad (3.166)$$

or, on rearranging

$$\frac{d^{(i)}n}{dh} = \left[\left(\frac{w}{v}\right) {}^{(i)}v - {}^{(i)}w\right] \frac{{}^{(i)}n \, g}{R_g T} \qquad (3.167)$$

which is the change in mol fraction of the ith species with a change in position in the column. Separating variables, assuming w/v is approximately constant, taking ${}^{(i)}n_0$ to be the mol fraction of component i at the base of the column, and integrating yields

$$^{(i)}n = {}^{(i)}n_0 e^{-({}^{(i)}w - w \,{}^{(i)}v/v)gh/R_g T} \qquad (3.168)$$

from which we see that the concentration of the solution decreases with increasing height in the column. We now consider the case in which the column is filled with partly or fully saturated soil. Soil water at saturations below field capacity, which is the value of $^{(1)}\theta$ at which gravity drainage ceases, is held in the soil against gravity by capillary action and has solute distributions described by (3.168). The only modification to (3.161) needed for this case is to add the additional work term

$$dW = -\Psi\, d(V\beta\,^{(1)}\theta) \tag{3.169}$$

to account for values of soil saturation below field capacity but large enough so that an equilibrium concentration of each solute can be reached throughout the column and to drop the term $(\partial^{(i)}\mu/\partial P)\, dP$ Thus, for a soil column having a degree of saturation below field capacity, we write in (3.181) in place of the term $(\partial\,^{(i)}\mu/\partial h)\, dh$ the term $(\partial\,^{(i)}\mu/\partial\Psi)\frac{d\Psi}{\partial h}\, dh$. Rearranging terms, separating variables, and integrating as in (3.167) and (3.168) yields

$$^{(i)}n = \,^{(i)}n_0 \exp\left[-\left(w_i + \frac{^{(i)}v}{g}\frac{\partial\Psi}{\partial h} - \frac{w\,^{(i)}v}{v}\right)\frac{gh}{R_g T}\right] \tag{3.170}$$

where it has been assumed that $\partial\Psi/\partial h$ is the same in all parts of the column (constant). In the more general case $\partial\Psi/\partial h$ would be found by making thermocouple psychrometer measurements along the column and fitting the data to an assumed function prior to integrating.

In Chapter 7 these ideas are applied to the centrifugation of soil cores to remove the water from them for chemical analysis. It can be seen from Eq. (3.170) that this can be accomplished with only limited success due to the effects of pressure on the equilibrium chemistry of water in soil.

Up to now, we have considered the chemistry of nonionic soil water solutions. We now consider the case in which the solute particles are ions, which is the topic of electrochemistry.

3.6   SELECTED ASPECTS OF ELECTROCHEMISTRY

Substances that become ionized in solution are called *electrolytes*. Strong electrolytes dissociate into ions in solution more completely than do weak electrolytes. According to this definition, NaCl, sodium chloride, would be viewed as a strong electrolyte while $HgCl_2$, mercuric chloride, would be viewed as a weak electrolyte. A solution of an electrolyte, an electrolytic solution, differs in a fundamental way from the dilute solutions previously discussed in that the solute particles interact with each other via Coulomb forces so that the simplifying ideal solution assumption of no solute particle interactions other than collisions is no longer viable.

The aspects of electrochemistry that are important in unsaturated flow are (1) the lowering of the soil solution vapor pressure, p_{0v}, which affects matric potential via Eq. (1.1a), (2) electrostatic interactions of the ions in solution with the possibly charged pore/capillary walls of the porous medium which would not be present if the solute species in the soil solution were not such as to dissociate into charged particles and which affects flow through porous media, and (3) the alteration in the flow regime that can be caused by applying an external electric field to the soil–water system.

We begin with the derivation of the relation between the vapor pressure of the solvent, which we identify with p_{0v} for a non–volatile solute in a dilute binary electrolytic solution, and the activity of the solute, which can be calculated directly from the Debye–Hückel theory of dilute electrolytic solutions, which we shall develop as well. The chemical potential, $^{(i)}\mu$, of a component of species i of a solution can be written as

$$^{(i)}\mu = {}^{(i)}\mu^0 + R_g T \ln\left(^{(i)}a\right) \tag{3.171}$$

where $^{(i)}\mu^0$ is the chemical potential of the species i in a standard state with respect to which $^{(i)}\mu$ is measured and in fact only the differences in chemical potential from those of the standard states are experimentally accessible. Further, $^{(i)}\mu^0$ depends on the species only and so is independent of whether the species i is in solution, in the vapor phase, etc. The symbol R_g is the universal gas constant, T is absolute temperature, $^{(i)}a$ is the activity of species i, and can be written as

$$^{(i)}a = {}^{(i)}x \, {}^{(i)}f \tag{3.172}$$

where $^{(i)}x$ is the mol fraction of species i, defined previously, $^{(i)}f$ is called the activity coefficient of species i, and $\ln()$ indicates logarithms to base e. The activity has the important property of having the same value for a given species in all phases in which it may occur in a system at equilibrium– the same property as chemical potential. Further, substitution of (3.172) into (3.171) shows this latter to be a generalization of (3.77) in that the mol fraction in (3.77), appropriate for an ideal dilute solution, is replaced by activity for a concentrated and/or electrolytic solution. From (3.45), the Gibbs–Duhem relation, we have for a binary solution (a binary solution consists of two components–the solvent and one solute) at constant temperature

$$^{(1)}x \, d\,^{(1)}\mu + {}^{(2)}x \, d\,^{(2)}\mu = 0 \tag{3.173}$$

in which species 1 is taken to be the solvent, species 2 is taken to be the solute, and the symbols are as previously defined. Substituting (3.171) into

(3.173) and noting that the $d^{(i)}\mu^0 = 0$, yields

$$^{(1)}x \, d\ln\left(^{(1)}a\right) = -\,^{(2)}x \, d\ln\left(^{(2)}a\right) \qquad (3.174)$$

Taking $^{(1)}a$ to be the activity of the solvent in the solution relative to the activity of the pure solvent for which $^{(1)}a = {}^{(1)}a^0 = 1$, $^{(2)}a$ to be the activity of the solute, which will be calculated shortly from Debye–Hückel theory, and rearranging, yields

$$d\ln\left(^{(1)}a\right) = -\frac{^{(2)}x}{^{(1)}x} \, d\ln\left(^{(2)}a\right) \qquad (3.175)$$

Replacing $^{(1)}a$ with $^{(1)}p/^{(1)}p^0$, where $^{(1)}p$ is the vapor pressure of the solvent in the solution just above a flat surface of solution, and $^{(1)}p^0$ is the vapor pressure of the pure solvent at the same temperature (Harned, 1930, p. 777), substituting (3.172) with i set equal to 2 into the right-hand side of (3.175), expanding the total differentials, and indicating integration yields

$$\int_{^{(1)}p^0}^{^{(1)}p} \frac{d\,^{(1)}p'}{^{(1)}p'} = -\frac{1}{^{(1)}x}\int_0^{^{(2)}x} d\,^{(2)}x' - \frac{1}{^{(1)}x'}\int_0^{\ln[\,^{(2)}f]} {}^{(2)}x'd\,\ln\left(^{(2)}f'\right)$$

$$(3.176)$$

where $^{(1)}p^0$ and $^{(1)}x$ have been assumed to be constant in the differentiations and indicated integrations. In order to evaluate the integral over $\ln\left(^{(2)}f'\right)$, we require its functional dependence on $^{(2)}x$. This can be derived from the theory of Debye and Hückel, to which we now turn.

3.6.1   Debye–Hückel Theory

We begin the development of the theory of Debye and Hückel by noting that the Coulomb force of attraction between ions of opposite sign in a solution is reduced from the value it would have in air by the dielectric properties of the solvent as follows. The magnitude of the Coulomb force F_c between two charged particles in vacuum is given by

$$F_c = \frac{1}{4\pi\varepsilon_o}\frac{q_1 q_2}{d^2} \qquad (3.177a)$$

where d is the distance between the two particles, their charges are q_1 and q_2, respectively, and $\varepsilon_o = 8.85 \times 10^{-12} \, C^2/N\,m^2$, is the permittivity of free space. The Coulomb is a quantity of electric charge and is abbreviated C, (1 Faraday is 96,519 C and is an Avogadro's number–1 mol–of electron charges), the Newton, abbreviated N, is a unit of force (1 N $= 1$ kg meter/sec^2), and it is assumed that the proximity of one charged particle to another does not significantly alter the distribution of charge on either particle. The effect of such alteration of charge, whether by redistribution or induction, would be to enhance the attraction of the charged particles for each other if q_1 and q_2 were of opposite sign and decrease the repulsive force between the particles if they were of the same sign. When the charges q_1 and q_2 are immersed in a dielectric of dielectric constant K_d, the magnitude of the force between them becomes

$$F_c = \frac{1}{4\pi\varepsilon_o K_d} \frac{q_1 q_2}{d^2} \tag{3.177b}$$

The higher the dielectric constant of the solvent, the more separated and mobile the ions become. The lowest-level approximation is to regard the average in-solvent interactions as electrostatic and consider the Coulomb interaction of a charged ion surrounded by a spherically symmetric solvation sheath, which is composed of solvent molecules polarized as depicted schematically in Figure 3.7. The fraction of central positive ions of species i per unit volume, n_i, having an electrostatic energy u due to a (pre-existing electrostatic) potential field Ψ_p at temperature T is given by the Boltzmann factor multiplied by the total number of central ions of species i per unit volume, n_{0i}, as

$$n_i = n_{0i} e^{-u/kT} \tag{3.178}$$

Electrostatic potential energy is defined as the work per unit charge required to bring a charge q from infinity to a position $\vec{r}$ in the presence of a potential field, Ψ_p, which itself may be due to a collection of static charges, which are assumed to be unaltered in position by the approach of charge q. Thus,

$$du = dq\,\Psi_p \tag{3.179}$$

so that (3.178) becomes

$$n_i = n_{0i} e^{-z_i \epsilon \Psi_p/kT} \tag{3.180}$$

where $z_i\epsilon$, which we identify with q, is the absolute value of the charge on one electron, $\epsilon = |-1.6 \times 10^{-19}|$ C, multiplied by z_i, the degree of ionization of positive ions of species i (singly ionized species have $|z_i| = 1$ etc.). Similarly, for m_{0i} negative solute ions of species i (negative values of z_i) in energy state u we have

$$m_i = m_{0i} e^{|z_i|\epsilon\Psi_p/kT} \tag{3.181}$$

where, as above, $||$ denotes absolute value. The total charge per unit volume due to solute ions, each having an electrostatic energy u, is thus

Figure 3.7: A central positive solute ion surrounded by solvent molecules polarized by it. The extended group of polarized solvent molecules surrounding the central solute ion is called a solvation sheath.

$$\rho_c = -\sum_i (z_i n_i - |z_i| m_i)\epsilon = \sum_i -2n_{0i}|z_i|\epsilon \sinh\left(\frac{|z_i|\epsilon\Psi_p}{kT}\right) \qquad (3.182)$$

where sinh is the hyperbolic sine, it has been assumed that $n_{0i} = m_{0i}$ so that $2n_{0i}$ is the total number of solute ions of species i per unit volume, and the rest of the quantities are as previously defined. The partial differential equation relating the potential Ψ_p to the volume density of the charge distribution to which it is due, in this case the charges on the solute ions and the induced polarization charge of the solvent, in an electrostatic (no moving charges) situation is called *Poisson's equation* and is given in MKS (meter-kilogram-second) units as

$$\nabla^2\Psi_p = -\frac{\rho_c}{\varepsilon_d} \qquad (3.183)$$

where

$$\varepsilon_d = K_d\varepsilon_0 \qquad (3.184)$$

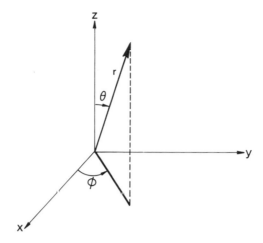

Figure 3.8: The quantities r, radial distance from the origin of coordinates, θ, the angle between r and the z-axis, and ϕ, the angle between the projection of r onto the $x - y$ plane and the x-axis, are shown.

where ε_0, the permittivity of free space, and K_d the dielectric constant of the medium in which the charges are located are as given previously (for vacuum $K_d = 1$), ε_d is the permittivity of the dielectric medium in which the charges are located, Ψ_p is calculated subject to (3.183) and appropriate boundary conditions, and ∇^2 is the usual Laplacian differential operator, given in three dimensions in spherical coordinates by

$$\nabla^2 \Psi_p = \frac{1}{r^2} \frac{\partial}{\partial r} \left(r^2 \frac{\partial \Psi_p}{\partial r} \right) + \frac{1}{r^2 \sin(\theta)} \frac{\partial}{\partial \theta} \left[\sin(\theta) \frac{\partial \Psi_p}{\partial \theta} \right] + \frac{1}{r^2 \sin^2(\theta)} \frac{\partial^2 \Psi_p}{\partial \phi^2}$$

(3.185)

where r, θ, and ϕ are the usual spherical coordinates shown in Figure 3.8. We now note that if the induced solvent charge distribution is spherically symmetric around the central ion, then the potential depends on r only, and (3.184) becomes in this case

$$\frac{1}{r^2} \frac{\partial}{\partial r} \left(r^2 \frac{\partial \Psi_p}{\partial r} \right) = -\frac{\rho_c}{\varepsilon_d} = \sum_i \frac{2 n_{0i} |z_i| \epsilon}{\varepsilon_d} \sinh \left(\frac{|z_i| \epsilon \Psi_p}{kT} \right)$$

(3.186)

Equation (3.186) is an ordinary (one independent variable) inhomogeneous (nonzero right-hand side) differential equation for Ψ_p as function of r. If we now expand $\sinh\left(|z_i|\epsilon\Psi_p/kT\right)$ in a Maclaurin series and retain only the first term, which is the argument of the sinh, (3.186) has been linearized in Ψ_p and becomes, after rearrangement,

$$\frac{\partial^2 \Psi_p}{\partial r^2} + \frac{2}{r}\frac{\partial \Psi_p}{\partial r} - \kappa^2 \Psi_p = 0 \qquad (3.190)$$

where

$$\kappa^2 = \sum_i \frac{2n_{0i}z_i^2\epsilon^2}{\varepsilon_d kT} = \frac{2\epsilon^2}{\varepsilon_d kT}\sum_i n_{0i}z_i^2 \qquad (3.188)$$

Equation (3.187) can be solved in various ways. Perhaps the simplest is to note that setting $\Psi_p(r) = \xi(r)/r$ in (3.187) yields

$$\frac{\partial^2 \xi(r)}{\partial r^2} - \kappa^2 \xi(r) = 0 \qquad (3.189)$$

and the two linearly independent solutions of (3.189) for $\xi(r)$ are $e^{+\kappa r}$ and $e^{-\kappa r}$ so that

$$\Psi_p(r) = \frac{Ae^{-\kappa r}}{r} + \frac{Be^{+k\kappa r}}{r} \qquad (3.190)$$

where A and B are constants to be determined. We first set $B = 0$ so that Ψ_p does not become infinite as $r \to \infty$. Thus

$$\Psi_p = \frac{Ae^{-\kappa r}}{r} \qquad (2.191)$$

We now determine the constant A as follows. The total polarization charge due to a central positive ion of charge $z_i\epsilon$ is $-|z_i|\epsilon$ due to the requirement for electroneutrality. Thus,

$$2\epsilon\sum_i n_{0i}|z_i| = -\int_{a_i}^{\infty} \rho_c 4\pi r^2 \, dr \qquad (3.192)$$

where the element of volume in spherical coordinates is $4\pi r^2 \, dr$, a_i is the effective radius of the central ion, and the rest of the quantities are as previously defined. From (3.186), (3.187) and (3.191) we have that

$$-\frac{\rho_c}{\varepsilon_d} = \kappa^2 \Psi_p = \kappa^2 \frac{Ae^{-\kappa r}}{r} \qquad (3.193)$$

or

$$\rho_c = \frac{-\varepsilon_d \kappa^2 A e^{-\kappa r}}{r} \qquad (3.194)$$

so that (3.192) becomes

$$2\epsilon \sum_i n_{0i}|z_i| = 4\pi\varepsilon_d\kappa^2 A \int_{a_i}^{\infty} re^{-\kappa r}\, dr \qquad (3.195)$$

Upon integrating, solving for A, which is found to be

$$A = \frac{2\epsilon e^{\kappa a_i} \sum_i n_{0i}|z_i|}{4\pi\varepsilon_d(1 + \kappa a_i)} \qquad (3.196)$$

and substituting into (3.191) we have for $\Psi_p(r)$,

$$\Psi_p(r) = \frac{2\epsilon e^{-\kappa(r-a_i)} \sum_i n_{0i}|z_i|}{4\pi\varepsilon_d(1 + \kappa a_i)r} \qquad (3.297)$$

for $r \geq a_i$. For a single ion of degree of ionization z_i, the potential becomes

$$\psi_p = \frac{|z_i|\epsilon e^{-\kappa(r-a_i)}}{4\pi\varepsilon_d(1 + \kappa a_i)r} \qquad (3.198)$$

which at $r = a_i$ can be written as

$$\psi_p = \frac{|z_i|\epsilon}{4\pi\varepsilon_d a_i} - \frac{|z_i|\epsilon\kappa}{4\pi\varepsilon_d(1 + \kappa a_i)} \qquad (3.199)$$

The first term in (3.199) is the potential due to the central ion, while the second term is the average potential due to the induced polarization of the solvent and the remaining solute ions. The energy per ion is, from comparison of (3.179) with (3.199),

$$du = -\frac{|z_i|^2\epsilon^2\kappa}{4\pi\varepsilon_d(1 + \kappa a_i)} \qquad (3.200)$$

Identifying du with $kT \ln(f_i)$ by inspection of (3.171) yields

$$\ln(f_i) = -\frac{|z_i|\epsilon^2\kappa}{4\pi\varepsilon_d(1 + \kappa a_i)} \qquad (3.201)$$

We now consider κ as follows. From (3.188) we have

$$\kappa = \left(\frac{2\epsilon^2}{\varepsilon_d kT}\sum_i n_{0i}z_i^2\right)^{\frac{1}{2}} = \left(\frac{2\epsilon^2}{\varepsilon_d kT}\sum_i \frac{c_i L}{1000}z_i^2\right)^{\frac{1}{2}} \qquad (3.202)$$

where c_i is the solute concentration in mols per liter, $L = 6.02 \times 10^{23}$ molecules per mol, Loschmidt's (Avogadro's) number, and

$$n_{0i} = \frac{c_i L}{1000} \tag{3.203}$$

for n_{0i} in units of particles per cm^3 of solution. The ionic strength, I, of an electrolytic solution is, by definition,

$$I = \frac{1}{2} \sum_i c_i z_i^2 \tag{3.204}$$

so that κ becomes

$$\kappa = \left(\frac{4\epsilon^2 L I}{1000 \, \varepsilon_d kT} \right)^{\frac{1}{2}} = \frac{\alpha I^{\frac{1}{2}}}{kT} \tag{3.205}$$

where

$$\alpha = \left(\frac{4\epsilon^2 L}{1000 \, \varepsilon_d} \right)^{\frac{1}{2}} \tag{3.206}$$

so that $\ln({}^{(i)}f_+)$ becomes

$$\ln({}^{(i)}f_+) = -\frac{|z_i|^2 \epsilon^2 \alpha I^{\frac{1}{2}}}{4\pi\varepsilon_d (kT)^{\frac{3}{2}} \left[1 + \alpha a_i I^{\frac{1}{2}}/kT \right]} \tag{3.207}$$

We now note that the mean activity coefficient of a binary electrolyte (a binary electrolyte is one that dissociates into two components in solution) of species i in solution, ${}^{(i)}f$, is related to the activity coefficients of the positive and negative ions of solute species i, ${}^{(i)}f_+$ and ${}^{(i)}f_-$, respectively, by

$$ {}^{(i)}f^{2n_{0i}} = {}^{(i)}f^{n_{0i}} \, {}^{(i)}f^{m_{0i}} \tag{3.208}$$

so that for $n_{0i} = m_{0i}$ as assumed previously and in (3.208)

$$\ln \left({}^{(i)}f \right) = \ln \left({}^{(i)}f_+^{\frac{1}{2}} \, {}^{(i)}f_-^{\frac{1}{2}} \right) \tag{3.209}$$

We set

$$ {}^{(i)}f_- = \frac{|z_i|^2 \epsilon^2 \alpha I^{\frac{1}{2}}}{4\pi\varepsilon_d (kT)^{\frac{3}{2}} \left[1 + \alpha a_i I^{\frac{1}{2}}/kT \right]} \tag{3.210}$$

which is the usual Debye–Hückel relation for the mean ionic activity coefficient. In order to use (3.210) in (3.176), our goal, we express $\ln({}^{(1)}f)$ in terms of mol fractions instead of concentrations as follows. Taking ${}^{(i)}n$ to be the number of mols of solute of species i, ${}^{(1)}n^0$ to be the number of mols of solvent in the solution, $^{(1)}v^0$ to be the volume per mol of solvent, we have

$$^{(i)}n = c_i \,^{(1)}n^0 \,^{(1)}v^0 \tag{3.211}$$

and the mol fraction of component 2 of the solution, $^{(2)}x$, is

$$^{(2)}x = \frac{^{(2)}n}{^{(2)}n^0} = c_2 \,^{(1)}v^0 \tag{3.212}$$

Thus,

$$c_2 = \frac{^{(2)}x}{^{(1)}v^0} \tag{3.213}$$

and we have for I in terms of $^{(2)}x$ in this case

$$I = \frac{^{(2)}x z_2^2}{2 \,^{(1)}v^0} = \beta^2 \,^{(2)}x \tag{3.214}$$

with $\beta^2 = z_2^2/2 \,^{(1)}v^0$ so that $\ln\left(^{(2)}f\right)$ becomes

$$\ln\left(^{(2)}f\right) = -\frac{z_2|z_2|\epsilon^2\alpha\beta[\,^{(2)}x]^{\frac{1}{2}}}{4\pi\varepsilon_d(kT)^{\frac{3}{2}}\left[1 + \alpha\beta a_i \,^{(2)}x^{\frac{1}{2}}/kT\right]} = \frac{A^0 \,^{(2)}x^{\frac{1}{2}}}{1 + B^0 \,^{(2)}x^{\frac{1}{2}}} \tag{3.215}$$

where

$$A^0 = -\frac{z_2|z_2|\epsilon^2\alpha\beta}{4\pi\varepsilon_d(kT)^{\frac{3}{2}}} \tag{3.216a}$$

$$B^0 = \frac{\alpha\beta a_i}{kT} \tag{3.216b}$$

Taking the total differential of (3.215) yields

$$d\ln\left(^{(2)}f\right) = \frac{A^0}{2}\left(\frac{B^0}{\left[1 + B^0 \,^{(2)}x^{\frac{1}{2}}\right]^2} + \,^{(2)}x^{-\frac{1}{2}}\right)d\,^{(2)}x \tag{3.217}$$

Substituting (3.217) into (3.176), carrying out the indicated integrations, and rearranging terms yields

$$^{(1)}p = \,^{(1)}p^0 \exp\left[-\frac{^{(2)}x}{^{(1)}x} - \frac{A^0}{^{(1)}x B^0}\left(\frac{(1 + B^0\sqrt{^{(2)}x})^2}{2} - 3(1 + B^0\sqrt{^{(2)}x})\right.\right.$$

$$\left.\left. +3\ln|1 + B^0\sqrt{^{(2)}x}| + \frac{1}{1 + B^0\sqrt{^{(2)}x}}\right) - \frac{3A^0}{2\,^{(2)}x B^0} + \frac{A^0 \,^{(2)}x^{\frac{3}{2}}}{3\,^{(1)}x}\right] \tag{3.218}$$

where $^{(1)}p$ is identified with p_{0v} of eq.(1.1) for an electrolytic solution. For dilute solutions the fractional lowering of the value of $^{(i)}p^0$ over the flat surface of a solvent due to both electrolytic and non-electrolytic solutes is the product of the fractional $(^{(i)}p_v/\,^{(i)}p^0)$ lowering due to each.

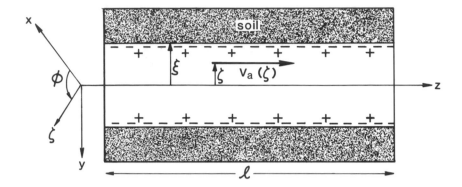

Figure 3.9: Cross-section of a capillary of showing the negative charge per unit area, σ, the surface of the clay walls of the capillary, and the induced positive charge per unit area, σ_i, on the water in the capillary adjacent to the walls.

3.6.2 Electro-Osmosis

Electro-osmosis in this context refers to the transport of water in a porous medium under the influence of a gradient of electrical potential (difference in voltage) applied to electrodes emplaced in the porous medium. We consider a section of a capillary of length ℓ, circular cross section, and radius ξ, as shown in Figure 3.9. A surface charge on the pore/capillary walls, negative in the case of clays, is required for electro-osmosis to occur since the actual force exerted on the water by the potential gradient is on the polarization charge induced in the water by the pre-existing charges on the clay surfaces with dissolved electrolytes entrained in the fluid enhancing this effect. Electro-osmosis is not observed to occur in media lacking these charges–sands, for example. The charge induced in the water is actually a dipole moment per unit volume, or polarization, $\vec{P}$, and has as its manifestation the induced surface charge per unit area, σ_i, given by

$$\sigma_i = \sigma \left(1 - \frac{1}{K_d} \right) \tag{3.219}$$

where σ is of the order of 9.64×10^{-6} C/cm^2 for montmorillonite, and K_d, as defined in connection with Eq. (3.177), is approximately equal to 81 for pure water at room temperature (K_d for most dielectric materials decreases as the temperature of the material increases) and hence for soil solution is an approximate value only. Note that electroneutrality has been assumed in the Debye–Hückel theory just discussed, so that the effects of ions in solution and the polarization they induce in the solute can be neglected to first approximation. The force $\vec{F}$ on the water in the capillary is

$$\vec{F} = -2\pi\xi\ell\sigma_i(\vec{\nabla}V \cdot \hat{n})\hat{n} \tag{3.220}$$

where V is the electrical potential applied to the soil as a function of position in the soil, $\hat{n}$ is a unit vector along the direction of the capillary, and the rest of the quantities are as previously defined. If it is now assumed that the potential gradient along the capillary is constant and given by $\partial V/\partial h$, where h is the distance along the capillary, as shown in Figure 3.9, then (3.220) corresponds to an additional pressure term, P_0, in Poiseuille's law of (2.111), and is

$$P_0 = -\sigma_i \frac{\partial V}{\partial h} \tag{3.221}$$

This corresponds to replacing A_1 in (2.121a) by $A_0 + A_1$, where A_0 is given by

$$A_0 = \frac{P_0\xi^2}{8\eta} \tag{3.222}$$

so that h_0 is replaced by h_2, where h_2 is given by

$$h_2 = \frac{A_0 + A_1}{B_1} = \frac{1}{\rho g \cos(\theta)} \left(\frac{2\gamma \cos(\theta)}{\xi} + P_0 \right) \tag{3.223}$$

and t_0 by t_2, where

$$t_2 = \frac{A_0 + A_1}{B_1^2} \tag{3.224}$$

$$\frac{1}{t_2} = \frac{\xi^3 d_0}{1 + \xi d_0/e_0} \tag{3.225}$$

where

$$d_0 = \frac{\rho^2 g^2 \cos^2(\theta)}{16\eta\gamma\cos(\phi)} \qquad (3.226)$$

$$e_0 = \frac{\rho^2 g^2 \cos^2(\theta)}{8P_0\eta} \qquad (3.227)$$

$$\frac{d_0}{e_0} = \frac{P_0}{2\gamma\cos(\phi)} \qquad (3.228)$$

so that if the gradient of electrical potential is zero, $P_0 = 0$, and $t_2 = t_0$ so that (2.128) for the average (across a cross section of the tube) velocity of fluid rising in a capillary tube of radius ξ due to a combination of capillary action and electro-osmosis, $v_{ae}(\xi)$, becomes

$$v_{ae}(\xi) = \frac{h_2}{t_2} \sum_{n=1}^{\infty} \frac{e^{-nt/t_2}n^n e^{-n}}{n!} \qquad (3.229)$$

For a bundle of capillaries having a pore radius frequency distribution $f(\xi)$, the Darcy velocity becomes, in place of (2.133),

$$J = \frac{\frac{h_2}{t_2}\sum_{n=1}^{\infty}\frac{n^n e^{-n}}{n!}\int_{r_m}^{r}\xi^4 f(x)e^{-nt/t_2}\,d\xi}{\int_{r_m}^{R}\xi^2 f(\xi)\,d\xi} \qquad (3.230)$$

Since the applied voltage gradient can be such that P_0 is in the same direction as P_2 of (2.113) or opposite to it, we now examine the relative magnitudes of P_0 and P_2 as follows. Assuming a constant gradient of electrical potential of -0.5 V/cm, since larger gradients in practice produce heating of the soil, a value of σ_i of 9.52×10^{-6} C/m^2, found by substituting the values for σ and K_d into (3.219), yields $P_0 \simeq 47.6$ dynes/cm^2, where the conversions 1 J/C = 1 V, 1 J = 10^7 ergs, and 1 erg = 1 dyn/cm have been used. By contrast, for values of $\gamma = 72$ dyn/cm, $\xi = 10^{-5}$, and $\varphi = 0$, $P_2 \simeq 1.44 \times 10^7$ dyn/cm. Thus, for unsaturated flow in an arid region, the contribution of electro-osmosis is generally small.

3.6.3 Streaming Potential

The streaming potential is the potential gradient developed when water is forced through materials having charged pore/capillary walls by an externally applied pressure and so is the converse of electro-osmosis. The streaming current is given for a capillary by

$$I = v_a(2\pi\xi)\sigma_i \qquad (3.231)$$

where v_a is given by (2.112) and the streaming potential V_s is given by

$$V_s = \frac{I\ell}{\pi\xi^2\rho_s} = \frac{2v_a\sigma_i\ell}{\rho_s\xi} \tag{3.232}$$

where ρ_s is the effective electrical resistivity of the capillary, a quantity to be determined experimentally, so that the streaming potential gradient is

$$\frac{V_s}{\ell} = \frac{2v_a\sigma_i}{\rho_s\xi} \tag{3.233}$$

Osmotic potential is often included as a part of the total potential and then ignored. It is certainly real when soil water chemistry varies from place to place, as noted previously, however, and acts in such a way as to minimize variations in soil water solution concentration from place to place. Osmotic pressure as such exists after osmosis has occurred. For dilute, nonelectrolytic soil solutions, this pressure is given by an equation of the same form as the ideal gas law, as noted previously.

An example of vapor transport, which might be viewed broadly as osmosis, is the evolution of water vapor from larger menisci and its condensation onto the water surfaces of smaller menisci with the intervening porous medium containing vapor(s) playing the role of the semipermeable membrane. Soil water chemistry and temperature affect this, as has been discussed, as well as in the next chapter, which has vapor transport as its subject.

SELECTED REFERENCES

Callen, H. B., *Thermodynamics*, 1st ed., John Wiley & Sons, N.Y., 1961.

Courant, R., *Differential and Integral Calculus*, Vol. II, Interscience Publishers, Inc., N.Y., 1947.

Eisenberg, D., and W. Kauzmann, *The Structure and Properties of Water*, Oxford University Press, N.Y., 1969.

Harned, H. S., "The Electrochemistry of Solutions," in *A Treatise on Physical Chemistry*, H. S. Taylor, ed., D. Van Nostrand Co., Inc., N.Y., 1931, p. 777.

Hill, T. E., "Three New Partition Functions in Statistical Mechanics," *J. Chemical Physics*, 29, 1423–24, 1958.

MacInnes, D. A., *The Principles of Electrochemistry*, Dover Publications Inc., N.Y., 1961.

Zemansky, M. W., *Heat and Thermodynamics*, McGraw–Hill Book Co., Inc., N.Y., 1957.

PROBLEMS

3.1 Compute the Legendre transform on S of $dU = T\,dS - P\,dV$ as $d(U - TS) = dF$ to yield $dF = -S\,dT - P\,dV$.

3.2 Given the total differential of F in Problem 3.1, show that the Maxwell relation that follows from it is

$$\left(\frac{\partial P}{\partial T}\right)_V = \left(\frac{\partial S}{\partial V}\right)_T$$

3.3 Considering S to be a function of T and A (surface area), compute $T\,dS$ in a form analogous to Eq. (3.53b) of the text. Next, write C_A in a form analogous to Eq. (3.54). Next, write the total differential of the Helmholtz free energy for a system in which the work term is given by Eq. (3.3) as

$$dF = -S\,dT + \gamma\,dA$$

and show that this leads to the Maxwell relation

$$\left(\frac{\partial \gamma}{\partial T}\right)_A = \left(\frac{\partial S}{\partial A}\right)_T$$

Use the results to arrive at Eq. (3.58) of the text.

3.4 Construct a mnemonic diagram for a system in which the work term is $dW = \gamma\,dA$ instead of $dW = -\Psi\,d^{(i)}\theta$ as was done in the text (i.e., replace Ψ by $-\gamma$ and $^{(i)}\theta$ by A) and use it to write the differentials of the Helmholtz free energy, Gibbs free energy, enthalpy, and internal energy at constant mol numbers as

$$dF = -S\,dT + \gamma\,dA$$

$$dG = -S\,dT - A\,d\gamma$$

$$dH = T\,dS - A\,d\gamma$$

$$dU = T\,dS + \gamma\,dA$$

3.5 Using either the mnemonic diagram of Problem 3.4 or the appropriate total differentials of the thermodynamic potential functions given there, arrive at the Maxwell relations

$$\left(\frac{\partial A}{\partial T}\right)_\gamma = \left(\frac{\partial S}{\partial \gamma}\right)_T$$

$$\left(\frac{\partial T}{\partial \gamma}\right)_S = -\left(\frac{\partial A}{\partial S}\right)_\gamma$$

$$\left(\frac{\partial T}{\partial A}\right)_S = \left(\frac{\partial \gamma}{\partial S}\right)_A$$

3.6 Carry out the derivations of Eqs. (3.99) and (3.105) of the text.

4

VAPOR TRANSPORT
IN A POROUS MEDIUM

4.1 POROUS MEDIUM CLAPEYRON EQUATION

The vapor available for transport in a partly saturated porous medium is that which is evolved from the concave menisci forming the surfaces of the water contained in the pores/capillaries in the medium. Since the vapor pressure over a concave meniscus is lower than that over a flat surface, the effective latent heat of vaporization is correspondingly higher, and the slope of the vapor-pressure–temperature curve for liquid in a porous medium, a parameter of interest in the calibration of thermocouple psychrometers, is correspondingly different from that for liquid having a flat surface and depends on the pore radius frequency distribution (a two-parameter frequency distribution would be required for capillaries of noncircular cross section) as well. For diffusion of vapor through ambient air in the pores/capillaries, transport is due to vapor concentration gradients that may be expressed as gradients of vapor pressure or density. We thus begin consideration of vapor transport in a porous medium by examining the evolution of vapor from the meniscus surfaces in the parts of the porous medium containing liquid. The slope of the usual vaporization curve of water over a flat water surface, shown in Figure 4.1, is given by the Clausius–Clapeyron equation, which we now derive as follows. We consider state A in the liquid phase and corresponding state B in the vapor phase just across the curve from A, as shown in Figure 4.1. In equilibrium we have

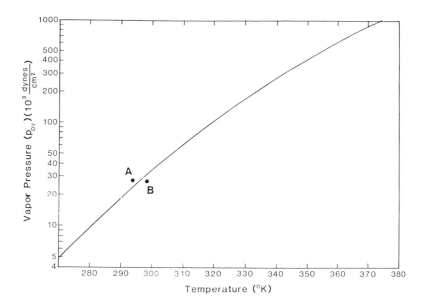

Figure 4.1: The vapor-pressure–temperature curve for pure water having a flat surface. Liquid water and its vapor are in equilibrium for values of pressure and temperature on the curve.

$$d\mu_A = d\mu_B \tag{4.1}$$

where μ_A is the chemical potential in the liquid at point A and μ_B the chemical potential of the vapor at point B. The total differentials of the Gibbs free energy per mol for points A and B are

$$d\mu_A = -s_A \, dT + v_A \, dp \tag{4.2a}$$

$$d\mu_B = -s_B \, dT + v_B \, dp \tag{4.2b}$$

where s_A is the entropy per mol of the liquid at point A, v_A is the volume per mol of the liquid at point A, and s_B and v_B are the entropy and volume per mol, respectively at the corresponding point in the vapor. Equating (4.2a) and (4.2b) via (4.1) and rearranging yields

$$\frac{dp}{dT} = \frac{s_B - s_A}{v_B - v_A} \tag{4.3}$$

where dp/dT is the slope of the vapor pressure curve, $v_B - v_A$ is the change in volume per mol that occurs upon condensation at the pressure and temperature common to points A and B, and $s_B - s_A$ is the change in entropy per mol that occurs upon condensation. Further, the heat evolved upon condensation, in this case when the system undergoes a transition from phase B to phase A, is

$$l_{BA} = T(s_B - s_A) \tag{4.4}$$

where l_{BA} equals the latent heat of vaporization at temperature T. The Clapeyron equation becomes

$$\frac{dp}{dT} = \frac{l_{BA}}{T(v_B - v_A)} \tag{4.5}$$

where all the quantities appearing in (4.5) are understood to be evaluated at temperature T. Equation (4.5) can be integrated to various degrees of accuracy depending on the assumptions made. The simplest assumptions are that $v_A << v_B$, l_{BA} is approximately constant (from Table 4.1 we see that there is approximately a 6% variation of l_{BA} with temperature in the range $20 - 80^{\circ}\mathrm{C}$), and that the water vapor is an ideal gas so that

$$v_B = \frac{R_g T}{p_{0v}} \tag{4.6}$$

Substituting into (4.5) and indicating integration yields

$$\int_{p_{0e}}^{p_{0v}} \frac{dp'_{0v}}{p'_{0v}} = \frac{l_{BA}}{R_g} \int_{T_0}^{T} \frac{dT'}{T'^2} \tag{4.7}$$

where p_{0e} is the vapor pressure at temperature T_0 and the primed quantity is a dummy variable of integration. Integrating and rearranging yields for the equation of the vapor pressure curve as a function of temperature, to lowest approximation,

$$p_{0v}(T) = p_e e^{-l_{BA}/R_g T} \tag{4.8a}$$

where

$$p_e = p_{0e} e^{l_{BA}/R_g T_0} \tag{4.8b}$$

Table 4.1: Latent heat of vaporization of water as a function of temperature.

Temp.($^\circ$C)	l_{BA}(cal/g)
0	595.9
10	590.4
20	584.9
30	579.5
40	574.0
50	568.5
60	563.2
70	557.5
80	551.7
90	545.8
100	539.55

If $p_{0v}(T)$, given by (4.8a), is the vapor pressure of water at temperature T when the water surface has an infinite radius of curvature, the vapor pressure over a water surface having the form of a concave meniscus of average radius of curvature r_a is given approximately by

$$p_v(r_a, T) = p_{0v}(T)e^{-A_0/r_a} \tag{4.9a}$$

which is the usual Kelvin relation given previously for a partly hemispherical meniscus surface in a capillary of circular cross section with radius ξ, and $r_a = \xi/\cos(\phi)$, where ϕ is the angle between the water meniscus and the soil surface with which it is in contact, so that

$$p_v(\xi, T) = p_{0v}(T)e^{-A_0 \cos(\phi)/\xi} \tag{4.9b}$$

where

$$A_0 = \frac{2\gamma m}{\rho k T} \tag{4.9c}$$

with ρ the density of liquid water, as previously defined. Differentiating (4.9b) with respect to T yields

$$\frac{\partial p_v(\xi, T)}{\partial T} = \frac{1}{p_{0v}(T)} \left(\frac{\partial p_{0v}(T)}{\partial T} \right) p_v(\xi, T)$$

$$+ \frac{A_0 \cos(\phi)}{\xi} p_v(\xi, T) \left(\frac{1}{T} + \frac{1}{\rho} \frac{\partial \rho}{\partial T} - \frac{1}{\gamma} \frac{\partial \gamma}{\partial T} \right) \tag{4.10}$$

where $\partial p_{0v}(T)/\partial T$ is the slope of the vapor pressure curve appearing in the Clapeyron equation.

Averaging (4.10) over the radii of the partially liquid–filled capillaries yields the expression for the equivalent of the Clapeyron equation for soil solution in a porous medium, namely the average slope of the vapor-pressure–temperature curve for water in a porous medium

$$\frac{\partial \bar{p}_v(r,T)}{\partial T} = \int_{r_m}^{r} f(\xi) \frac{\partial p_v(\xi,T)}{\partial T} d\xi$$

$$= \frac{1}{p_{0v}} \left[\frac{\partial p_{0v}(T)}{\partial T} \right] \int_{r_m}^{r} p_v(\xi,T) f(\xi) d\xi$$

$$+ A_0 \left[\frac{1}{T} + \frac{1}{\rho}\frac{\partial \rho}{\partial T} - \frac{1}{\gamma}\frac{\partial \gamma}{\partial T} \right] \int_{r_m}^{r} \frac{f(\xi) p_v(\xi,T) \cos(\phi)}{\xi} d\xi \quad (4.11)$$

An essentially equivalent result, which will be given in Section 4.5, can be derived using phase barrier theory. In that development the increase in the latent heat of vaporization for water forming a concave meniscus over that of water having a flat surface is given and leads to a qualitative discussion of the origins of capillary hysteresis in porous media.

The capillary-level description of the movement of water vapor moving from locations where a given concentration exists to locations of initially lower concentration depends on the ratio of average effective capillary radius to a quantity called the *mean free path* which is the average distance a vapor molecule travels before it undergoes a collision with another vapor molecule. If the capillary radius is large enough so that most of the collisions a vapor molecule makes are with other molecules instead of the walls of the capillary, then the concepts of Poiseuille's law and molecular diffusion apply. "Large" in this context means radii 8–10 times larger than the mean free path. An average value for the mean free path for air at room temperature and one atmosphere pressure is 6.40×10^{-6} cm. Capillaries of circular cross section and radius 6.4×10^{-5} cm would correspond approximately to those assumed to be present in a 2.25 bar ceramic. (The "bar" designation for a ceramic is the gas pressure, if applied to one side of a water-saturated ceramic sample, required to begin to push water out of the ceramic against the action of the capillary forces holding it in place.) If, on the other hand, the capillary radius is comparable to or less than the mean free path, then a more detailed scattering mechanism must be considered to describe vapor transport in it. Capillaries of radius 6.40×10^{-6} cm correspond approximately to those assumed to be present in what would be called a 22.5 bar ceramic. The intermediate range of capillary radii corresponding approximately to ceramics having air entry pressures in the 3.0–20.0 bar range is not amenable to either of the limiting case treatments mentioned but can be dealt with using a semiempirical approximation due to Kennard

(1938). The level of approximation of these approaches to the transport of water vapor in a porous medium, assumed to be made up of some form of collections of capillaries, is that of the classical kinetic theory of gases, to which we now turn.

4.2 VAPOR TRANSPORT IN CAPILLARIES

4.2.1 Introduction to the Kinetic Theory of Gases

The Maxwell–Boltzmann distribution for the velocities of n_v molecules per unit volume of an ideal gas, the molecules having no internal degrees of freedom, is, in three dimensions,

$$f(v_v) = n_v \left(\frac{m}{2\pi kT}\right)^{3/2} e^{-mv_v^2/2kT} \tag{4.12a}$$

where

$$v_v^2 = v_x^2 + v_y^2 + v_z^2 \tag{4.12b}$$

is the square of the magnitude of the velocity (the magnitude of the velocity of a molecule is its speed) of a molecule having Cartesian velocity components v_x, v_y, and v_z, m is the mass of a gas molecule, n_v is the number of molecules per unit spatial volume, $f(v_v)$ is the fraction of the total number of molecules per unit volume having velocities in the range v_x to $v_x + dv_x$, v_y to $v_y + dv_y$, and v_z to $v_z + dv_z$ in the volume element $dv_x\, dv_y\, dv_z$ of velocity space, and the rest of the symbols are as previously defined. If $f(v_v)$ is integrated over all angles in spherical coordinates in velocity space there results $n(v_v)$, the number of molecules per unit volume having a range of speeds $v_v + dv_v$. Thus,

$$n(v_v) = \int_0^{2\pi} \int_0^{\pi} f(v_v) \frac{\partial(v_x, v_y, v_z)}{\partial(v_v, \theta_v, \phi_v)}\, dv_v\, d\theta_v\, d\phi_v \tag{4.13}$$

where the transformation from the Cartesian volume element $dv_x\, dv_y\, dv_z$ of velocity space to the volume element in spherical velocity coordinates, denoted by the subscript v, is given by the Jacobian determinant

$$\frac{\partial(v_x, v_y, v_z)}{\partial(v_v, \theta_v, \phi_v)} = \det \begin{bmatrix} \frac{\partial v_x}{\partial v_v} & \frac{\partial v_x}{\partial \theta_v} & \frac{\partial v_x}{\partial \phi_v} \\[2mm] \frac{\partial v_y}{\partial v_v} & \frac{\partial v_y}{\partial \theta_v} & \frac{\partial v_y}{\partial \phi_v} \\[2mm] \frac{\partial v_z}{\partial v_v} & \frac{\partial v_z}{\partial \theta_v} & \frac{\partial v_z}{\partial \phi_v} \end{bmatrix} \tag{4.14}$$

multiplied by $dv_v\,d\theta_v\,d\phi_v$. The relations between the Cartesian and spherical velocity components are given by

$$v_x = v_v \sin(\theta_v)\cos(\phi_v) \tag{4.15a}$$

$$v_y = v_v \sin(\theta_v)\cos(\phi_v) \tag{4.15b}$$

$$v_z = v_v \cos(\theta_v) \tag{4.15c}$$

Evaluating the derivatives and determinant of (4.14) yields

$$\frac{\partial(v_x, v_y, v_z)}{\partial(v_v, \theta_v, \phi_v)} = v_v^2 \sin(\theta_v) \tag{4.16}$$

Substituting (4.16) and (4.15a) into (4.13) and performing the indicated integrations over θ_v and ϕ_v yields

$$n(v_v)\,dv_v = 4\pi n_v \left(\frac{m}{2\pi kT}\right)^{3/2} v_v^2 e^{-mv_c^2/2kT}\,dv_v \tag{4.17}$$

Note in passing that

$$\int_0^\infty n(v_v)\,dv_v = n_v \tag{4.18}$$

The most probable speed, v_m, where the subscript m denotes the mode of the distribution of molecular speeds, is found by solving

$$\left.\frac{dn(v_c)}{dv_c}\right|_{v_c=v_m} = 0 \tag{4.19}$$

for v_m. This yields

$$v_m = \left(\frac{\rho_g kT}{m}\right)^{1/2} \tag{4.20}$$

where $\rho_g = mn_v$ is the mass density of the gas. To compute the average speed $\bar{v}_v$ as an average over the distribution given by (4.17), we first notice that the quantity $n(v_v)/n_v$ is a distribution giving the fraction of molecules per unit volume having speeds between v_v and $v_v + dv_v$, that is, a distribution normalized to one. Thus,

$$\bar{v}_v = \int_0^\infty \frac{v_v\,n(v_v)}{n_v}\,dv_v = \left(\frac{8kT}{\pi m}\right)^{1/2} \tag{4.21}$$

when (4.17), rearranged, is substituted into (4.21) and the indicated integration carried out. The integration can be performed with the help of the gamma function previously described in (3.153) after the dummy variable of integration in (3.153) is changed from y to y^2 and y is identified with v_v of (4.21). Similarly it may be shown that the average of the square of v_v is given by

$$\overline{v_v^2} = \int_0^\infty \frac{v_v^2 n(v_v)}{n_v} dv_v = \frac{3kT}{m} \tag{4.22a}$$

where

$$\sqrt{\overline{v_v^2}} = v_{\text{rms}} = \left(\frac{3kT}{m}\right)^{1/2} \tag{4.22b}$$

is called the root-mean-square (rms) speed of the molecule.

We now consider the number of collisions per unit time a gas molecule moving with velocity $\vec{v}_1$ makes with other molecules of a gas moving as a whole with velocity $\vec{v}_g$. We consider each molecule to have an effective diameter, so far as collisions are concerned, of d_g, and it is further assumed that the molecules are spherically symmetric in this regard. If we have n_v molecules per unit volume, then the volume per molecule is $1/n_v$. Further, the shape of the volume in space swept out per unit time by a molecule between collisions is a "cylinder" of effective diameter d_g and volume $\pi (d_g/2)^2 |\vec{v}_1 - \vec{v}_g|$, where $\vec{v}_1 - \vec{v}_g$ is the velocity of the molecule relative to that of the gas as a whole and the vertical bars denote absolute value. A collision will occur if the molecules have a center-to-center distance d_g or less, and since collisions can occur anywhere on the circumference of the molecule we have approximately the situation shown in Figure 4.2, where the arrows represent the directions of motion of molecules being scattering relative to the central scattering molecule. The cylindrical volume swept out per unit time by the equivalent "molecule" of diameter $2d_g$ as it moves to collide with the surrounding gas of "point" molecules is $\pi d_g^2 |\vec{v}_1 - \vec{v}_g|$. The ratio of the volume per unit time swept out by the equivalent molecule to the spatial volume per point particle, $1/n_v$, is the number of collisions per unit time that the original molecule undergoes. Further, since it has been assumed that the molecules of the gas have no internal degrees of freedom, no energy of motion (kinetic energy) is lost to the excitation of energy levels associated with internal degrees of freedom. Thus, at constant temperature the Maxwell–Boltzmann velocity distribution of the gas as a whole and its total kinetic energy remain constant. Hence the number of collisions per unit time, n_c, is

$$n_c = \frac{\pi d_g |\vec{v}_1 - \vec{v}_g|}{1/n_v} = \pi n_v d_g |\vec{v}_1 - \vec{v}_g| \tag{4.23}$$

However, by our previous assumption, both $\vec{v}_1$ and $\vec{v}_g$ have the Maxwell–Boltzmann velocity distribution, so that the average collision frequency, $\bar{n}_c$, is given by

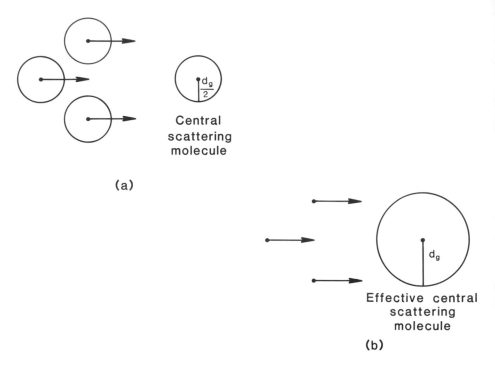

Figure 4.2: (a) "Spherical" molecules having their centers a distance d_g away from the central molecule potentially scattering one of them; (b) the situation equivalent to (a) of a "molecule" of diameter $2d_g$ scattering "point" molecules.

$$\bar{n}_c = \frac{\pi d_g^2 n_v \int \int |\vec{v}_1 - \vec{v}_g| e^{-m(v_1^2 + v_g^2)/2kT}\, d\vec{v}_1\, d\vec{v}_g}{\int \int e^{-m(v_1^2 + v_g^2)/2kT}\, d\vec{v}_1\, d\vec{v}_g} \qquad (4.24)$$

where the pre-exponential factors of the Maxwell–Boltzmann distribution have been divided out, $d\vec{v}_1 = dv_{1x}\, dv_{1y}\, dv_{1z}$, and similarly for $d\vec{v}_g$; $v_1^2 = \vec{v}_1 \cdot \vec{v}_1 = v_{1x}^2 + v_{1y}^2 + v_{1z}^2$, and similarly for v_g^2, and the sixfold integrations in the numerator and denominator are from $-\infty$ to ∞ over the components of $d\vec{v}_1\, d\vec{v}_g$. It is convenient at this point to change the variables of integration in (4.24) as follows.

$$\vec{x}_1 = \frac{\vec{v}_1 - \vec{v}_g}{\sqrt{2}} \qquad (4.25a)$$

$$\vec{x}_2 = \frac{\vec{v}_1 + \vec{v}_g}{\sqrt{2}} \qquad (4.25b)$$

The limits of integration are still $-\infty$ to ∞ over the components of $d\vec{x}_1$ and $d\vec{x}_2$ and $|\vec{v}_1 - \vec{v}_g| = \sqrt{2}|\vec{x}_1| = x_1$. Further, solving (4.25a) and (4.25b) for $\vec{v}_1$ and $\vec{v}_g$ yields

$$v_1^2 + v_g^2 = \frac{1}{2}\left[(\vec{x}_1 + \vec{x}_2)\cdot(\vec{x}_1 + \vec{x}_2) + (\vec{x}_1 - \vec{x}_2)\cdot(\vec{x}_1 - \vec{x}_2)\right]$$

$$= \frac{1}{2}\left(x_1^2 + 2\vec{x}_1\cdot\vec{x}_2 + x_2^2 + x_1^2 - 2\vec{x}_1\cdot\vec{x}_2 + x_2^2\right)$$

$$= x_1^2 + x_2^2 \tag{4.26}$$

The volume element in the new space is computed via the Jacobian determinant formalism as follows.

$$dv_{1x}\, dv_{1y}\, dv_{1z}\, dv_{gx}\, dv_{gy}\, dv_{gz} = \frac{\partial(\vec{v}_1, \vec{v}_g)}{\partial(\vec{x}_1, \vec{x}_2)}\, d\vec{x}_1\, d\vec{x}_2 \tag{4.27}$$

where

$$d\vec{x}_1\, d\vec{x}_2 = dx_{1x}\, dx_{1y}\, dx_{1z}\, dx_{2x}\, dx_{2y}\, dx_{2z} \tag{4.28}$$

Further, $\partial(\vec{v}_1, \vec{v}_g)/\partial(\vec{x}_1, \vec{x}_2)$ is understood to mean

$$\frac{\partial(\vec{v}_1, \vec{v}_g)}{\partial(\vec{x}_1, \vec{x}_2)} = \det\begin{bmatrix} \frac{\partial\vec{v}_1}{\partial\vec{x}_1} & \frac{\partial\vec{v}_1}{\partial\vec{x}_2} \\ \\ \frac{\partial\vec{v}_g}{\partial\vec{x}_1} & \frac{\partial\vec{v}_g}{\partial\vec{x}_2} \end{bmatrix} \tag{4.29}$$

in which (4.25a) and (4.25b) have been solved for $\vec{v}_1$ and $\vec{v}_g$ to yield

$$\vec{v}_1 = \frac{1}{\sqrt{2}}(\vec{x}_1 + \vec{x}_2) \tag{4.30a}$$

$$\vec{v}_g = \frac{1}{\sqrt{2}}(\vec{x}_2 - \vec{x}_1) \tag{4.30b}$$

In this particular case, the elements of (4.29) can be evaluated directly from (4.30a) and (4.30b) to yield

$$\frac{\partial(\vec{v}_1, \vec{v}_g)}{\partial(\vec{x}_1, \vec{x}_2)} = 1 \tag{4.31}$$

If instead the components of the $\vec{v}_1$, etc., were considered explicitly, then (4.30a) and (4.30b) would each be decomposed into three scalar equations, and a Jacobian of the form $\partial(v_{1x}, \cdots, v_{gx}, \cdots)/\partial(x_{1x}, \cdots, x_{2x}, \cdots)$ would be used. The use of (4.29) instead of the 36-element determinant is an example of the technique of partitioning matrices into contiguous groups of elements–submatrices–which are then treated as elements of what has thereby become a lower-rank matrix. Substituting (4.31) into (4.27) and thence into (4.24) along with (4.26) and (4.25) yields

$$\bar{n}_c = \frac{\pi d_g^2 n_v \int_{-\infty}^{\infty} \cdots \int_{-\infty}^{\infty} \sqrt{2}|\vec{x}_1|^{-m(x_1^2+x_2^2)/2kT} \, d\vec{x}_1 \, d\vec{x}_2}{\int_{-\infty}^{\infty} \cdots \int_{\infty}^{-\infty} e^{-m(x_1^2+x_2^2)/2kT} \, d\vec{x}_1 \, d\vec{x}_2} \tag{4.32}$$

or

$$\bar{n}_c = \frac{\sqrt{2}\pi d_g^2 n_v \int_{-\infty}^{\infty} \int_{-\infty}^{\infty} x_1 e^{-mx_1^2/2kT} \, dx_{1x} \, dx_{1y} \, dx_{1z}}{\int_{-\infty}^{\infty} \int_{-\infty}^{\infty} \int_{-\infty}^{\infty} e^{-mx_1^2/2kT} \, dx_{1x} \, dx_{1y} \, dx_{1z}} \tag{4.33}$$

We now change variables from Cartesian to spherical coordinates in velocity space as previously by setting

$$x_{1x} = v_v \sin(\theta_v) \cos(\phi_v) \tag{4.34a}$$

$$x_{1y} = v_v \sin(\theta_v) \sin(\phi_v) \tag{4.34b}$$

$$x_{1z} = v_v \cos(\theta_v) \tag{4.34c}$$

and carrying out the resulting integrals over v_v, θ_v, and ϕ_v to yield

$$\bar{n}_c = \sqrt{2}\pi d_g^2 n_v \bar{v}_v \tag{4.35}$$

where $\bar{v}_v$ is given by (4.21). The average distance a gas molecule travels between collisions is called the mean free path, l_f, and is equal to its average speed, $\bar{v}$, divided by its average collision frequency, $\bar{n}_c$, so that, from (4.35)

$$l_f = \frac{\bar{v}}{\bar{n}_c} = \frac{1}{\sqrt{2}\pi d_g^2 n_v} \tag{4.36}$$

From the ideal gas law we have for n_v

$$n_v = \frac{p_v}{kT} \tag{4.37}$$

where p_v is the vapor pressure as previously, so that (4.36) becomes

$$l_f = \frac{kT}{\sqrt{2}\pi d_g^2 p_v} \tag{4.38}$$

For water in a capillary of circular cross-section and radius ξ forming a meniscus, the pressure of water vapor just above it is given approximately by the Kelvin relation

$$p_v = p_{ov} e^{-2\gamma m \cos(\phi)/\rho kT\xi} \tag{4.39}$$

where all the symbols have been previously defined. Thus, $\bar{p}_v$, the average of Eq. (4.39) over a capillary pore radius frequency distribution, as given in Eq. (1.3), would be substituted into Eq. (4.38) to compute average mean free paths of vapor molecules in porous media.

Up to now we have been treating the vapor as a homogeneous gas. We now consider the mean free paths of the two species of a binary mixture of gases. We begin by considering the collision frequency of a molecule traveling with speed v_v through a Maxwellian gas of molecules of a different type. We write the differential collision frequency, $d\nu'$, as

$$d\nu' = \sigma_{12}\bar{v}_r n(v_g)\, dv_g \qquad (4.40)$$

where σ_{12} is the scattering cross-section for a collision between the incoming molecule of species 1 and the molecules of species 2, in this case gas molecules, so that in this case σ_{12} could have been written as σ_{1g}, and similarly ν' could have been written as ν_{1g} (recall Figure 4.2, in which σ_{12} would become $\sigma_{11} = \pi d_g^2$), $n(v_g)dv_g$ is the number of gas molecules per unit volume that have speeds in the range v_g to $v_g + dv_g$ and is given by (4.17) with v_v set equal to v_g, and v_r is the magnitude of the average velocity of the incoming molecule, $\vec{v}_1$, relative to the velocity of a gas molecule,

$$v_r = |\vec{v}_1 - \vec{v}_g| \qquad (4.41)$$

We now consider a special case of (4.40), namely one in which the magnitudes of $\vec{v}_1$ and $\vec{v}_g$ are fixed and the variable differences between them arise from varying differences in their relative directions only. This is depicted graphically for an arbitrary angle in Figure 4.3. We can average v_r over the available values of θ by writing, in spherical coordinates

$$v_r = \frac{1}{2}\int_0^\pi v_r \sin(\theta)\, d\theta = \frac{1}{2}\int_0^\pi [v_1^2 + v_g^2 - 2v_1 v_g \cos(\theta)]^{1/2} \sin(\theta)\, d\theta \quad (4.42)$$

where the law of cosines has been used to express the magnitude of v_r in terms of v_1, v_g, and θ, and the averaging kernel is the fraction $2\pi \sin(\theta)d\theta$ of the solid angle 4π steradians subtended by the unit sphere. Carrying out the integration indicated in (4.42) yields

$$v_r = \frac{1}{6v_1 v_g}\left\{[(v_1 + v_g)^2]^{3/2} - [(v_1 - v_g)^2]^{3/2}\right\}$$

$$= \frac{1}{6v_1 v_g}\left[(v_1 + v_g)^3 - |v_1 - v_g|^3\right] \qquad (4.43)$$

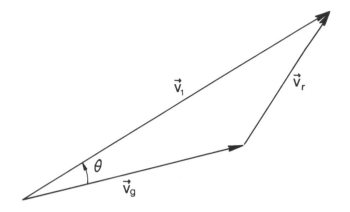

Figure 4.3: The vector sum of $\vec{v}_g$, and $\vec{v}_r$ equals $\vec{v}_1$. The instantaneous direction of $\vec{v}_1$ makes an angle θ with the direction of $\vec{v}_g$.

where $|v_1 - v_g|$, the absolute value of the difference between the speeds of the molecule traversing the gas and the gas molecules, is used to retain the positive nature of the quantity $[(v_1 - v_g)^2]^{3/2}$. Further, we can write this absolute value as

$$|v_1 - v_g| = v_1 - v_g, \qquad \text{for} \qquad v_1 > v_g \qquad (4.44a)$$

$$|v_1 - v_g| = v_g - v_1, \qquad \text{for} \qquad v_g > v_1 \qquad (4.44b)$$

Thus, on substituting (4.44a) and (4.44b) into (4.43) we have

$$v_r = v_1 + \frac{v_g^2}{3v_1}, \qquad \text{for} \qquad v_1 > v_g \qquad (4.45a)$$

$$v_r = v_g + \frac{v_1^2}{3v_g}, \qquad \text{for} \qquad v_g > v_1 \qquad (4.45b)$$

We now substitute (4.17), (4.45a), and (4.45b) into (4.40) and integrate to find the collision frequency. Thus,

$$\int_0^\nu d\nu' = 4\pi n_v \left(\frac{m}{2\pi kT}\right)^{3/2} \sigma_{12}$$

$$\times \left[\int_0^{v_1} \left(v_1 + \frac{v_g^2}{3v_1}\right) v_g^2 e^{-mv_g/2kT}\, dv_g\right.$$

$$\left. + \int_{v_1}^\infty \left(v_g + \frac{v_1^2}{3v_g}\right) v_g^2 e^{-mv_g^2/2kT}\, dv_g\right] \quad (4.46)$$

where now v_g is a dummy variable of integration. Setting $m/2kT = a$ for ease of manipulation, the integrals within the brackets, denoted collectively by I, may be rearranged as follows

$$I = v_1 \int_0^{v_1} v_g^2 e^{-av_g^2}\, dv_g + \frac{1}{3v_1} \int_0^{v_1} v_g^4 e^{-av_g^2}\, dv_g$$

$$+ \int_{v_1}^\infty v_g^3 e^{-av_g^2}\, dv_g + \frac{v_1^2}{3} \int_{v_1}^\infty v_g e^{-av_g^2}\, dv_g \quad (4.47)$$

The first two integrals may be reduced by successive integrations by parts using, in the integration by parts formula,

$$\int u\, dv = [uv] - \int v\, du \quad (4.48a)$$

with

$$dv = v_g e^{-av_g^2}\, dv_g \quad (4.48b)$$

so that

$$v = \frac{e^{-av_g^2}}{-2a} \quad (4.48c)$$

and

$$u = v_g \quad (4.48d)$$

until all powers of v_g in the integrand(s) have been removed. The third and fourth integrals can be evaluated by changing the variable of integration from v_g to $x = v_g^2$ and evaluating the result using integral tables. These manipulations yield for the integrals in Eq. (4.47)

$$I = \left(\frac{-v_1^2}{2a^2} - \frac{v_1^2}{6a^2} - \frac{1}{4a^4} + \frac{1}{2a^4} + \frac{v_1^2}{2a^2} + \frac{v_1^2}{6a^2}\right) e^{-a^2 v_1^2}$$

$$+ \left(\frac{v_1}{2a^2} + \frac{1}{4a^4 v_1}\right) \int_0^{v_1} e^{-a^2 v_g^2}\, dv_g \quad (4.49)$$

Setting

$$x = \sqrt{a} v_g \qquad (4.50)$$

in the remaining integral and rearranging terms yields

$$I = \frac{1}{4a^4} e^{-av_1^2} + \left(\frac{v_1}{2a^2} + \frac{1}{4a^4 v_1} \right) \frac{1}{a} \int_o^{av_1} e^{-x^2} \, dx \qquad (4.51)$$

Substituting (4.50) into (4.51), replacing a by $m\, 2kT$, and recalling that the error function, erf(), is defined as

$$\text{erf}(z) = \frac{2}{\sqrt{\pi}} \int_0^z e^{-x^2} \, dx \qquad (4.52)$$

yields for the collision frequency, ν, of a molecule having velocity v_1 encountering a hard-sphere Maxwellian gas composed of molecules of mass m at temperature T

$$\nu = \frac{n_v \sigma_{12}}{\sqrt{\pi} c} \left[e^{-mv_1^2/2kT} + \sqrt{\pi} \left(v_1 c + \frac{1}{2v_1 c} \right) \text{erf}\,(v_1 c) \right] \qquad (4.53a)$$

where

$$c = \left(\frac{m}{2kT} \right)^{1/2} \qquad (4.53b)$$

The mean free path of the molecule, l_1, is thus

$$l_1 = \frac{v_1}{\nu} \qquad (4.54)$$

If the molecule encounters a number of intermixed Maxwellian gases, such as might be assumed to comprise air, the resultant collision frequency is the sum of the collision frequencies computed using (4.53a) and (4.53b) for each component gas, while the resultant mean free path is $\bar{v}_v$ of Eq. (4.21) divided by the composite collision frequency.

    We now consider the collision frequency of the molecules of mass m_1 of a hard-sphere Maxwellian gas having n_{1v} molecules per unit volume colliding with another Maxwellian gas composed of molecules of mass m_2 and having n_{2v} molecules per unit volume. We begin by denoting m in Eq. (4.53a) and (4.53b) by m_2 and n_v by n_{2v} to yield

$$\nu = \frac{n_{2v} \sigma_{12}}{\sqrt{\pi} c_2} \left[e^{-m_2 v_1^2/2kT} + \sqrt{\pi} \left(v_1 c_2 + \frac{1}{2v_1 c_2} \right) \text{erf}\,(v_1 c_2) \right] \qquad (4.55a)$$

where

$$c_2 = \left(\frac{m_2}{2kT} \right)^{1/2} \qquad (4.55b)$$

The number of collisions per unit time, $d\nu'_{12}$, made by molecules of the first gas, having speeds in the range v_1 to $v_1 + dv_1$, with those of the second gas is

$$n_{1v} \, d\nu'_{12} = 4\pi n_{1v} \left(\frac{m_1}{2kT} \right)^{3/2} v_1^2 e^{-m_1 v_1^2/2kT} \nu \, dv_1 \tag{4.56}$$

where (4.17), with n_v written as n_{1v}, m set equal to m_1, and v_c identified with v_1, has been used. Indicating integration of (4.56) and dividing both sides by n_{1v} yields

$$\int_0^{v_{12}} d\nu'_{12} = \frac{4}{\pi} c_1^3 \sigma_{12} n_{2v}$$

$$\times \int_0^\infty c_2^{-1} \left[e^{-m_2 v_1^2/2kT} + \sqrt{\pi} \left(v_1 c_2 + \frac{1}{2v_1 c_2} \right) \mathrm{erf} \left(v_1 c_2 \right) \right]$$

$$\times e^{-m_1 v_1^2/2kT} v_1^2 \, dv_1 \tag{4.57a}$$

where

$$c_1 = \left(\frac{m_1}{2kT} \right)^{1/2} \tag{4.57b}$$

The three integrals appearing on the right-hand side of Eq. (4.57a) are of the forms

$$I_1 = \int_0^\infty v_1^2 e^{-a_2^2 v_1^2} \, dv_1 \tag{4.58a}$$

$$I_2 = \int_0^\infty v_1^3 e^{-a_1^2 v_1^2} \, \mathrm{erf}(bv_1) \, dv_1 \tag{4.58b}$$

$$I_3 = \int_0^\infty v_1 e^{-a_1^2 v_1^2} \, \mathrm{erf}(bv_1) \, dv_1 \tag{4.58c}$$

where

$$a_2^2 = \frac{m_1 + m_2}{2kT} \tag{4.59a}$$

$$a_1^2 = \frac{m_1}{2kT} \tag{4.59b}$$

$$b = \left(\frac{m_2}{2kT} \right)^{1/2} \tag{4.59c}$$

The first integral can be performed using the gamma function, which can be written as

$$\int_0^\infty e^{-c^2 x^2} x^{2z-1} \, dx = \frac{\Gamma(z)}{2c^2} \tag{4.60}$$

with $z = \frac{3}{2}$ and $c = a_2$. The second integral can be deduced by applying integration by parts to the third integral, which is

$$\int_0^\infty v_1 e^{-a_1^2 v_1^2} \, \mathrm{erf}(bv_1) \, dv_1 = \frac{b}{2a_1^2} (a_1^2 + b^2)^{-1/2} \tag{4.61}$$

where the conditions $\mathrm{Re}(a_1^2) > \mathrm{Re}(b^2)$ and $\mathrm{Re}(a_1^2) > 0$, where $\mathrm{Re}()$ means "the real part of" are required for the integral to converge. In the present case both a_1 and b are taken to be real and positive and the stated conditions mean that $m_1/2kT > m_2/2kT$ or $m_1 > m_2$ If the roles of the gases are reversed, we have the condition $m_2 > m_1$ and since the final result will be symmetric between the two gases, these conditions do not restrict it. Finally we have upon integrating (4.61) by parts

$$\int_0^\infty v_1^3 e^{-a_1^2 v_1^2} \, \mathrm{erf}(bv_1) \, dv_1 = \frac{b}{2a_1^2(a_1^2 + b^2)^{1/2}} \left(\frac{1}{a_1^2} + \frac{1}{2(a_1^2 + b^2)} \right) \quad (4.62)$$

Applying these results to Eq. (4.61), substituting Eqs. (4.58) and (4.59) into Eq. (4.57), and rearranging yields

$$\nu_{12} = \sigma_{12} n_{2v} \left(\frac{8kT(m_1 + m_2)}{\pi m_1 m_2} \right)^{1/2} \quad (4.63)$$

Recalling that the average speed of a molecule in a Maxwellian gas is

$$\bar{v} = \left(\frac{8kT}{\pi m} \right)^{1/2} \quad (4.21)$$

we have for the binary collision frequency, ν_{12},

$$\nu_{12} = \sigma_{12} n_{2v} (\bar{v}_1^2 + \bar{v}_2^2)^{1/2} \quad (4.64)$$

where $\bar{v}_1$ and $\bar{v}_2$ are the average speeds of the molecules having masses m_1 and m_2, respectively. Writing ν_i for the collision frequency of the molecules having mass m_i of a hard-sphere Maxwellian gas with those of a mixture of hard-sphere Maxwellian gases, the molecules of which have masses m_j and numbers of molecules per unit volume n_{jv}, where $j = 1, \ldots, m$, and factoring out the average speed of the molecules having mass m_i yields

$$\nu_i = \sum_{j=1}^n \nu_{ij} = \bar{v}_i \sum_{j=1}^n \sigma_{ij} n_{jv} \left(1 + \frac{m_i}{m_j} \right)^{1/2} \quad (4.65)$$

We now write for the mean free path of the molecules of mass m_i, l_{if},

$$l_{if} = \frac{\bar{v}_i}{\nu_i} = \frac{1}{\sum_{j=1}^n \sigma_{ij} n_{jv} \left(1 + \frac{m_i}{m_j} \right)^{1/2}} \quad (4.66)$$

The expressions given for collision frequency and mean free path account for the collisions of the molecules of a given gas with each other as well as with the molecules of the other gases that may be present.

We now consider the simplest theory available for the viscosity of a mixture of Maxwellian gases. We begin by deriving an expression for the viscosity of a single-component Maxwellian gas. We consider gas flowing with average macroscopic velocity v_{0z} along the lower of two parallel plates viewed from their edges. We presume that a velocity gradient exists along the x direction and further that $dv_{0z}/dx > 0$, that is, that as x increases, v_{0z} increases as well. We consider a volume element dV which is an average distance $\bar{x}$ away from the lower plate and assume that the average velocity of the molecules scattered out of the element of volume dV is equal to the average macroscopic velocity of gas flow at that point which in this case can be written $v_{0z} + \bar{x}\, \partial v_{0z}/\partial x$. Further, since the chance for a molecule crossing the lower plate to be scattered by another molecule is the same as if it had just been scattered at the lower plate, $\bar{x}$ can be viewed as the projection of $-l_f$ onto the x axis, or

$$\bar{x} = -l_f \cos(\theta) \tag{4.67}$$

We now compute the number of molecules per unit area per unit time impinging on the lower plate from below. In spherical coordinates in velocity space the angular and speed distribution, f_r, of the molecule of a Maxwellian gas is

$$f_r = \left(\frac{m}{2\pi kT}\right)^{3/2} e^{-mv_v^2/2kT} v_v^2 \sin(\theta_v)\, d\theta_v\, d\phi_v\, dv_v \tag{4.68}$$

Integrating (4.68) over ϕ_v ($0 \le \phi_v \le 2\pi$) and θ_v ($0 \le \theta_v \le \pi$) yields the number of molecules moving radially in all directions in velocity space having speeds in the range v_v to $v_v + dv_v$, $n(v_v)\, dv_v$, as given previously,

$$n(v_v)\, dv_v = 4\pi n_v \left(\frac{m}{2\pi kT}\right)^{3/2} e^{-mv_v^2/2kT} v_v^2\, dv_v \tag{4.17}$$

The fraction of the number of molecules contained in the solid angle $d\Omega$ steradians having speeds in the range v_v to $v_v + dv_v$ is $n(v_v)\, dv_v\, d\Omega/4\pi$, where 4π steradians, the result of the integrals over θ_v and ϕ_v of (4.68) leading to (4.17), is one complete solid angle. The number of molecules impacting an area A of the lower plate in a time dt that approach it from an angle θ_v with its normal is the number contained in a cylinder of length $v_v\, dt$ and projected (onto the lower plate) cross-sectional area $A \cos(\theta_v)$. Thus, the number of molecules per unit area per unit time impinging on the lower plate from an angle θ_v is, on using the substitution $d\Omega = \sin(\theta_v)\, d\theta_v\, d\phi_v$,

$$\int_0^\infty \frac{n(v_v)v_v\,dv_v}{4\pi} \int_0^{2\pi} \int_0^{\pi/2} \sin(\theta_v)\cos(\theta_v)d\theta_v\,d\phi_v$$

$$= \pi n_v \left(\frac{m}{2\pi kT}\right)^{3/2} \int_0^\infty e^{-mv_v^2/2kT}v_v^3\,dv_v$$

$$= \frac{n_v}{2}\left(\frac{2kT}{\pi m}\right)^{1/2} = \frac{1}{4}n_v\bar{v} \qquad (4.69)$$

where (4.60) has been used and the upper limit of $\pi/2$ of the θ_v integration reflects the fact that the molecules impinging on the lower plate come from the "upper" solid angle of 2π steradians.

We can now continue with the evaluation of $\bar{x}$ by averaging it over the "upper" solid angle weighted by the factor $\cos(\theta_v)$ as previously to yield

$$\bar{x} = \frac{\int_0^{2\pi}\int_0^{\pi/2}[-l_f\cos(\theta_v)]\sin(\theta_v)\cos(\theta_v)\,d\theta_v\,d\phi_v}{\int_0^{2\pi}\int_0^{\pi/2}\sin(\theta_v)\cos(\theta_v)\,d\theta_v\,d\phi_v} = \frac{-2}{3}l_f \qquad (4.70)$$

The momentum per molecule transferred from the volume element dV to the moving gas stream as a result of scattering molecules reaching it from the bottom plate is

$$m\left(v_{0z} + \bar{x}\frac{dv_{0z}}{dx}\right) = m\left(v_{oz} - \frac{2l_f}{3}\frac{dv_{0z}}{dx}\right) \qquad (4.71a)$$

while for molecules reaching dV from the upper plate for which the sign of $\bar{x}$ is reversed we have

$$m\left(v_{0z} + \bar{x}\frac{dv_0}{dx}\right) = m\left(v_{0z} + \frac{2l_f}{3}\frac{dv_{0z}}{dx}\right) \qquad (4.71b)$$

Subtracting (4.71b) from (4.71a) yields $-\frac{4}{3}ml_f\,dv_{0z}/dx$ for the net z component of momentum per molecule transferred from dV to $+x$ via shear. Multiplying this result by the right-hand side of (4.69), which is the number of molecules per unit area per unit time that arrive from each side of dV to be scattered [Eqs. (4.71a) and (4.71b) could each have been so multiplied before subtracting] yields for the total z component of momentum per unit area (area parallel to the $x$axis) per unit time transferred to $+x$ via viscous shear.

$$P_{xz} = -\frac{1}{3}mn_v\bar{v}_vl_f\frac{dv_{0z}}{dx} \qquad (4.72)$$

where $\bar{v}_v$ is given by (4.21). However,

$$P_{xz} = -\eta\frac{dv_{0z}}{dx} \qquad (4.73)$$

where η is the viscosity of the gas and the minus sign arises from the fact that momentum is being transferred in the positive x direction. Equating (4.72) and (4.73) yields

$$\eta = \frac{1}{3}mn_v\bar{v}l_f = \frac{1}{3}\rho_g\bar{v}_vl_f \qquad (4.74)$$

where

$$\rho_g = mn_v \qquad (4.75)$$

is the mass density of the gas as noted in connection with Eq. (4.21). To a first approximation, the viscosity of a mixture of Maxwellian gases of species i is

$$\eta = \frac{1}{3}\sum_{i=1}^{m_i}\rho_i\bar{v}_il_{if} \qquad (4.76)$$

where l_{if} is given by (4.66) and the sum is over species of gas. We now introduce the notation $\bar{l}_{fi}$ to signify special cases of (4.66) where only the $i = j$ term in the sum is selected. This term is of course the mean free path of a species i of gas with no other gases present and is

$$\bar{l}_{fi} = \frac{1}{\sqrt{2}n_{iv}\sigma_{ii}} \qquad (4.77)$$

We may now express the viscosity of an individual species of gas as

$$\eta_i = \frac{1}{3}\rho_i\bar{v}_v\bar{l}_{fi} \qquad (4.78)$$

and the total viscosity as

$$\eta = \sum_{i=1}^{n}\frac{\eta_il_{if}}{\bar{l}_{fi}} \qquad (4.79)$$

which has the form of a weighted sum of the viscosities of the individual gases. The concept of viscosity for vapor transport in capillaries has meaning only for capillary radii greater than or equal to ten times the mean free path of the gas molecules at the pressure and temperature being considered. Poiseuille's law with slip, which we will now consider, holds approximately in this regime containing as it does the viscosity and the gradient of velocity.

4.2.2   Vapor Transport in Large-Radius Capillaries

Poiseuille's law has been discussed for the nonturbulent flow of liquid through a tube in Section 2.9. For flow in a filled tube of circular cross section of radius ξ and length ℓ along the z axis of a ρ-θ-z polar coordinate system, we have, upon writing the steady-state equation of motion,

$$\frac{dv_z}{d\zeta} = \frac{-(p_{v2} - p_{v1})\zeta}{2\eta\ell} \tag{2.103}$$

where the symbols are as previously defined and the subscript v stands for vapor. The boundary condition at the wall of the tube, which was given as $v_z = 0$ for liquid flow, is now

$$v_z\big|_{\zeta=\xi} = -K_s \frac{dv_z}{d\zeta}\bigg|_{\zeta=\xi} \tag{4.80}$$

where K_s is the coefficient of slip of gas along the wall. The velocity of slip of the gas past the tube wall evaluated at the wall is given by combining (2.148) and (4.80) to yield

$$v_z\big|_{\zeta=\xi} = \frac{(p_{v2} - p_{v1})\xi K_s}{2\eta\ell} \tag{4.81}$$

Separating variables in (2.103) and integrating yields

$$v_0 - v_z = \frac{\zeta^2(p_{v2} - p_{v1})}{4\eta\ell} \tag{2.105}$$

Combining (4.81) and (2.105) yields at the wall of the tube where $\zeta = \xi$

$$v_0 = \frac{(p_{v2} - p_{v1})}{4\eta\ell}(\xi^2 + 2K_s\xi) \tag{4.82}$$

and combining (2.105) and (4.82) yields

$$v_z = \frac{p_{v2} - p_{v1}}{4\eta\ell}(\xi^2 - \zeta^2 + 2K_s\xi) \tag{4.83}$$

which is the velocity of the gas flowing through a tube corrected for slip along the wall of the tube. We now, in anticipation of developments shortly to be carried out, replace $(p_{v2} - p_{v1})/\ell$ by $-dp_v/dz$, where the minus sign expresses the fact that p decreases as z increases, to obtain

$$v_z = \frac{-1}{4\eta}(\xi^2 - \zeta^2 + 2K_s\xi)\frac{dp_v}{dz} \tag{4.84}$$

We now consider K_s in more detail. We assume that a fraction f_m of the molecules of the gas striking the inner surface of the tube give up some part of the component of their total momentum tangent to the wall and thus that the observed viscous drag of the gas is due to the change in tangential momentum that the molecules undergo upon diffuse reflection from the wall. Molecules reflected with no change in their momentum are said to have been specularly reflected. Momentum is brought to the wall in the amount $(-f_m/2)\,\eta\,(dv_z/d\zeta)$ per unit area per unit time by the aggregate of gas molecules approaching it (half the total viscous drag multiplied by the fraction, f, of molecules losing some tangential momentum to the wall) and also in amount $\frac{1}{4}n_v\bar{v}_v m\, v_z|_{\zeta=\xi}$ due to viscous slip along the wall, where we recall from (4.69) that $\frac{1}{4}n_v\bar{v}_v$ is the number of molecules per unit area per unit time impinging on the wall and $mv_z|_{\zeta=\xi}$ is the momentum per molecule of mass m due to its viscous slip velocity $v_z|_{\zeta=\xi}$ along the wall. The total momentum transferred to the wall per unit area per unit time [compare (4.73)] is $-\eta\,(dv_z/d\zeta)$ so that

$$f_m\left(\frac{-\eta}{2}\frac{dv_z}{d\zeta} + \frac{1}{4}n_v m\bar{v}\,v_z|_{\zeta=\xi}\right) = -\eta\frac{dv_z}{d\zeta} \tag{4.85}$$

or, on eliminating $v_z|_{\zeta=\xi}$ between (4.85) and (4.80), we have, upon solving for K_s,

$$K_s = \frac{2\eta(2 - f_m)}{f_m n_v m\bar{v}_v} \tag{4.86}$$

Substituting

$$n_v = \frac{p_v}{kT} \tag{4.21}$$

from the ideal gas law and

$$\bar{v}_v = \left(\frac{8kT}{\pi m}\right)^{1/2}$$

from (4.21) yields

$$K_s = \frac{\eta(2 - f_m)}{f_m}\left(\frac{\pi kT}{2m}\right)^{1/2}\frac{1}{p_v} = \frac{C_s}{p_v} \tag{4.87a}$$

where

$$C_s = \frac{\eta(2 - f_m)}{f_m}\left(\frac{\pi kT}{2m}\right)^{1/2} \tag{4.87b}$$

from which we see that K_s is inversely proportional to the gas pressure, p, at each point in the tube. In a porous medium a reasonable assumption is that $f_m = 1$. In that case K_s becomes

$$K_s = \eta\left(\frac{\pi kT}{2m}\right)^{1/2}\frac{1}{p_v} \tag{4.88}$$

Substituting (4.87a) into (4.84) yields

$$v_z = \frac{-1}{4\eta} \left(\xi^2 - \zeta^2 + \frac{2C_s\xi}{p_v} \right) \frac{dp_v}{dz} \tag{4.89}$$

In the development of Poiseuille's law for liquid flowing through a tube of circular cross-section, the volume per unit time passing through the tube was given in (2.111). Multiplying this quantity by the density of the liquid would yield the mass per unit time flowing through the tube since the fluid is assumed to be essentially incompressible. For the case of gas flowing through a tube, the volume per time is not a particularly meaningful quantity due to its high compressibility, and a more refined approach must be employed. The mass per unit time flowing through the tube is a much more meaningful quantity, and will now be calculated. The volume of gas per unit time flowing through the tube is given by (2.111) with q set equal to q_g, as

$$q_g = \int_0^\xi v_z(\zeta) 2\pi\zeta \, d\zeta \tag{4.90}$$

Multiplying q_g by the density of the gas, ρ_g, as given using the ideal gas law,

$$\rho_g = mn_v = \frac{mp_v}{kT} \tag{4.91}$$

and substituting into (4.90) yields

$$M_g = \rho_g q_g = \frac{-mp_v}{4\eta kT} \frac{dp_v}{dz} \int_0^\xi \left(\xi^2 - \zeta^2 + \frac{2C_s}{p_v} \right) 2\pi\zeta \, d\zeta \tag{4.92}$$

where M_g is the mass of gas per unit time flowing through the tube. Since p_v is assumed to be uniform across any cross-section of the tube, it does not depend on ζ, and we carry out the integration with C_s/p_v held constant to yield

$$M_g = \frac{-\pi m\xi^4 p_v}{8\eta kT} \frac{dp_v}{dz} \left(1 + \frac{4C_s}{p_v\xi} \right) \tag{4.93}$$

For steady flow in the tube, at any given time, M_g is a constant and we can separate variables, p_v and z, and write

$$M_g \int_0^l dz = \frac{-\pi m\xi^4}{8\eta kT} \int_{p_{v2}}^{p_{v1}} \left(p_v + \frac{4C_s}{\xi} \right) dp_v \tag{4.94}$$

where, as before, $p_{v2} > p_{v1}$, thus causing flow of gas from $z = 0$ to l. Integrating and rearranging yields

$$M_g = \frac{\pi m\xi^4}{8\eta \ell kT} \left(\frac{p_{v2}^2 - p_{v1}^2}{2} + \frac{4C_s}{\xi}(p_{v2} - p_{v1}) \right) \tag{4.95}$$

It can further be shown that when a temperature gradient, dT/dz, exists along the tube, M_g can be written

$$M_g = \frac{-\pi m \xi^4 p_v}{8\eta kT} \frac{dp_v}{dz} \left(1 + \frac{4C_s}{p_v \xi}\right) - \frac{3\pi\eta\xi^2}{4T} \frac{dT}{dz} \qquad (4.96)$$

where the sign convention is that temperature decreases as z increases and that $T_2 > T_1$. Separating variables and integrating yields

$$M_g = \frac{\pi m \xi^4}{8\eta k \overline{T}} \left(\frac{p_{v2}^2 - p_{v1}^2}{2} + \frac{4C_s}{\xi}(p_{v2} - p_{v1})\right) + \frac{3\pi\eta\xi^2}{4} \ln\left(\frac{T_2}{T_1}\right) \qquad (4.97)$$

where T_2 and T_1 are the temperatures at the ends of the tube corresponding to $z = 0$ and ℓ, respectively and $\overline{T} = \frac{1}{2}(T_2 + T_1)$ has been substituted for T in the first term on the right-hand side of (4.97) and assumed constant, making (4.97) an approximate result. Equation (4.94) holds for the movement of gas as a whole through a tube of radius large compared to the mean free path of the gas in it. If the gas consists of a mixture of species, then the mean free paths of the components are given by (4.66), and the viscosity of the mixture, η in Eq. (4.97), is given by (4.76) or (4.79).

4.2.3 Vapor Transport in Medium-Radius Capillaries

When the density of the vapor being transported is sufficiently low as to result in the mean free path of the gas molecules being longer than the diameter of the tube along which flow is taking place, the ordinary concepts of viscosity and a hydrodynamic approach to vapor transport, such as would include Poiseuille's law, fail and a mathematical description based on molecular scattering must be used instead. We first consider a tube long compared to its diameter, so that end effects may be neglected, as depicted in cross-section in Figure 4.4. We consider the molecules that leave an element of area dA_1 on the wall of the tube after impinging on it that subsequently pass through dA. Assuming, as above, that the fraction, f_m, of molecules undergoing a change in their tangential momentum upon being scattered from the wall is equal to 1 for earth materials means that molecules leaving dA_1 have no memory of their original velocities. Thus, we assume that the molecules leaving dA_1 do so as if dA_1 were a source of molecules having a Maxwellian distribution of velocities. In spherical coordinates in velocity space centered on dA_1, we write for the number, n_1, of molecules leaving dA_1 per unit time per unit volume in momentum space to pass through dA in the fraction of the element of solid angle $d\Omega_1/4\pi$ where $d\Omega_1$ is the element of solid angle subtended by dA at the location of

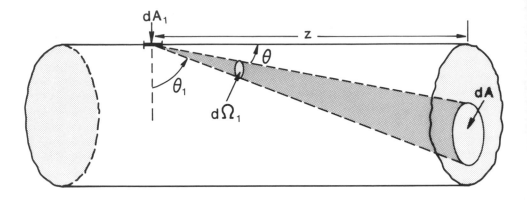

Figure 4.4: A capillary tube of uniform, though not necessarily circular, cross section with elements of area dA and dA_1 on planes perpendicular to the cross section of the tube and the wall, respectively.

dA_1,

$$n_1 = \left(\frac{m}{2\pi kT}\right)^{3/2} n_{v_1} v_1^2 e^{-v^2/2kT} \left(\frac{d\Omega}{4\pi}\right) \cos(\theta_1) v_1 \, dv_1 \, dA_1 \qquad (4.98)$$

where $\cos(\theta_1) v_1 \, dA_1$ is the volume traversed in unit time by a molecule leaving dA_1 at an angle θ_1 with respect to the normal to dA_1, as shown in Figure 4.4, m is the mass of a gas molecule, and n_1 is the average number of gas molecules per unit volume, in the vicinity of dA_1. Integrating (4.98) over v_1 from 0 to ∞ yields,

$$f_1 = \int_0^{\infty} n_1 \, dv_1 = n_{v_1} \bar{v}_1 \frac{d\Omega}{4\pi} \cos(\theta_1) \, dA_1 \qquad (4.99)$$

where f_1 is the number of molecules per unit solid angle passing through the element of area dA, and $\bar{v}_1$ is the average speed of a molecule as given in (4.21).

Figure 4.5 shows the same tube with the positive z axis directed vertically downward. The distance between dA and the tube wall at a point vertically below dA_1 is r_2, while a perpendicular to the tube wall at that point,

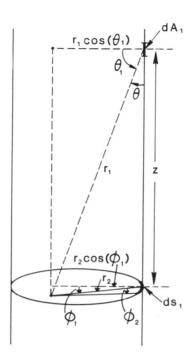

Figure 4.5: A tube of constant cross-section with the z axis directed vertically downward. The element of area dA is taken not to lie on a diameter of the cross section of the tube.

$r_2 \cos(\phi_1)$, is the base of the right triangle having r_2 as the hypotenuse, while ϕ_2 is the angle with its vertex in dA subtending an element of arc length ds on the wall of the cylinder, and r_1 is drawn from dA_1 to the projection of $r_2 \cos(\phi_1)$ through dA as shown. From Figure 4.5 we thus see that

$$z = r_1 \cos(\theta) = r_2 \cot(\theta) \tag{4.100a}$$

$$r_1 \cos(\theta_1) = r_2 \cos(\phi_1) \tag{4.100b}$$

$$r_1^2 = r_2^2 + z^2 \tag{4.101}$$

We recall that the element of solid angle $d\Omega$ can be expressed as

$$d\Omega_1 = \frac{dA \cos(\theta)}{r_1^2} \tag{4.102}$$

Substituting (4.101) and (4.102) into (4.99) and indicating integration yields

$$N_1 = \frac{\bar{v}}{4\pi} \int dA \int \frac{n_{v_1} \cos(\theta_1) \cos(\theta)}{(r_2^2 + z^2)} dA_1 \qquad (4.103)$$

Writing dA_1 as $ds_1\, dz$, using (4.100a,b), and noting that the change in n_{v_1} along the tube can be written as

$$n_{v_1} - n_{v_0} = z \frac{dn_{v_1}}{dz} \qquad (4.104)$$

where n_{v_0} is the molecular distribution function at the location of the cross-section of the tube containing dA, the expression for N_1, the number of molecules per unit time passing through the cross-section of the tube containing dA, becomes

$$N_1 = \frac{\bar{v}}{4\pi} \int dA \int r_2 \cos(\phi_1)\, ds_1 \int_{-\infty}^{\infty} \frac{z}{(r_2^2 + z^2)^2} \left(n_{v_0} + z \frac{dn_{v_1}}{dz} \right) dz \qquad (4.105)$$

where the long tube approximation, $-\infty \le z \le \infty$, has been used. Assuming that n_{v_0} is an even function, and that the pressure gradient is uniform along the tube so that dn_{v_1}/dz is a constant yields, when the indicated integrations over z are carried out,

$$N_1 = \frac{dn_{v_1}}{dz} \frac{\bar{v}_v}{8} \int dA \int \cos(\phi_1)\, ds_1 \qquad (4.106)$$

Using $\bar{v}_v$ as given in (4.21), the ideal gas law to cast M_{v_1} in terms of pressure, and writing $ds_1 \cos(\phi_1) = r_2\, d\phi_2$ (compare Figure 4.5) yields for N_1

$$N_1 = \frac{dp_v}{dz} \left(\frac{m}{8\pi kT} \right)^{1/2} \int dA \int_0^{2\pi} r_2\, d\phi_2 \qquad (4.107)$$

The mass of gas per unit time moving through the tube, M_m, is given by mN_1. If we assume a constant and uniform gradient along a length ℓ of tube, dp_v/dz becomes

$$\frac{dp_v}{dz} = \frac{p_{v2} - p_{v1}}{\ell} \qquad (4.108)$$

and we have for M_m

$$M_m = \frac{(p_{v2} - p_{v1})}{\ell} \left(\frac{m}{8\pi kT} \right)^{1/2} \int dA \int_0^{2\pi} r_2\, d\phi_2 \qquad (4.109)$$

Equation (4.109) can now be integrated over the circumference of the tube and over its cross-sectional area for tubes of various cross-sectional shapes.

For a tube of circular cross section and radius a, carrying out the indicated integrals results in $16a^3\pi/3$ so that for this case the mass per unit area per unit time moving through the tube, Q_m, becomes

$$Q_m = \frac{8a}{3\ell}\left(\frac{m}{2\pi kT}\right)^{1/2}(p_{v2} - p_{v1}) \qquad (4.110)$$

If f_m, the fraction of molecules undergoing a change in the component(s) of their momenta tangential to the wall were not equal to 1, then (4.110) would become

$$Q_m = \frac{2 - f_m}{f_m}\left(\frac{8\pi a^3(p_{v2} - p_{v1})}{3\ell(2\pi mkT)^{\frac{1}{2}}}\right) \qquad (4.111)$$

The case of effusion of molecules through a hole smaller in diameter than the mean free path and placed in a thin wall is the opposite of the long tube case just considered. We now consider effusion.

4.2.4   Molecular Effusion

We consider a Maxwellian distribution of molecular speeds and write for the number of molecules moving in all directions in velocity space with speeds in the range v to $v + dv$ as in (4.17)

$$n(v)\,dv = 4\pi n_{v_1}\left(\frac{m}{2\pi kT}\right)^{3/2}e^{-mv^2/2kT}v^2\,dv \qquad (4.112)$$

The fraction of the number of molecules passing through the solid angle $d\Omega$ per unit area per unit time having speeds in the range v to $v + dv$ is

$$\frac{n(v)v\cos(\theta)\,d\Omega}{4\pi} = n_{v_1}\left(\frac{m}{2\pi kT}\right)^{3/2}e^{-mv^2/2kT}v^3\,dv\sin(\theta)\cos(\theta)\,d\theta\,d\phi \qquad (4.113)$$

where the volume the molecules traverse in time dt is $v\,dt\cos(\theta)\,dA$, where dA is the element of area subtending $d\Omega$ that the molecules pass through in time dt and is oriented at an angle θ with respect to the normal to the opening through which effusion is occurring. Integrating (4.113) over θ, ϕ, and v yields the total number of molecules effusing through the hole per unit area per unit time, N_e,

$$N_e = \frac{n_{v_1}}{4}\bar{v} \qquad (4.114)$$

Multiplying (4.109) by m and using the ideal gas law to express n_{v_1} in terms of net pressure difference across the hole gives for the effusive mass flux

$$Q_e = \frac{m\bar{v}}{4kT}(p_{v2} - p_{v1}) = \left(\frac{m}{2\pi kT}\right)^{1/2}(p_{v2} - p_{v1}) \qquad (4.115)$$

We see that Q_m in general is much less that Q_e

4.2.5   Vapor Transport in Medium-Length Capillaries

For cases of tube length intermediate between (4.111) and (4.115) no closed-form expression is available but a semiempirical form that reduces to the two cases given has been developed by Kennard (1938, p. 308) for the mass flux (mass per unit area per unit time) through a tube of circular cross-section of radius a in the notation used here and is

$$Q_s = \frac{20 + \frac{8\ell}{a}}{20 + \frac{19\ell}{a} + 3\left(\frac{\ell}{a}\right)^2} \left(\frac{m}{2\pi kT}\right)^{1/2} (p_{v2} - p_{v1}) \qquad (4.116)$$

where p_{v2} is taken to be larger than p_{v1}. We see that for $\ell/a \ll 1$ (4.116) reduces to (4.115) while for $\ell/a \gg 1$ (4.116) reduces to (4.110).

4.2.6   Vapor Transport in a Bundle of Capillaries

In order to apply these results to a porous medium, either to the transport of vapor through it or to evaporation from a soil surface, some assumed model of the medium is required. One such model in common use is that of a bundle of capillary tubes having a frequency distribution of capillary radii as has been used for illustrative purposes in Chapters 2 and 3. We write the mass flux, $^{(i)}j$, for a section of porous medium idealized as a bundle of parallel capillaries through which pure vapor of species i is moving as

$$^{(i)}j = \frac{N'\,^{(0)}\epsilon \int_{r_m}^{r} {}^{(i)}Q_s \pi \xi^2 f(\xi)\, d\xi}{A'} = \frac{\int_{r_m}^{r} {}^{(i)}Q_s \xi^2 f(\xi)\, d\xi}{\int_{r_m}^{R} \xi^2 f(\xi)\, d\xi} \qquad (4.117)$$

where A' is given by

$$A' = \frac{N'}{^{(0)}\epsilon} \int_{r_m}^{R} \pi \xi^2 f(\xi)\, d\xi \qquad (4.118)$$

as in (2.130) and is the area of the section of porous medium, N' is the total number of capillary openings in the section, $^{(0)}\epsilon$ is the effective porosity of the section, $f(\xi)$ is the frequency distribution of the capillary radii, r_m is the smallest capillary radius of interest, while R is the largest, and $Q_s(\xi)$ is written as

$$^{(i)}Q_s = \frac{20 + \frac{8\ell}{\xi}}{20 + \frac{19\ell}{\xi} + 3\left(\frac{\ell}{\xi}\right)^2} \left(\frac{^{(i)}m}{2\pi kT}\right)^{1/2} (p_{v2} - p_{v1}) \qquad (4.119)$$

Models of the porous medium consisting of combinations of capillaries in series and parallel could also be considered, but this development will not be carried out here.

We now consider the binary diffusion of two species of gas in a single capillary in the case in which the capillary radius is large enough for the concepts of Poiseuille's law and viscosity to be valid, keeping in mind that the later developments of this chapter and the next show that the usual (partial differential) diffusion equation does not actually describe transport phenomena.

4.3   BINARY MOLECULAR DIFFUSION

The transport of one species of vapor molecule through an atmosphere of vapor molecules of another type due to spatial variations in concentration of the first species is called binary (two species present) molecular diffusion. The mass flux, $^{(1)}\vec{j}$, of species 1 through species 2 due to gradients in its concentration is given by

$$^{(1)}\vec{j} = -D_{12}\vec{\nabla}C_1 \tag{4.120}$$

where C_1 is the concentration of species 1 in units of mass per unit volume and D_{12} is the binary diffusion coefficient, which is given to first approximation in kinetic theory by (Kennard, 1938, p. 194)

$$D_{12} = \frac{3}{8}\left(\frac{2}{\pi}\right)^{1/2}\frac{1}{n_v\sigma_{12}}\left(\frac{m_1+m_2}{m_1 m_2}kT\right)^{1/2} \tag{4.121}$$

where σ_{12} is the effective scattering cross-section for species 1 and 2 relative to each other, m_1 and m_2 are the molecular masses for species 1 and 2, respectively, k is Boltzmann's constant, and

$$n_v = n_{1v} + n_{2v} \tag{4.122}$$

where n_{1v} and n_{2v} are the numbers of molecules of species 1 and 2, respectively, per unit volume. If the vapors of species and 1 and 2 are assumed to be ideal, then from the ideal gas law

$$n_{iv} = \frac{p_i}{kT} \tag{4.123}$$

where p_{vi} is the partial vapor pressure of the ith species of vapor. The expression for D_{12} was derived for C_i in units of n_{iv}. If we now consider the diffusion of water vapor through air, we can find an average value for m_2, for air, from the measured binary diffusion coefficient for water vapor in air of $D_{12} = 0.239\,\mathrm{cm^2/sec}$ if the effective scattering coefficient and the mass of a water molecule, m_1, are assumed known. Alternatively, if a value is chosen for the average mass of an air molecule, a value for σ_{12} can be found from (4.121).

Equation (4.120) is an approximation to a more general, though still approximate, equation that can be derived from considerations of the molecular distribution function, which is the number of molecules in a given volume element of momentum space and coordinate space at any time, and molecular scattering. This approximate generalization of (4.120) is

$$^{(1)}\vec{j}(\vec{r}, t) = -D_{12}\vec{\nabla}\rho_1(\vec{r}, t) - \frac{D_{12}}{v^2}\frac{\partial^{(1)}\vec{j}(\vec{r}, t)}{\partial t} \qquad (4.124)$$

where $\vec{r}$ is a position vector, t is time, v is the magnitude of the velocity of the moving edge of the "front" of water vapor diffusing through the surrounding air, and $\rho_1(\vec{r}, t)$ is the mass density of the water vapor. The associated equation of continuity is

$$\frac{\partial\rho_1(\vec{r}, t)}{\partial t} + \vec{\nabla} \cdot {}^{(1)}\vec{j}(\vec{r}, t) = \kappa\rho_1(\vec{r}, t) \qquad (4.125)$$

where the quantity on the right side of (4.125) is a source (if $\kappa > 0$) or sink (if $\kappa < 0$) term for water vapor molecules. Taking the divergence of (4.124) and the partial derivative with respect to time of (4.125) and combining the results yields

$$\nabla^2\rho_1(\vec{r}, t) = \left(\frac{v^2 - \kappa D_{12}}{v^2 D_{12}}\right)\frac{\partial\rho_1(\vec{r}, t)}{\partial t} + \frac{1}{v^2}\frac{\partial^2\rho_1(\vec{r}, t)}{\partial t^2} - \frac{\kappa}{D_{12}}\rho_1(\vec{r}, t) \quad (4.126)$$

where, to lowest approximation, D_{12}, v, and κ are assumed to be constants independent of time and concentration. Equation (4.126) has the form of the equation of telegraphy (Webster, 1955, p. 173 et seq.; Morse and Feshbach, 1953, p. 865 et seq.). The rationale for using the telegrapher's equation rather than the usual diffusion equation to discuss the relative movement of water vapor and air is that the usual diffusion equation would be obtained by letting $v \to \infty$ in (4.124), which means that solutions of the diffusion equation are all of such a form that the disturbances they describe reach the boundaries of the region being studied, even if the boundaries are located at infinity, instantaneously. This is of course unphysical and also means, for example, that the concept of "radius of influence" of a pumping well in saturated groundwater hydrology is fundamentally meaningless since the "influence"–drawdown–is at infinity, in the case of the Theis equation, say, as soon as pumping begins. (The partial saving grace here is that v may be taken in this case to be the speed of sound in the aquifer, which puts the true radius of influence at distances large compared with the usual distance of an observation well from the pumping well at times of a few seconds.) Further, and more important from the standpoint of describing contaminant transport, the diffusion equation does not describe a moving front, even one becoming less concentrated with time. The movement of such a front with no change in impurity concentration would be described by solutions of the equation that would result from setting the first and third terms on the right of (4.126) equal to zero. (This particular partial differential equation is called the *wave equation.*) Solutions of the telegrapher's equation have both diffusion-equation-like properties–namely change in disturbance in a given location with time–and wave-equation-like properties–namely the propagation of a disturbance, such as concentration of a contaminant being transported, with a definite front moving at a finite velocity. Setting

$$\frac{v^2 - \kappa D_{12}}{v^2 D_{12}} = \frac{1}{D_e} \tag{4.127}$$

$$\kappa_1 = \frac{\kappa}{D_{12}} \tag{4.128}$$

for convenience (note that for $\kappa = 0$, D_e would reduce to D_{12}) and assuming D_e is constant (not entirely correct since v is expected to change with time as the migration process proceeds), we now consider, by way of example, the following initial-value problem for one-dimensional time-dependent movement of water vapor in a capillary oriented along the z axis having a radius large compared to the mean free path of the gas molecules it contains. The differential equation and initial conditions are

$$\frac{\partial^2 \rho_1(z,t)}{\partial z^2} = \frac{1}{D_e}\frac{\partial \rho_1(z,t)}{\partial t} - \kappa_1 \rho_1(z,t) + \frac{1}{v^2}\frac{\partial^2 \rho_1}{\partial t^2} \tag{4.129a}$$

$$\rho_1(0,t) = \rho_0 \tag{4.129b}$$

$$\rho_1(z,0) = \rho_0 e^{-\kappa_2 z} \tag{4.129c}$$

$$\left.\frac{\partial \rho_1}{\partial t}\right|_{t=0} = \rho_1'(z,0) = 0 \tag{4.129d}$$

$$\left.\frac{\partial \rho_1}{\partial z}\right|_{z=0} = \rho_1'(0,t) = 0 \tag{4.129e}$$

where ρ_0 is the density of water vapor at $z = 0$–just above a capillary meniscus, for example. We employ, for convenience, integral transform techniques to solve this equation set for $\rho_1(z,t)$ as follows. The infinite cosine transform of a second derivative term is

$$C_\infty \{f''(x)\} = \left(\frac{2}{\pi}\right)^{\frac{1}{2}} \int_0^\infty f''(x) \cos(kx) \, dx$$

$$= -\left(\frac{2}{\pi}\right)^{\frac{1}{2}} f'(0) - k^2 C_\infty \{f(x)\}$$

$$= -\left(\frac{2}{\pi}\right)^{\frac{1}{2}} f'(0) - k^2 \bar{f}(k) \tag{4.130}$$

where

$$C_\infty \{f(x)\} = \left(\frac{2}{\pi}\right)^{\frac{1}{2}} \int_0^\infty f(x) \cos(kx) \, dx = \bar{f}(k) \tag{4.131}$$

and k is called the transform variable. Equation (4.131) along with

$$f(x) = \left(\frac{2}{\pi}\right)^{\frac{1}{2}} \int_0^\infty \bar{f}(k) \cos(kx) \, dk \tag{4.132}$$

form an integral transform–inversion pair. Similarly, the Laplace transforms of the first and second derivatives of $f(t)$ are

$$L\{f'(t)\} = p\bar{f}(p) - f(0) \tag{4.133}$$

and

$$L\{f''(t)\} = p^2 \bar{f}(p) - pf(0) - f'(0) \tag{4.134}$$

respectively, where p is the transform variable in this case,

$$L\{f(t)\} = \int_0^\infty e^{-pt} f(t) \, dt = \bar{f}(p) \tag{4.135}$$

and

$$f(t) = L^{-1}\{\bar{f}(p)\} = \frac{1}{2\pi i} \int_{r-i\infty}^{r+i\infty} e^{zt} \bar{f}(z) \, dz \tag{4.136}$$

Equations (4.135) and (4.136) form a Laplace transform–inverse pair. Equation (4.136) with $z = x + iy$, a complex variable, $i = \sqrt{-1}$, and r a value of $x > 0$ such that it is to the right of any singularities of $\bar{f}(z)$, represents an integral in the complex plane. Laplace transforming (4.129a) to remove the derivatives with respect to time yields

$$\frac{\partial^2 \bar{\rho}_1(z,p)}{\partial z^2} = \frac{1}{D_e} \left[p\bar{\rho}_1(z,p) - \rho_1(z,0) \right]$$

$$+ \frac{1}{v^2} \left[p^2 \bar{\rho}_1(z,p) - p\rho_1(z,0) - \rho_1'(z,0) \right] - \kappa_1 \bar{\rho}_1(z,p) \quad (4.137)$$

where all operations have been carried out with respect to t only. Substituting (4.129c) and (4.129d) into (4.137) and applying the infinite cosine transform to the result yields

$$-k^2 \bar{\bar{\rho}}_1(k,p) = \frac{p}{D_e} \bar{\bar{\rho}}(k,p) - \left(\frac{2}{\pi}\right)^{\frac{1}{2}} \frac{\rho_0}{D_e} \left(\frac{\kappa_2}{\kappa_2^2 + k^2}\right)$$

$$+ \frac{p^2}{v^2} \bar{\bar{\rho}}_1(k,p) - \left(\frac{2}{\pi}\right)^{\frac{1}{2}} \frac{p\rho_0}{v^2} \left(\frac{\kappa_2}{\kappa_2^2 + k^2}\right) - \kappa_1 \bar{\bar{\rho}}_1(k,p) \quad (4.138)$$

where

$$C_\infty \left\{ L \left\{ \rho_1(z,t) \right\} \right\} = L \left\{ C_\infty \left\{ \rho_1(z,t) \right\} \right\} = \bar{\bar{\rho}}_1(k,p) \quad (4.139)$$

$$\bar{\rho}_1(k,0) = C_\infty \left\{ \rho_0 e^{-\kappa_2 z} \right\} = \left(\frac{2}{\pi}\right)^{\frac{1}{2}} \rho_0 \frac{\kappa_2}{\kappa_2^2 + k^2} \quad (4.140)$$

and the result of Laplace transforming (4.129e), which yields

$$\bar{\rho}_1'(0,p) = 0 \quad (4.141)$$

has been used. Solving algebraically for $\bar{\bar{\rho}}_1(k,p)$ yields

$$\bar{\bar{\rho}}(k,p) = \frac{\sqrt{2}\rho_0 \kappa_2}{\sqrt{\pi} D_e \left(\kappa_2^2 + k^2\right) \left[A(p) + k^2\right]} \left(\frac{1}{D_e} - \frac{p}{v^2}\right) \quad (4.142)$$

where

$$A(p) = \frac{1}{v^2} \left[\left(p + \frac{v^2}{2D_e}\right)^2 - \frac{v^4}{4D_e^2} - \kappa_1 v^2 \right] = \frac{(p+a)^2 - c^2}{v^2} \quad (4.143a)$$

where

$$a = \frac{v^2}{2D_e} \quad (4.143b)$$

$$c^2 = \frac{v^4}{4D_e^2} - \kappa_1 v^2 \quad (4.143c)$$

Applying (4.132) to (4.142) yields

$$C_\infty^{-1}\left\{\bar{\bar{\rho}}_1(k,p)\right\} = \bar{\rho}_1(z,p)$$

$$= \left(\frac{1}{D_e} - \frac{p}{v^2}\right)\frac{2\rho_0\kappa_2}{\pi}\int_0^\infty \frac{1}{(\kappa_2^2+k^2)\,[A(p)+k^2]}\cos(kz)\,dk \quad (4.144)$$

Carrying out the indicated integration and rearranging terms yields

$$\bar{\rho}(z,p) = \frac{\rho_0}{D}\left(\sqrt{A(p)}e^{-z\kappa_2} - \kappa_2 e^{-z\sqrt{A(p)}}\right)\left(\frac{1}{D_e} - \frac{p}{v^2}\right) \quad (4.145)$$

where

$$D = \sqrt{A(p)}\,[A(p) - \kappa_2^2] \quad (4.146)$$

We now indicate the Laplace inversion as follows.

$$L^{-1}\left\{\bar{\rho}_1(z,p)\right\} = \rho_1(z,t)$$

$$= \frac{\rho_0\kappa_2 v^2 e^{-\kappa_2 z}}{D_e}L^{-1}\left\{\frac{1}{(p+a)^2-b^2}\right\}$$

$$-\frac{\rho_0 v^3}{D_e}L^{-1}\left\{\frac{pe^{-(z/a)\sqrt{p^2-c^2}}}{\sqrt{(p+a)^2-c^2}\,[(p+a)^2-b^2]}\right\}$$

$$+\rho_0 e^{-\kappa_2 z}L^{-1}\left\{\frac{p}{(p+a)^2-b^2}\right\}$$

$$-\rho_0\kappa_2 v^3 L^{-1}\left\{\frac{pe^{-(z/v)\sqrt{(p+a)^2-c^2}}}{\sqrt{(p+a)^2-c^2}\,[(p+a)^2-b^2]}\right\} \quad (4.147)$$

where

$$A(p) - \kappa_2^2 = \frac{(p+a)^2-b^2}{v^2} \quad (4.148a)$$

with

$$b^2 = c^2 + \kappa_2^2 v^2 \quad (4.148b)$$

$$\sqrt{A(p)} = \frac{1}{v}\left[(p+a)^2-c^2\right]^{\frac{1}{2}} \quad (4.148c)$$

have been used. It is assumed that b^2 and c^2 are real and positive and thus that $c^2 \geq \kappa_2^2 v^2$. This assumption depends for its validity on the relative magnitudes of v^2/D_e, κ_1, and κ_2. If, in fact, $c^2 < \kappa_2^2 v^2$ so that b^2 is negative, then the Laplace inversions will yield other results. To carry out the Laplace inversions indicated we recall that

$$L^{-1}\left\{\bar{f}(p+a)\right\} = e^{-at}L^{-1}\left\{\bar{f}(p)\right\} \quad (4.149)$$

Applying this result to (4.147) yields

$$L^{-1}\{\bar{\rho}_1(z,p)\} = \rho_1(z,t)$$

$$= \frac{\rho_0 \kappa_2 v^2 e^{-\kappa_2 z}}{D_e} e^{-v^2 t/2D_e} L^{-1}\left\{\frac{1}{p^2 - b^2}\right\}$$

$$- \frac{\rho_0 v^3}{D_e} e^{-v^2 t/2D_e} L^{-1}\left\{\frac{e^{-(z/v)\sqrt{p^2-c^2}}}{\sqrt{p^2-c^2}\,(p^2-b^2)}\right\}$$

$$+ \rho_0 e^{-\kappa_2 z} e^{-v^2 t/2D_e} L^{-1}\left\{\frac{p - \frac{v^2}{2D_e}}{p^2 - b^2}\right\}$$

$$- \rho_0 \kappa_2 v^3 e^{-v^2 t/2D_e} L^{-1}\left\{\frac{\left(p - \frac{v^2}{2D_e}\right) e^{-(z/v)\sqrt{p^2-c^2}}}{\sqrt{p^2-c^2}\,(p^2-b^2)}\right\} \quad (4.150)$$

We now recall that

$$L^{-1}\left\{\frac{1}{p^2 - b^2}\right\} = \frac{1}{b}\sinh(bt) \quad (4.151)$$

which allows inversion of the first term and part of the third term on the right-hand side of (4.150). To invert the second term on the right-hand side of (4.150), we recall that (Roberts and Kaufman, 1966, p. 154)

$$L^{-1}\left\{\frac{e^{-c_1(p^2-c^2)^{\frac{1}{2}}}}{(p^2-c^2)^{\frac{1}{2}}}\right\} = I_0\left[c(t^2 - c_1^2)\right] H(t - c_1) \quad (4.152)$$

where $H(t - c_1)$ is the Heaviside step function written here as

$$H(t - c_1) = 1, \qquad t \geq c_1$$

$$= 0, \qquad t < c_1 \quad (4.153)$$

I_0 is the modified Bessel function of the first kind of order zero, and we set $c_1 = z/v$. We further recall the convolution theorem for Laplace transforms,

$$L^{-1}\{\bar{f}_1(p)\bar{f}_2(p)\} = \int_0^t L^{-1}\{\bar{f}_1(p)\}\big|_{t=t-\tau}\; L^{-1}\{\bar{f}_2(p)\}\big|_{t=\tau}\, d\tau \quad (4.154)$$

which is of course unaltered if the roles of f_1 and f_2 are interchanged. Applying (4.153) to the indicated inversion in the second term of the right-hand side of (4.150) with the identifications

$$\bar{f}_1(p) = \frac{1}{p^2 - b^2} \quad (4.155a)$$

$$\bar{f}_2(p) = \frac{e^{-\sqrt{(p+c)(p-c)}\frac{z}{v}}}{\sqrt{(p+c)(p-c)}} \tag{4.155b}$$

so that

$$L^{-1}\left\{\frac{e^{-\sqrt{p^2-c^2}z/v}}{\sqrt{p^2-c^2}\,(p^2-b^2)}\right\} = \frac{1}{b}\int_0^t I_0\left(c\left(\tau^2 - \frac{z^2}{v^2}\right)\right)$$

$$\times H\left(\tau - \frac{z}{v}\right)\sinh\left[b(t-\tau)\right]d\tau \tag{4.156}$$

where the Heaviside step function guarantees the transport of the diffusing water vapor with a finite velocity. Completing the inversion of (4.150) yields for the transport of water vapor through air in a large-radius capillary tube

$$\rho_1(x,t) = \rho_0 v^4 e^{-\kappa_2 z - v^2 t/2D_e}\left(\frac{\sinh(bt)}{D_e b}\right)$$

$$-\frac{\rho_0 v^2}{D_e b}e^{-v^2 t/2D_e}\int_0^t I_0\left(c\left(\tau^2 - \frac{z^2}{v^2}\right)\right) H\left(\tau - \frac{z}{v}\right)\sinh\left[b(t-\tau)\right]d\tau$$

$$+\rho_0 e^{-\kappa_2 z - v^2 t/2D_{12}}\left(\cosh(bt) - \frac{v^2}{2D_e b}\sinh(bt)\right)$$

$$+\frac{\rho_0\kappa_2 v^3 a}{b}e^{-v^2 t/2D_e}\left[\int_0^t I_0\left[c\left(\tau^2 - \frac{z^2}{v^2}\right)\right] H\left(\tau - \frac{z}{v}\right)\sinh\left[b(t-\tau)\right]\right]d\tau$$

$$-\rho_0\kappa_2 v^3 e^{-at}\int_0^t I_0\left(c\left(\tau^2 - \frac{z^2}{v^2}\right)\right) H\left(\tau - \frac{z}{v}\right)\cosh\left[b(t-\tau)\right]d\tau \tag{4.157}$$

where a is given by (4.143b) and the result $L^{-1}\{p/(p^2-b^2)\} = \cosh(bt)$ has been used.

4.4   MACROSCOPIC VAPOR TRANSPORT

In the diffusion calculation just discussed, it will be noticed that neither the viscosity nor the equation of state, even at the level of the ideal gas law, was present. This, along with the fact that (4.126) is a somewhat ad hoc approximation, is symptomatic of the deeper problem there alluded, to which is that diffusionlike formalism, even with the extension of (4.124), is unsatisfactory for the description of transport in porous media. The correct partial differential equations, which will be discussed in more detail in Chapter 5, are essentially those of fluid mechanics with the densities of the interpenetrating continuae, one of which is the porous medium itself, given in the form of the bulk density for each medium as discussed in

Section 1.5. Writing Eq. (1.27) in tensor notation (see Appendix A for a brief discussion of tensor formalism) yields

$$\frac{\partial\left({}^{(i)}\rho_b\,{}^{(i)}v^l\right)}{\partial t} = -\,{}^{(i)}v^l\,{}^{(i)}j^m_{,m} - {}^{(i)}j^k\,{}^{(i)}v^l_{,k} + {}^{(i)}\tau^{ln}_{,n} + {}^{(i)}\rho_b\,{}^{(i)}F^l \quad (4.158)$$

where the stress tensor for species i, ${}^{(i)}\tau^{ln}$, is given by

$$^{(i)}\tau^{ln} = {}^{(i)}\eta\left({}^{(i)}v_{\alpha,\beta} + {}^{(i)}v_{\beta,\alpha}\right)g^{\alpha l}g^{\beta n} + {}^{(i)}\eta'g^{ln}\,{}^{(i)}v^k_{,k} - {}^{(i)}pg^{ln} \quad (4.159)$$

and ${}^{(i)}\rho_b$ is the bulk density of species i, ${}^{(i)}v^l$ is the lth contravariant component of the velocity of the ith species, ${}^{(i)}j^m_{,m}$ is the divergence (contraction of the covariant derivative) of the mth component of the mass flux of species i, ${}^{(i)}F^l$ is the lth component of the body force per unit mass acting on species i, ${}^{(i)}\eta$ is the dynamic viscosity of the ith species as given in (4.74), also called the shear viscosity [the kinematic viscosity of species i, ${}^{(i)}\nu$, equals ${}^{(i)}\eta/{}^{(i)}\rho$ and for ${}^{(i)}\eta$ in poise(s), ${}^{(i)}\rho$ in g/cm^3, ${}^{(i)}\nu$ has units of cm^2/sec or stoke(s)], ${}^{(i)}\eta'$ is the second coefficient of viscosity of species i, also called the bulk, or dilational viscosity, and equals zero for a monatomic gas, ${}^{(i)}v_{\alpha,\beta}$ is the covariant derivative of the αth covariant component of the velocity of species i, and similarly for ${}^{(i)}v_{\beta,\alpha}$, g^{lm} is the inverse of the metric tensor, and ${}^{(i)}p$ is the pressure of the ith species and is related to the density and temperature of the ith species by the appropriate equation of state. The Einstein summation convention, in which a dummy index appearing once as a superscript and once as a subscript in a given term is assumed to be summed over, is used as well. For an ideal gas at constant temperature in a porous medium, the equation of state (relation between pressure, temperature, and density) is

$$^{(i)}p = \frac{{}^{(i)}_b\rho kT}{{}^{(i)}m} \quad (4.160)$$

where ${}^{(i)}m$ is the mass of a gas molecule, ${}^{(i)}p$ is understood to be the partial pressure due to species i, in the sense of Dalton's law of partial pressures, and the rest of the symbols are as previously defined. Writing Fick's law of diffusion, as in (4.120), as

$$^{(i)}j^m = -D\,{}^{(i)}c_{,q}\,g^{qm} = -D\,{}^{(i)}\rho_b\,g^{qm} \quad (4.161)$$

where the gradient (covariant derivative) of the concentration of species i is ${}^{(i)}c_{,q}$, and we identify ${}^{(i)}c$ with ${}^{(i)}\rho_b$. We now write, as in (1.22),

$$^{(i)}j^m = {}^{(i)}\rho_b\,{}^{(i)}v^m \quad (4.162)$$

so that from (4.163) and (4.164)

$$^{(i)}v^m = -\frac{D\,g^{mq}}{^{(i)}_b\rho}\,^{(i)}\rho_{b,q} \tag{4.163}$$

or, in covariant components,

$$^{(i)}v_m = -\frac{D}{^{(i)}\rho_b}\,^{(i)}\rho_{b,m} \tag{4.164}$$

Assuming that the effects of the porous medium are characterized by D, which is assumed to be a constant in this development, considering a single species, $i = 1$, in one dimension, and assuming $^{(i)}\eta' = 0$, yields for (4.158)

$$\frac{\partial}{\partial t}\left(-Dg^{lq}\,^{(i)}\rho_{b,q}\right) = -\frac{D}{^{(i)}\rho_b}\left(g^{lq}\,^{(i)}\rho_{b,q}\right)D\left[g^{qm}\,^{(i)}\rho_{b,q}\right]_{,m}$$

$$-D\left(^{(i)}\rho_{b,q}g^{kq}\right)D\left[g^{lq}\frac{^{(i)}\rho_{b,q}}{^{(i)}\rho_b}\right]_{,k}$$

$$+\,^{(1)}\tau^{ln}_{,n} + \,^{(i)}\rho_b\,^{(1)}F^l \tag{4.165}$$

We now substitute (4.163) into (4.159) to yield for the stress tensor

$$^{(1)}\tau^{ln} = \,^{(1)}\eta\left(\left[\frac{-D}{^{(1)}\rho_b}\,^{(1)}\rho_{b,\alpha}\right]_{,\beta} + \left[\frac{-D}{^{(1)}\rho_b}\,^{(1)}\rho_{b,\beta}\right]_{,\alpha}\right)g^{\alpha l}g^{\beta n} -\,^{(1)}pg^l \tag{4.166}$$

where g^l is the lth component of the acceleration due to gravity, which is assumed to be the only body force per unit mass acting in this case and the rest of the symbols are as previously defined. Noting that the covariant derivative of a scalar is equal to the usual partial derivative, assuming a cylindrical polar coordinate system with $x^1 = r$, $x^2 = \phi$, and $x^3 = z$, for which the metric tensor and its inverse are given by

$$g_{ij} = \begin{bmatrix} 1 & 0 & 0 \\ 0 & (x^1)^2 & 0 \\ 0 & 0 & 1 \end{bmatrix} \tag{4.167}$$

$$g^{ij} = \begin{bmatrix} 1 & 0 & 0 \\ 0 & (x^1)^{-2} & 0 \\ 0 & 0 & 1 \end{bmatrix} \tag{4.168}$$

respectively, and assuming that $^{(1)}_b\rho$ depends on x^3 only, yields for (4.165), after some reduction,

$$\frac{1}{D} g^{l3} \frac{\partial^2 \, ^{(1)}\rho_b}{\partial t \, \partial x^3} = \frac{1}{^{(1)}\rho_b} g^{l3} \frac{\partial \, ^{(1)}\rho_b}{\partial x^3} \left(g^{33} \frac{\partial^2 \, ^{(1)}\rho_b}{\partial x^3 \, \partial x^3} \right)$$

$$+ g^{k3} \frac{\partial \, ^{(1)}\rho_b}{\partial x^3} \left[\frac{1}{^{(1)}\rho_b} g^{l3} \frac{\partial \, ^{(1)}\rho_b}{\partial x^3} \right]_{,k}$$

$$- \frac{1}{D^2} \, ^{(1)}\tau^{ln}_{,n} + \frac{1}{D^2} \, ^{(1)}\rho_b \, g^l \qquad (4.169)$$

Setting $l = 3$ and carrying out the indicated covariant derivative with respect to x^k yields

$$\frac{1}{D} \frac{\partial^2 \, ^{(1)}\rho_b}{\partial t \, \partial x^3} = \frac{2}{^{(1)}\rho_b} \frac{\partial \, ^{(1)}\rho_b}{\partial x^3} \frac{\partial^2 \, ^{(1)}\rho_b}{\partial x^3 \, \partial x^3} - \frac{1}{\left(^{(1)}\rho_b \right)^2} \left(\frac{\partial \, ^{(1)}\rho_b}{\partial x^3} \right)^3$$

$$- \frac{1}{D^2} \, ^{(1)}\tau^{ln}_{,n} + \frac{1}{D^2} \, ^{(1)}\rho_b \, g^l \qquad (4.170)$$

We now consider a similar expansion of $^{(1)}\tau^{ln}$ to yield

$$^{(1)}\tau^{ln} = \frac{2D \, ^{(1)}\eta}{^{(1)}\rho_b} \left(\frac{1}{^{(1)}\rho_b} \frac{\partial \, ^{(1)}\rho_b}{\partial x^\alpha} \frac{\partial \, ^{(1)}\rho_b}{\partial x^\beta} - \frac{\partial^2 \, ^{(1)}\rho_b}{\partial x^\alpha \, \partial x^\beta} \right) g^{\alpha l} g^{\beta n} - \, ^{(1)}p \, g^{ln} \qquad (4.171)$$

which becomes, on carrying out the summations over α and β,

$$^{(1)}\tau^{ln} = \frac{2D \, ^{(1)}\eta \, g^{3l} g^{3n}}{^{(1)}\rho_b} \left[\frac{1}{^{(1)}\rho_b} \left(\frac{\partial \, ^{(1)}\rho_b}{\partial x^3} \right)^2 - \frac{\partial^2 \, ^{(1)}\rho_b}{\partial x^3 \, \partial x^3} \right] - \, ^{(1)}p \, g^{nl} \qquad (4.172)$$

We now carry out the divergence of $^{(1)}\tau^{ln}$ to yield

$$^{(1)}\tau^{ln}_{,n} = \frac{2D \, ^{(1)}\eta \, g^{3l}}{^{(1)}\rho_b} \left[\frac{3}{^{(1)}\rho_b} \frac{\partial \, ^{(1)}\rho_b}{\partial x^3} \frac{\partial^2 \, ^{(1)}\rho_b}{\partial x^3 \, \partial x^3} - \frac{2}{\left(^{(1)}\rho_b \right)^2} \left(\frac{\partial \, ^{(1)}\rho_b}{\partial x^3} \right)^3 \right.$$

$$\left. - \frac{\partial^3 \, ^{(1)}\rho_b}{\partial x^3 \, \partial x^3 \, \partial x^3} \right] - g^{3l} \frac{\partial \, ^{(1)}p}{\partial x^3} \qquad (4.173)$$

Combining (4.173) and (4.160) with (4.170) yields

$$\frac{\partial^2 \,^{(1)}\rho_b}{\partial t \, \partial x^3} = - \frac{2}{^{(1)}\rho_b} \frac{\partial \,^{(1)}\rho_b}{\partial x^3} \frac{\partial^2 \,^{(1)}\rho_b}{\partial x^3 \, \partial x^3} \left(\frac{3 \,^{(1)}\eta}{^{(1)}\rho_b} - D \right)$$

$$+ \frac{1}{\left(^{(1)}\rho_b \right)^2} \left(\frac{\partial \,^{(1)}\rho_b}{\partial x^3} \right)^3 \left(\frac{4 \,^{(1)}\eta}{^{(1)}\rho_b} - D \right)$$

$$+ \frac{2 \,^{(1)}\eta}{^{(1)}\rho_b} \frac{\partial^3 \,^{(1)}\rho_b}{\partial x^3 \partial x^3 \, \partial x^3} + \frac{1}{D} \frac{\partial \,^{(1)}p}{\partial x^3} + \frac{1}{D} \,^{(1)}\rho_b \, g \quad (4.174)$$

Equation (4.174) is the partial differential equation describing either binary molecular diffusion along a single capillary or one-dimensional vapor transport in a porous medium, depending on the choice of D. The equation of continuity for species i in a porous medium corresponding to (4.158) (Eulerian rather than Lagrangian formulation of the equations of motion) is

$$\frac{\partial \,^{(i)}\rho_b}{\partial t} = - \,^{(i)}j^m_{,m} \quad (4.175)$$

which becomes in the present instance

$$\frac{\partial \,^{(1)}\rho_b}{\partial t} = - \,^{(1)}\rho_b \frac{\partial \,^{(1)}v^3}{\partial x^3} - \frac{\partial \,^{(1)}\rho_b}{\partial x^3} \,^{(1)}v^3 \quad (4.176)$$

In addition to the equations of motion, continuity, state, and flux laws for each species, a complete equation set for isothermal gas transport requires the equation of energy flow, which is

$$\frac{\partial \left(^{(m)}E \,^{(m)}\rho_b \right)}{\partial t} = - \vec{\nabla} \cdot \left(^{(m)}\rho_b \,^{(m)}\vec{v} \,^{(m)}E \right) - \vec{\nabla} \cdot \vec{q} + ^{(m)}\vec{v} \cdot \vec{q} - \vec{\nabla} \cdot \left(^{(m)}\overleftrightarrow{\chi} \cdot \,^{(m)}\vec{v} \right)$$
$$(4.177a)$$

with $\vec{q}$, the heat current due to an applied temperature gradient, set equal to zero due to the isothermal assumption, $^{(m)}E$ given by

$$^{(m)}E = \frac{1}{2} \,^{(m)}\vec{v} \cdot \,^{(m)}\vec{v} + \,^{(m)}U \quad (4.177b)$$

where $^{(m)}U$ is the internal energy of species m, $\vec{g}$ is the acceleration due to gravity, the stress tensor for a viscoelastic system, $\overleftrightarrow{\chi}$, reduces in the present case to the expression given in (4.172), and the other symbols are as previously defined. In this development it is explicitly assumed that no coupled effects, in the sense of nonequilibrium thermodynamics, are present. Consideration of these, which will be given in Chapter 6, would require additional gradient (generalized force) terms in the expressions for the $^{(m)}\vec{v}$, one such for each of the flows being considered.

4.5 PHASE BARRIER THEORY

We now consider the increase in the effective latent heat of vaporization when the water surface is concave and the contribution of this increase to the hysteresis of the moisture characteristic. This is most conveniently done using phase barrier formalism (see, for example, Lewis and Randall, 1923, pp.181–184 for an analogous discussion of a porous barrier separating the liquid and vapor phases of mercury, which would apply to mercury injection porosimetry). In the present case the phase barrier between liquid at one pressure and its vapor at a lower pressure is the ambient atmosphere pressing on the liquid surface, and we thus proceed as follows. We define p_{0l} as the pressure of the liquid phase, p_{0v} as the pressure of the vapor phase when the liquid phase has a surface of infinite radius of curvature (flat), v_A as the volume per mol of liquid, v_B as the volume per mol of vapor, μ_A as the chemical potential of the liquid and μ_B as the chemical potential of the vapor over the flat liquid surface, s_A as the entropy per mol of the liquid, and s_B as the entropy per mol of the vapor. Thus

$$d\mu_A = \left(\frac{\partial \mu_A}{\partial T}\right)_T dp_{0l} + \left(\frac{\partial \mu_A}{\partial T}\right)_{p_{0l}} dT = v_A \, dp_{0l} - s_A \, dT \qquad (4.178a)$$

$$d\mu_B = \left(\frac{\partial \mu_B}{\partial p_{0v}}\right)_T dp_{0l} + \left(\frac{\partial \mu_B}{\partial T}\right)_{p_{0v}} dT = v_B \, dp_{0v} - s_B \, dT \qquad (4.178b)$$

when equilibrium exists between the liquid and vapor phases, in this case in the presence of the phase barrier,

$$d\mu_A = d\mu_B \qquad (4.179)$$

Substituting (4.178a) and (4.178b) into (4.179) and rearranging terms yields

$$v_A \, dp_{0l} - v_B \, dp_{0l} = (s_A - s_B) \, dT = \frac{-l_{BA}}{T} \, dT \qquad (4.180)$$

where l_{BA} is the usual latent heat of vaporization of the liquid in units of energy per mol. If p_{0l} is constant, $dp_{0l} = 0$ and (4.180) becomes

$$\left(\frac{\partial p_{0v}}{\partial T}\right)_{p_{0l}} = \frac{l_{BA}}{T v_B} \qquad (4.181)$$

When the liquid surface is concave, for example, when it wets the walls of a capillary tube of radius ξ, the pressure of its vapor, p_{0v}, is given approximately by the Kelvin relation as

$$p_v = p_{0v} e^{-A_0 \cos(\phi)/\xi} \qquad (4.182)$$

where A_0 is given by Eq. (4.9c) and (4.181) becomes

$$\left(\frac{\partial p_v}{\partial T}\right)_{p_l} = \frac{l_1}{T v'_B} \tag{4.183}$$

where l_1 is the effective latent heat of vaporization of the water at its surface in the capillary, not counting the binding energy between the water and the inner surfaces of the porous medium, p_l is the pressure of the liquid just below the capillary meniscus, and v'_B, is the volume per mol of the liquid vapor above the curved surface. From Eqs. (4.181) and (4.181) we have, neglecting the temperature variation of γ and ρ,

$$\left(\frac{\partial p_v}{\partial T}\right)_{p_l} = p_{0v}e^{-A_0 \cos(\phi)/\xi}\left[\frac{A_0 \cos(\phi)}{T\xi}\right]$$

$$+ e^{-A_0 \cos(\phi)/\xi}\left(\frac{\partial p_{0v}}{\partial T}\right)_{p_l}$$

$$= p_v \frac{A_0 \cos(\phi)}{T\xi} + \frac{p_v}{p_{0v}}\frac{l_{BA}}{T v_B} = \frac{l_1}{T v'_B} \tag{4.184}$$

Assuming that the vapor above the flat and curved surfaces is ideal, we have

$$v_B = \frac{R_g T}{p_{0v}} \tag{4.185a}$$

$$v'_B = \frac{R_g T}{p_v} \tag{4.185b}$$

where $R_g = 8.3143 \times 10^7$ ergs/mol °K. Combining (4.185a) and (4.185b) with (4.184) and rearranging terms yields

$$l_1 = l_{BA} + \frac{A_0 R_g T \cos(\phi)}{\xi} \tag{4.186}$$

or, since $R_g = Lk$, where L is Avogadro's number, 6.02252×10^{23} particles per mol, and k is Boltzmann's constant, (4.186) becomes

$$l_1 = l_{BA} + \frac{2\gamma m L \cos(\phi)}{\rho\xi} \tag{4.187}$$

For a surface with infinite radius of curvature, $\xi \to \infty$ and $l_1 = l_{BA}$ as is observed, even if air is present. Substituting numerical values for the constants appearing in (4.187) yields

$$l_1 = l_{BA} + \frac{2.602 \times 10^3 \cos(\phi)}{\xi}\frac{\text{ergs}}{\text{mol}} \tag{4.188a}$$

or

$$l_1 = l_{BA} + \frac{1.446 \times 10^2 \cos(\phi)}{\xi} \frac{\text{ergs}}{\text{g}} \qquad (4.188\text{b})$$

The contribution of (4.186) to hysteresis arises as follows. When a mole of water molecules is absorbed on the curved water surfaces at the ends of capillaries having average effective radii ξ, in the porous medium, l_{BA} units of heat (energy) are liberated. In order to evaporate the same mole of water molecules from the same surfaces, however, l_1 units of energy are required because of the curvature of the water surfaces. This difference between l_1 and l_{BA} contributes to the area enclosed by the main branches of the matric potential hysteresis loop. The average value of $l_1, \bar{l}_1$, is given by

$$\bar{l}_1 = l_{BA} \int_{r_m}^{r} f(\xi)\, d\xi + A_0 R_g T \cos(\phi) \int_{r_m}^{r} \xi^{-1} f(\xi)\, d\xi \qquad (4.189)$$

Thus, the average contribution to the area of the hysteresis loop is

$$\Delta A = \bar{l}_1 - l_{BA} = A_0 R_g T \cos(\phi) \int_{r_m}^{r} \xi^{-1} f(\xi)\, d\xi \qquad (4.190)$$

Comparison of values of l_{BA} in Table 4.1 with (4.188b) shows that the "hysteresis correction" is significant. Equation (4.186) can be generalized in two ways. First, derivatives of γ and ρ in (4.184) can be taken with respect to temperature, and (2.95) can be used in place of (1.2) which was used previously. We do not carry out these generalizations here but note that from (4.185a) and (4.185b) we have

$$\frac{\partial p_v}{\partial T} = \frac{l_1 p_v}{R_g T^2} \qquad (4.191\text{a})$$

$$\frac{\partial p_{0v}}{\partial T} = \frac{l_{DA} p_{0v}}{R_g T^2} \qquad (4.191\text{b})$$

We now consider the origins of capillary hysteresis partly in terms of Eq. (4.190).

4.6   CAPILLARY HYSTERESIS

The contribution to the area enclosed by the hysteresis loop in the $\Psi - {}^{(i)}\theta$ plane from the difference between the heat per mol liberated during the adsorption of water molecules from the vapor onto the concave water menisci in the porous medium and the energy per mol required to desorb it due only to the fact that the menisci are concave is, as previously,

$$\Delta A = \bar{l}_1 - l_{BA} \qquad (4.192)$$

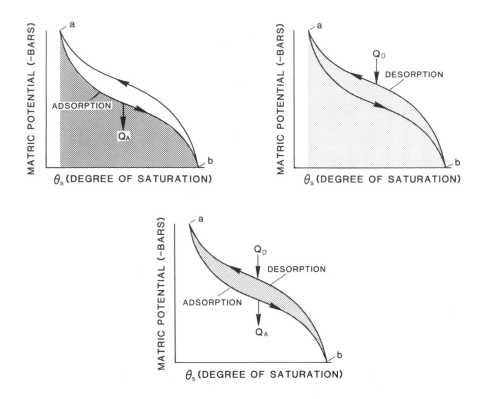

Figure 4.6: Heat Q_A is released on adsorption of water by soil. Energy Q_D is needed to remove the adsorbed water. $Q_A - Q_D$ is the net energy input and is the area enclosed by the main branches of the hysteresis loop.

We now discuss briefly the physical origins of the nonzero area enclosed by the main wetting–drying branches of the moisture characteristic. The work done when adsorption occurs is represented by the entire area under the adsorption branch of the moisture characteristic, as shown in Figure 4.6. The area has units of energy per unit bulk volume and represents the heat liberated when water is taken up by an initially unsaturated soil sample. The heat of adsorption has three components. The first is the heat evolved when water adsorbs on soil surfaces and represents the difference in energy in the soil–water system before the water is adsorbed and the lower energy of the soil–water system after adsorption has occurred. This part of the heat evolved is essentially reversible, and the energy it represents is replaced when the water is removed from the soil sample by the act of breaking the water–soil bonds. The second contribution to the shaded area under the adsorption curve is from the heat of adsorption per mol (or gram) of water vapor adsorbed onto existing water meniscus surfaces. This adsorption can lead to heat evolution from water–soil surface interactions as the capillaries fill due to adsorption of water on their ends. Further, water–water heat can be liberated as capillary action fills capillaries already containing water adsorbed on their inner surfaces. While these contributions to the area under the adsorption curve are certainly time dependent, the spontaneous redistribution that occurs, the third component, after the initial adsorption has taken place, proceeds more slowly, and if sufficient time for equilibration is allowed is indistinguishable in its effects from the principal adsorption process. When desorption of water from the interior surfaces of the sample takes place, more energy per mol on the average is required to desorb a molecule from concave menisci than that released when adsorption occurred, as shown in Eq. (4.190). Also, in incompletely filled pores/capillaries, the change in shape of the macroscopic–on the pore/capillary scale–water surfaces can often be such, particularly if a capillary is being filled from the side inward, as to form surfaces of smaller average radius of curvature, and hence lower average vapor pressure and greater energy are required for desorption than were present/evolved when the adsorption process began. Both these causes, in addition to spontaneous redistribution, depending on the time involved, require an energy of desorption for the water in the soil sample per unit bulk volume equal to the entire area under the desorption branch of the moisture characteristic. The difference between the two areas is the hysteresis loop. The area enclosed by the loop represents the difference between the energy liberated as heat when adsorption occurs, Q_A, and the energy required to remove the water from the soil sample, Q_D, principally due to the concave nature of the water surfaces in the sample pores/capillaries. As noted in Figure 4.6, the difference between Q_D and Q_A is given by

$$Q_A - Q_D = \int_a^b \Psi_A \, d^{(i)}\theta - \int_a^b \Psi_D \, d^{(i)}\theta \qquad (4.193)$$

Note that our original thermodynamic convention, by which the work done on a system is positive and the work done by a system is negative, means that Q_A, since it represents work done by the soil–water system, is negative. This accords with the fact that the area under the adsorption curve in the $\Psi - {}^{(i)}\theta$ plane is actually negative since $\Psi < 0$. Further, Q_d, plotted as the area enclosed by the desorption branch of the moisture characteristic, though it represents work done on the system, and hence a positive quantity, is plotted as a larger negative area, and hence the area enclosed by the hysteresis loop is Q_A, a negative quantity, minus Q_D, a larger negative quantity, to yield a positive quantity, the net work done on the system during one adsorption–desorption cycle.

SELECTED REFERENCES

Ames, W. F., *Nonlinear Partial Differential Equations in Engineering*, Academic Press, New York, 1965.

Ames, W. F., *Nonlinear Partial Differential Equations*, Academic Press, New York, 1967.

Courant, R., and D. Hilbert, *Methods of Mathematical Physics*, Vol. II, Interscience Publishers, New York, 1962.

Davis, H. T., *Introduction to Nonlinear Differential and Integral Equations*, Dover Publications, Inc., New York, 1962.

Desloge, E. A., *Statistical Physics*, Holt, Rinehart and Winston, Inc., New York, 1966.

Kennard, E. H., *Kinetic Theory of Gases*, McGraw-Hill Book Co., Inc., New York, 1938.

Lewis, G. N., and M. Randall, *Thermodynamics*, McGraw-Hill Book Co., Inc., New York, 1923.

Loeb, L. B., *The Kinetic Theory of Gases*, McGraw-Hill Book Co., Inc., New York, 1934.

Morse, P. M., and H. Feshbach, *Methods of Theoretical Physics*, McGraw-Hill Book Co., New York, 1953.

Roberts, G.E., and H. Kaufman, *Table of Laplace Transforms*, W. B. Saunders Co., Philadelphia, PA, 1966.

Webster, A. G., *Partial Differential Equations of Mathematical Physics*, Dover Publications Inc., 1955.

PROBLEMS

4.1 Integrate the Clapeyron equation to show that the vapor pressure over a flat liquid surface is given approximately by

$$p_{0v} = p_e e^{-l_{BA}/R_g T}$$

where

$$p_e = p_{0e} e^{l_{BA}/R_g T_0}$$

4.2 Show that the Clapeyron equation, the slope of the vapor-pressure–temperature curve, for liquid in a capillary is given by Eq. (4.10).

4.3 Evaluate Eq. (4.14) of the text with the help of Eqs. (4.15a) and (4.15c) to yield (4.16).

4.4 Beginning with Eq. (4.17), derive Eq. (4.18) and interpret the result physically.

4.5 Show explicitly that the most probable speed, v_m, of a molecule of a Maxwellian gas is

$$v_m = \left(\frac{\rho_g kT}{m}\right)^{1/2}$$

where $\rho_g = mn_v$ is the mass density of the gas.

4.6 Show explicitly that the root-mean-square (rms) speed of a molecule of a Maxwellian gas is

$$v_{\text{rms}} = \left(\frac{3kT}{m}\right)^{1/2}$$

4.7 Using the definite integral

$$\int_0^\infty \frac{\cos(\alpha z)}{[\beta^2 + z^2][\gamma^2 + z^2]} dz = \frac{\pi\left(\beta e^{-\alpha\gamma} - \gamma e^{-\alpha\beta}\right)}{2\beta\gamma\left(\beta^2 - \gamma^2\right)}$$

show that

$$C_\infty^{-1}\left\{\frac{1}{[\kappa_2^2 + k^2]\,(A(p) + k^2)}\right\} = \frac{\pi\left[\kappa_2 e^{-z\sqrt{A(p)}} - \sqrt{A(p)}e^{-z\kappa_2}\right]}{2\kappa_2\sqrt{A(p)}\,[\kappa_2^2 - A(p)]}$$

4.8 Compute numerically the mean free path of water vapor molecules above a capillary meniscus in a vacuum bell jar (water vapor only is present) as a function of temperature in ten degree intervals ranging from 10 to 80 °C assuming that the water vapor is a Maxwellian gas. Assume that the density of the water and its surface tension are independent of temperature, but assume that the Kelvin relation as given in Eq. (4.39) holds with the angle of capillarity equal to zero. Assume that $d_g \simeq 4.5 \times 10^{-8}$ cm, that p_{0v} is given by Eq. (4.8a) and that p_{0e} and T_0 of Eq. (4.8b) are $1.2277 \times 10^4\,\text{dyn/cm}^2$ and $283.16\,°\text{K}$, respectively.

5

MULTISPECIES TRANSPORT
IN UNSATURATED
DEFORMABLE POROUS MEDIA

5.1 INTRODUCTION

The mathematical description of the flow of water and the impurities it may contain in an unsaturated porous medium can be approached on a number of different levels. The most basic of these contains the assumptions that (1) the medium and fluid(s) are isothermal, (2) the porosity of each distinct soil horizon is constant, (3) gradients of matric potential and gravitation are the driving forces for liquid and vapor transport, (4) unsaturated hydraulic conductivity, even if it is assumed for convenience in specific applications to have a prescribed mathematical form, is treated as a phenomenological coefficient to be measured, as is the moisture characteristic, and (5) the equation of continuity holds in the form in which sources and sinks are absent. Mathematically these ideas are expressed as

$$\frac{\partial(\epsilon\,^{(i)}\theta)}{\partial t} = \epsilon\frac{\partial(\,^{(i)}\theta)}{\partial t} = -\vec{\nabla}\cdot\,^{(i)}\vec{J} \tag{5.1}$$

for the equation of continuity. Here the assumption of constant porosity, which means physically that the porous medium is assumed to be rigid, and assumed not to expand, contrary to the behavior of a swelling clay, for example, is explicitly displayed.

In (5.1), $^{(i)}\theta$ is the degree of saturation of an elementary volume of porous medium by a medium of species i, and $^{(i)}\vec{J}$ is the volume flux of species i leaving the elementary volume of the porous medium. The flux, $^{(1)}\vec{J}$, which is the volume per unit time of water, taken by convention to be

160

species 1, flowing through a unit area perpendicular to the direction of flow and having dimensions of velocity (volume per unit area per unit time, or Darcy velocity) is assumed to be given by the phenomenological relation

$$^{(i)}\vec{J} = -^{(1)}\overleftrightarrow{K}(^{(1)}\theta)\ \vec{\nabla}\left(\frac{^{(1)}\Psi(^{(1)}\theta)}{^{(1)}\rho g} + z\right) \tag{5.2}$$

in which $^{(1)}\overleftrightarrow{K}(^{(1)}\theta)$ is the second-rank tensor for unsaturated hydraulic conductivity, $^{(1)}\rho$ is the density of medium 1, and the rest of the symbols are as previously defined. Both $^{(1)}\overleftrightarrow{K}(^{(1)}\theta)$, the unsaturated hydraulic conductivity, and $^{(1)}\Psi(^{(1)}\theta)$, the matric potential, as functions of degree of liquid water saturation, $^{(1)}\theta$, show hysteresis, which means in this context that the values of $^{(1)}\Psi(^{(1)}\theta)$ and $^{(1)}\overleftrightarrow{K}(^{(1)}\theta)$ are different, respectively, for a given value of $^{(1)}\theta$ depending on whether or not the soil is wetting or drying.

Often, however, it is assumed for computational simplicity at this level of approximation that hysteresis is absent and that $^{(1)}\Psi(^{(1)}\theta)$ and $^{(1)}\overleftrightarrow{K}(^{(1)}\theta)$ respectively, have the same values for a given degree of saturation whether the soil is wetting or drying. The assumption that the fluid and porous medium are isothermal and at the same temperature is often not correct—particularly when water movement near land surface is being considered and diurnal temperature changes occur in the soil. That this problem can be particularly severe is seen from the fact that the matric potential depends directly on temperature, as shown in Eqs. (1.1a) and (1.1b) from which we see that as temperature increases, $^{(1)}\Psi$ becomes less negative for a given soil water content. However, moisture tends to flow from warmer toward cooler locations in the soil, and, in general, drier soils mean smaller (more negative) values of $^{(1)}\Psi$ so, when water and vapor transport can occur, the effect of increasing temperature at a given location is to make $^{(1)}\Psi$ more negative. Also, fluid density and interfacial tension decrease with increasing temperature, and as the temperature increases the initial effect is to make the matric potential of water in soil become less negative.

Combining (5.1) and (5.2) and specifying $\Psi(^{(1)}\theta)$ and $^{(1)}\overleftrightarrow{K}(^{(1)}\theta)$ as functions of $^{(1)}\theta$ yield a nonlinear partial differential equation for $^{(1)}\theta$ as a function of position and time with boundary conditions and macroscopic sources and sinks, such as the locations and rates of transpiring plants, for example, yet to be specified.

Hitherto, the equation so derived has been taken to be an equation of motion for a fluid in a porous medium. As will be discussed shortly, however, partial differential equations derived as an application of the principle of conservation of momentum rather than conservation of mass are the appropriate equations to be applied to describe the transport of water and impurities in porous media. The relation between $^{(1)}\Psi$ and $^{(1)}\theta$ is called a *constitutive relation* and is actually a form of the equation of state for the soil–water system in the same sense that the ideal gas law, used in preceding chapters, is the equation of state of an ideal gas–namely the relation between fluid pressure, temperature, and volume in the system.

The porous material will initially be taken to be fixed in a coordinate system fixed in space, and water movement is measured relative to these coordinates, that is, relative to the medium. Since the flow field as a whole is described by solutions of this equation set, following standard usage we term the mathematical description *Eulerian*. The equation set employed to follow the movement of small "parcels" of fluid through the porous medium is called the *Lagrangian description of fluid flow*, though both modes of description are actually due to Euler.

In the following sections we will develop the equations of motion describing the movement of fluids through an unsaturated porous medium and will eventually consider the porous medium and the fluids flowing through it as nonisothermal, interpenetrating, viscoelastic continua, with the viscoelastic descriptions of the media given via fractional calculus.

5.2   SINGLE-SPECIES EQUATIONS OF MOTION

5.2.1   Elastic Medium Equation of Motion

We consider the following sequence of events. A finite sample of elastic medium originally at equilibrium and free from internal stresses (stress is defined as force per unit area and will be formulated more precisely later on) and strains (changes in length per unit length) is deformed by applying forces to the outside of the sample, which is then held in place. The distortions, strains, thus induced in the material give rise to reaction forces–the stresses–internal to the material, to the forces originally applied. The notion of internal stress may be quantified as follows. Consider a coordinate system in which the material is located before and after it is deformed, as shown in Figure 5.1.

The presence of stresses induced in the medium by the initial "infinitesimal" deformation gives rise to forces acting across imaginary surfaces drawn in the medium such as the force of magnitude F acting on the imaginary element of surface $\Delta x \, \Delta y$ as shown. The force of magnitude F acting across the element of surface of area $\Delta x \, \Delta y$ perpendicular to the z axis is thus due to the strains (displacements) set up in the medium by the initial deformation. The force $\vec{F}$ can be resolved into components along each of the coordinate axes. These components we will term F_x, F_y, and F_z, respectively. The component of $\vec{F}$ perpendicular to the plane across which

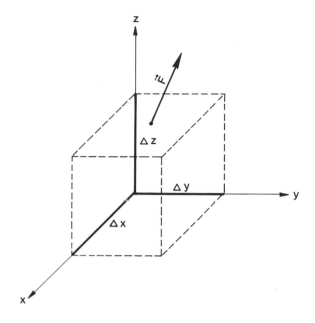

Figure 5.1: An element of volume in a distorted elastic medium.

it acts, F_z in this case, is a pure tension or compression. The two components of $\vec{F}$ parallel to the surfaces, ΔF_x and ΔF_y in this case, give rise to shear and rotation. If the magnitude of each force component is divided by the area across which it acts and the component of force and the plane across which it acts are indicated by a dual subscript (or superscript or mixed superscript and subscript) notation, we can write for the three components of a small change of $\vec{F}$ across the surface

$$f_{xz} = \lim_{\Delta A \to 0} \frac{\Delta F_x}{\Delta x\, \Delta y} \qquad (5.3a)$$

$$f_{yz} = \lim_{\Delta A \to 0} \frac{\Delta F_y}{\Delta x\, \Delta y} \qquad (5.3b)$$

$$f_{zz} = \lim_{\Delta A \to 0} \frac{\Delta F_z}{\Delta x\, \Delta y} \qquad (5.3c)$$

The quantities defined in (5.3a)–(5.3c) in the limit that $\Delta A = \Delta x\, \Delta y \to 0$ are three components of a second-rank tensor called the *stress tensor* (see

Appendix A for a brief discussion of tensor quantities and their manipulation–
as previously mentioned, the unsaturated hydraulic conductivity is another
example of a second-rank tensor). Consideration of elements of surface perpendicular to the x and y axes yields expressions similar to (5.3a)–(5.3c)
for the other six components of the stress tensor. Collectively these may
be summarized as

$$f_{ij} = \lim_{\Delta A_j \to 0} \frac{\Delta F_i}{\Delta A_j} \tag{5.4}$$

where the element of area ΔA_j is perpendicular to the direction denoted by
i, $1 \leq i \leq 3$, $1 \leq j \leq 3$ and the f_{ij} are taken to be the covariant components
of the stress tensor. The order of subscripts in (5.4) as to which represents
the component of force and which represents the axis perpendicular to the
element of area across which the force is acting varies from one reference
work to the next, so that care in reading is required. The fact that the
stress tensor is symmetric, that is,

$$f_{ij} = f_{ji} \tag{5.5}$$

does not affect this, as perusal of the literature will show.

We now consider the displacements of elements of the elastic medium
from equilibrium occurring as a result of the original deformation, that is,
the strains. We consider the coordinate system shown in Figure 5.2 and
the displacements of two points, p_1 and p_2, from their original equilibrium
positions resulting from the original deformation. In Figure 2(a) are shown
the displacements the points p_1 and p_2 embedded in the elastic medium
undergo when it is initially deformed, and we note the relations

$$\vec{u}_1 = \vec{r}_1' - \vec{r}_1 \tag{5.6}$$

$$\vec{u}_2 = \vec{r}_2' - \vec{r}_2 \tag{5.7}$$

in which $\vec{u}_1$ and $\vec{u}_2$ are the displacements the points p_1 and p_2 undergo on
deformation measured relative to their original locations in the unstrained
state of the medium. In Figure 2(b) the difference in relative displacement
between the two, $\vec{u}_a$, is displayed. Thus

$$\vec{u}_a = \vec{u}_2 - \vec{u}_1 \tag{5.8}$$

Writing $\vec{u}_a$ in terms of its x, y, and z components (denoted by 1–3, respectively), we have

$$\vec{u}_a = \hat{i} u_1(x, y, z) + \hat{j} u_2(x, y, z) + \hat{k} u_3(x, y, z) \tag{5.9}$$

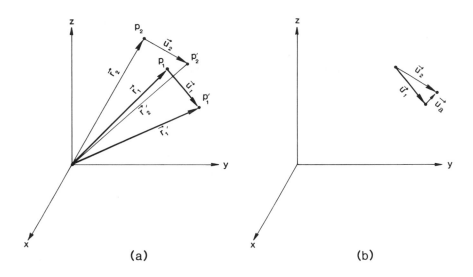

Figure 5.2: In (a) the element of matter at the point p_1 moves as a result of deformation and/or rotation to location p_1' while simultaneously the point p_2 has moved to location p_2'. In (b) is shown the incremental displacement, $\vec{u}_a$, between $\vec{u}_1$ and $\vec{u}_2$.

where $\hat{i}$, $\hat{j}$, and $\hat{k}$ are unit vectors in the x, y, and z directions, respectively. The total differential of $\vec{u}_a$, $d\vec{u}_a(x, y, z)$, is thus

$$d\vec{u}_a = \frac{\partial u_1}{\partial x^j} \, dx^j \, \hat{e}^1 + \frac{\partial u_2}{\partial x^j} \, dx^j \, \hat{e}^2 + \frac{\partial u_3}{\partial x^j} \, dx^j \, \hat{e}^3 = \frac{\partial u_i}{\partial x^j} dx^j \, \hat{e}^i \qquad (5.10)$$

where $\hat{e}^i$ is a unit vector in the i direction so that $\hat{e}^1 = \hat{i}$, $\hat{e}^2 = \hat{j}$, and $\hat{e}^3 = \hat{k}$, in this case, $x^1 = x$, $x^2 = y$, $x^3 = z$, and $1 \leq j \leq 3$. The indicated spatial coordinates are written with superscripts since the total differentials of the coordinates transform from one coordinate system to another as the components of a contravariant vector (actually, as noted in Appendix A, scalars are tensors of rank zero and vectors are tensors of rank one, so that no semantic distinction between tensors on the one hand and scalars and vectors on the other need actually be made, but the customary terminology will be adhered to here as elsewhere), and the Einstein summation convention, in which, unless specifically stated otherwise, covariant (sub-scripts) and contravariant (superscripts) indices repeated once in a given term (the same dummy index used for each) are assumed to be summed. In the case of (5.10) this last condition means that

$$d\vec{u}_a = \frac{\partial u_i}{\partial x^j} \, dx^j \, \hat{e}^i = \sum_{i=1}^{3} \sum_{j=1}^{3} \frac{\partial u_i}{\partial x^j} \, dx^j \, \hat{e}^i \qquad (5.11)$$

so that the total differential of $\vec{u}_a$ is, so far, the sum of nine terms. To facilitate later developments we now write $d\vec{u}_a$ as

$$d\vec{u}_a = \left[\frac{1}{2} \left(\frac{\partial u_i}{\partial x^j} + \frac{\partial u_j}{\partial x^i} \right) + \frac{1}{2} \left(\frac{\partial u_i}{\partial x^j} - \frac{\partial u_j}{\partial x^i} \right) \right] dx^j \, \hat{e}^i \qquad (5.12)$$

We now focus on the terms in square brackets in (5.12) and consider first the quantity $\frac{1}{2} \left(\partial u_i / \partial x^j + \partial u_j / \partial x^i \right)$. We see that for $i = j$ it reduces to a term having the significance of an extension or compression per unit length along each of the coordinate axes, namely along the x axis for $i = j = 1$, etc. For $i \neq j$, this term represents a shear, as we now show for the case in which $j = 1$ and $i = 2$. We consider Figure 5.3 showing the cross-section of an element of volume in the $x^1 - x^2$ plane. In Figure 5.3(a) we have that $AD = dx^2$, $DC = dx^1$, and $AA' = du_1$. We take $\theta/2$ to be a measure of the shear strain that has occurred so that in this case, for strains sufficiently small that $\theta \simeq \tan(\theta)$, we have

$$\frac{\theta}{2} = \frac{1}{2} \frac{\partial u_1}{\partial x^2} \qquad (5.13)$$

Similarly, if in Figure 5.3(b) we take $\phi/2$ to be a measure of the strain due to shear in the x^2 direction, we have

$$\frac{\phi}{2} = \frac{1}{2} \frac{\partial u_2}{\partial x^1} \qquad (5.14)$$

where $BB' = du_2$ and $DC = dx^1$. Thus, since the total strain about the x^3 axis is the sum of the two component strains just given, we have for the Cartesian tensor components of strain about the x^3 axis

$$\sigma_{12} = \sigma_{21} = \frac{1}{2} \left(\frac{\partial u_1}{\partial x^2} + \frac{\partial u_2}{\partial x^1} \right) \qquad (5.15)$$

Similar considerations for shear about the x^1 and x^2 axes and collection of the results into one equation yields for the Cartesian tensor components of the strain tensor

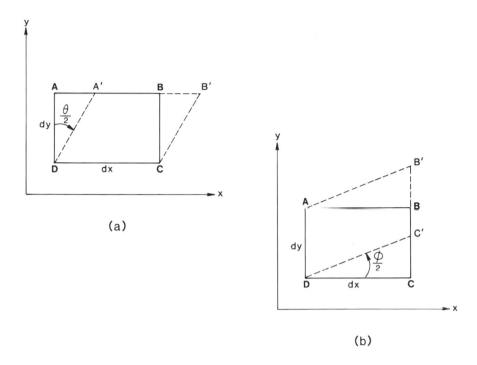

Figure 5.3: Cross-section of an initially unstressed volume element (solid lines) superimposed on the cross-section of the volume element after a small shear has occurred (dashed lines). In (a) is shown shear along the x^1 direction, while in (b) is shown shear along the x^2 direction.

$$\sigma_{ij} = \frac{1}{2}\left(\frac{\partial u_i}{\partial x^j} + \frac{\partial u_j}{\partial x^i}\right) \tag{5.16}$$

In a general curvilinear coordinate system, undeformed for the present since we are considering infinitesimal elastic distortions, the strain tensor is written in terms of its covariant components as

$$\sigma_{ij} = \frac{1}{2}\left(u_{i,j} + u_{j,i}\right) \tag{5.17}$$

where $u_{i,j}$ and $u_{j,i}$ are the usual covariant derivatives of u_i and u_j, respectively. [Since Christoffal symbols of the second kind appearing in the expressions for the covariant derivatives are zero in a Cartesian coordinate system, (5.16) reduces to (5.17) in that case.] In the case of pure rotation, for example, about the x^3 axis, rather than actual shear, the elements of mass in the elastic medium do not change position relative to each other, and so in that case, θ of Figure 5.3(a) would equal $-\phi$ of Figure 5.3(b) and σ_{12} would equal zero. The two terms in the second expression in square brackets of (5.12), which becomes in this case

$$\omega_{12} = \frac{1}{2}\left(\frac{\partial u_1}{\partial x^2} - \frac{\partial u_2}{\partial x^1}\right) \tag{5.18}$$

are seen to add and not go zero. We thus identify the second-rank anti-symmetric Cartesian tensor ω_{ij}, the components of which are given by

$$\omega_{ij} = -\omega_{ji} = \frac{1}{2}\left(\frac{\partial u_i}{\partial x^j} - \frac{\partial u_j}{\partial x^i}\right) \tag{5.19}$$

as representing rotations of elements of the sample. In a general curvilinear coordinate system ω_{ij}, the rotation tensor, is expressed in terms of its covariant components as

$$\omega_{ij} = \frac{1}{2}\left(u_{i,j} - u_{j,i}\right) \tag{5.20}$$

where $u_{i,j}$ and $u_{j,i}$ are the covariant derivatives noted previously and in Cartesian coordinates (5.20) reduces to (5.19). Since (5.20) describes pure rotations, we do not consider it further in our description of elasticity but use (5.17) instead as representing the full description of strain in an elastic medium.

The relations between stress and strain in an elastic medium can be expressed as

$$f_{ij} = C_{ij}^{kl}\sigma_{kl} \tag{5.21}$$

where C_{ij}^{kl} is called the *tensor of elasticity* and, from the sum of the number of superscripts and subscripts, is seen to be a fourth-rank tensor. We now write

$$\sigma_{kl} = \gamma_{kl}^{ij}f_{ij} \tag{5.22}$$

where γ_{kl}^{ij} is the inverse of C_{ij}^{kl} in the sense that

$$C_{ij}^{kl}\,\gamma_{kl}^{ij} = \frac{1}{2}\left(1 + \delta_i^j\delta_j^i\right) \tag{5.23a}$$

or, more generally,

$$C_{ij}^{kl}\gamma_{mn}^{pq} = \frac{1}{2}\left(\delta_i^p\delta_j^q + \delta_i^q\delta_j^p\right) = C_{mn}^{pq}\,\gamma_{ij}^{kl} \tag{5.23b}$$

where the Einstein summation convention has been used. The choice to write f_{ij} and σ_{ij} in terms of their covariant components is entirely arbitrary.

Any combination of components, including contravariant and/or mixed, for example, is possible and equally valid. For example, (5.21) could be written

$$f^{ij} = g^{im} g^{jn} C_{mn}^{kl} \sigma_{kl} \tag{5.24}$$

by raising the i and j indices using the inverse, g^{im}, for example, of the metric tensor, g_{im}, which describes in this case the space of the unde-formed coordinate system in which the displacements of the sample from its unstressed state are measured, to raise the index i from covariant to contravariant and g^{jn} to similarly raise the index j. (A discussion of rais-ing and lowering indices can be found in Appendix A.) The forms of the tensor of elasticity and its inverse depend on the symmetry of the elastic body being considered. For an elastic medium the tensor of elasticity has the symmetries

$$C_{ij}^{kl} = C_{ij}^{lk} = C_{ji}^{kl} = C_{ji}^{lk} \tag{5.25}$$

For a homogeneous medium, strain is in the same direction as stress so that if the covariant derivative (the covariant derivative is the generalization of the gradient operator from use on scalars to use on vectors and tensors) of the stress tensor is zero, the covariant derivative of the strain tensor is zero as well. Thus, using (5.21), we find

$$f_{ij,m} = C_{ij}^{kl} \sigma_{kl,m} + C_{ij,m}^{kl} \sigma_{kl} = 0 \tag{5.26}$$

where, as above, the comma followed by a subscript denotes covariant dif-ferentiation. Since $\sigma_{kl,m}$ is assumed to be zero in this case, we have for the covariant derivative of the tensor of elasticity for a homogeneous, isotropic medium,

$$C_{ij,m}^{kl} = 0 \tag{5.27}$$

Consideration of the elementary form of Hooke's law, the shear modulus, μ, and Poisson's ratio, σ, along with (5.25) leads eventually to the result for a homogeneous, isotropic elastic medium,

$$C_{ij}^{kl} = \mu \left(\delta_i^k \delta_j^l + \delta_i^l \delta_j^k \right) + \lambda g_{ij} g^{kl} \tag{5.28}$$

where the Kronecker delta functions, δ_i^k, are mixed, second-rank tensors and are equal to one if $i = k$ and equal to zero otherwise, μ is the shear modulus or first Lamé coefficient, to be discussed in more detail shortly, while λ is the second Lamé coefficient for the medium, g_{ij} is the metric tensor describing the original coordinate system embedded in the unstressed material, and g^{kl} is its inverse. Similarly, we have for γ_{kl}^{ij} for a homogeneous, isotropic medium,

$$\gamma_{kl}^{ij} = \frac{1}{4\mu} \left(\delta_k^i \delta_l^j + \delta_k^j \delta_l^i \right) - \frac{\lambda}{2\mu(3\lambda + 2\mu)} g_{kl} g^{ij} \tag{5.29}$$

which is the inverse of (5.28). The relation between the shear modulus, μ, Young's modulus, Y, and Poisson's ratio, σ, is

$$\mu = \frac{Y}{2(1 + \sigma)} \tag{5.30}$$

Similarly, we have for the definition of the second Lamé coefficient in terms of the other elastic constants for a homogeneous isotropic medium,

$$\lambda = B - \frac{2\mu}{3} = \frac{Y\sigma}{(1 + \sigma)(1 - 2\sigma)} \tag{5.31}$$

where B is called the *bulk modulus* of the medium.

We now derive directly the relation between the bulk modulus, Young's modulus, and Poisson's ratio for a homogeneous isotropic medium. We consider a parallelepiped (bar) of rectangular cross section having length ℓ_0, width w_0, and height h_0. If a tensile stress, in this case force per unit cross-sectional area of the body, is applied along its length, as shown in Figure 5.4, and we denote the cross-sectional area of the body after the stress has been applied by A, the resultant small change in length, $\Delta\ell$, can be written as

$$\frac{F}{A} = Y\frac{\Delta\ell}{\ell} \tag{5.32}$$

where Y is the ratio of tensile (or compressive) stress to strain, which is change in length per unit length and is called *Young's modulus*. As the bar is stretched, it becomes smaller in cross-sectional area as well. Specifically,

$$\frac{\Delta w}{w} = \frac{\Delta h}{h} = -\sigma\frac{\Delta\ell}{\ell} \tag{5.33}$$

where σ is called *Poisson's ratio* and is the ratio between the tensile and dilational strains. The minus sign describes explicitly the fact that as $\Delta\ell$ increases Δw and Δh decrease and vice-versa. If the bar is now subjected to a uniform external pressure p, causing it to change in volume by ΔV, we have

$$p = -B\frac{\Delta V}{V} \tag{5.34}$$

for a homogeneous isotropic elastic medium. Compressive strains, however, tend to cause dilation of the bar. Thus, for the compressive stress–pressure in this case–along the length of the bar, we have from (5.32) for what is now a component of the change in length of the bar, $\Delta\ell_1$,

$$-Y\frac{\Delta\ell_1}{\ell} = p \tag{5.35}$$

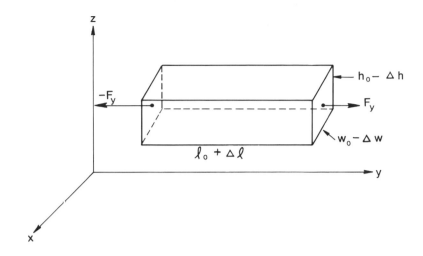

Figure 5.4: A bar of length ℓ and initial cross-sectional area A_0 oriented along the y axis is subjected to a force F_y. Thus, F_y/A is the tensile stress to which the bar is subjected.

Further, the pressure on a pair of faces on opposite sides of the bar, those of initial (prestress) area $w_0\ell_0$, for example, gives a component of the change in length of the bar, $\Delta\ell_2$, via

$$\frac{\Delta w}{w}\sigma = -\frac{\Delta\ell_2}{\ell} \qquad (5.36)$$

as an application of the cause and effect relations given in (5.33) and also

$$\frac{\Delta w}{w} = -\frac{p}{Y} \qquad (5.37)$$

as an application of (5.32) so that

$$\frac{p\sigma}{Y} = \frac{\Delta\ell_2}{\ell} \qquad (5.38)$$

while the compressive stress p on the pair of faces of initial area $h_0\ell_0$ gives rise to a component of the change in length of the bar, $\Delta \ell_3$, via

$$\frac{\Delta h}{h} \sigma = -\frac{\Delta \ell_3}{\ell} = -\frac{p\sigma}{Y} \qquad (5.39)$$

Thus, the change in length of the elastic bar subjected to uniform hydrostatic stress is

$$\Delta \ell = \Delta \ell_1 + \Delta \ell_2 + \Delta \ell_3 = -\frac{p\ell}{Y} + \frac{2p\sigma\ell}{Y} = -\frac{p\ell}{Y}(1 - 2\sigma) \qquad (5.40)$$

If we now begin with the initial width w_0, in place of ℓ_0, and consider the three components of its change in length as previously, we find

$$\Delta w = -\frac{pw}{Y}(1 - 2\sigma) \qquad (5.41)$$

Similarly, considering the three components of the change in height, h_0, yields

$$\Delta h = -\frac{ph}{Y}(1 - 2\sigma) \qquad (5.42)$$

as well. The volume of the bar after compression is

$$V = (\ell_0 - \Delta \ell)(w_0 - \Delta w)(h_0 - \Delta h)$$

$$\simeq V_0 - \ell_0 w_0 \Delta h - h_0 \ell_0 \Delta w - h_0 w_0 \Delta \ell \qquad (5.43)$$

so that in this case

$$\frac{V_0 - V}{V} \simeq \frac{\Delta V}{V_0} = \left(\frac{\Delta \ell}{\ell_0} + \frac{\Delta w}{w_0} + \frac{\Delta h}{h_0}\right) \simeq \frac{-3p}{Y}(1 - 2\sigma) = -\frac{p}{B} \qquad (5.44)$$

or,

$$B = \frac{Y}{3(1 - 2\sigma)} \qquad (5.45)$$

for the relation between the bulk modulus, Poisson's ratio, and Young's modulus for a homogeneous isotropic elastic solid. We now derive directly the relation between the shear modulus, Young's modulus, and Poisson's ratio given in (5.30) as follows. We consider the bar for the case in which $\ell_0 = h_0$ after a shear stress has been applied, as shown in Figure 5.5. The applied force causes elongation of the original diagonal, d_0, per unit length, $\Delta d/d_0$, where the line segment Δd, one leg of the small isosceles right triangle shown in Figure 5.5, makes an angle of $45°$ with the line segment that is the hypotenuse of the small triangle shown in Figure 5.5. We write the elongation per unit length as

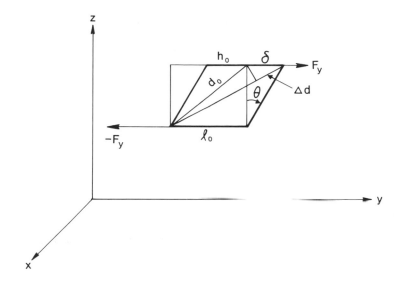

Figure 5.5: An elastic bar to which a shearing force F_y has been applied.

$$\frac{\Delta d}{d_0} = \frac{\delta}{\sqrt{2}} \times \frac{1}{\sqrt{2}\ell_0} = \frac{\theta}{2} \qquad (5.46)$$

We now examine the two contributions to Δd, Δd_1, from the component of F_y giving rise to linear extension along the diagonal, and Δd_2, the component of extension along the diagonal arising from the component of F_y acting perpendicular to the diagonal as follows. The component of F_y acting along the diagonal is $F_\parallel = F_y/\sqrt{2}$, so that the total force of $2F_\parallel$ is $\sqrt{2}F_y$ and the component of F_y perpendicular to the diagonal, $F_\perp$, is $F_\perp = F_y/\sqrt{2}$ so that the total force of $2F_\perp$ is $\sqrt{2}F_y$. Applying the definition of Young's modulus, (5.32), yields,

$$\frac{\sqrt{2}F_\parallel}{A_\parallel} = Y\frac{\Delta d_1}{d} \qquad (5.47)$$

where $A_\parallel$ is the area of the bar in the plane of the diagonal perpendicular to d_0 and is approximately $A_\parallel = w_0\sqrt{2}\ell_0 = \sqrt{2}A$, where A is the area of the upper face of the bar. Applying (5.32) along with (5.33) yields

$$\frac{\Delta d_2}{d} = \frac{\sqrt{2}\sigma F}{YA} \tag{5.48}$$

where A is also approximately $\sqrt{2}d_0\ell_0$. Approximating d by d_0, writing

$$\frac{\theta}{2} = \frac{\Delta d_1 + \Delta d_2}{d} \tag{5.49}$$

and inserting Δd_1 and Δd_2 from (5.47) and (5.48), respectively, yields

$$\frac{\theta}{2} = \frac{F_y}{AY}(1 + \sigma) \tag{5.50}$$

Writing the shearing stress f_y as F_y/A and using the shear modulus, μ, of (5.28) yields for the relation between shearing stress and shearing strain, θ,

$$f_y = \mu\theta \tag{5.51}$$

We now identify μ in terms of Young's modulus and Poisson's ratio by comparing (5.50) and (5.51) to yield (5.30).

The Lagrangian form, to be discussed in more detail shortly, of the equation of motion for an isothermal elastic medium is derived by equating the divergence of the stress tensor, which is the force per unit volume, at points interior to the medium, plus any body forces per unit volume, to the acceleration per unit volume of a point in the medium. Thus we have

$$f^{ij}_{,j} + \rho F^i = \rho a^i \tag{5.52}$$

where ρ is the mass density of the material, F^i is the total body force per unit mass acting on the elastic body (two examples are centrifugal force and gravity) written in terms of its contravariant components, a^i is the acceleration of the material comprising the elastic body, and $f^{ij}_{,j}$ is the divergence (contraction of the covariant derivative) of the stress tensor given in (5.24). Equation (5.52) is written in terms of the original coordinate system embedded in the elastic body prior to its being deformed. Substituting (5.24) into (5.52) and using (5.27) and (5.28) yields

$$\left[\mu g^{im} g^{jn}(\delta^k_m \delta^l_n + \delta^l_m \delta^k_n) + \lambda g^{im} g^{jn} g_{mn} g^{kl}\right]\sigma_{kl,j} + \rho F^i = \rho a^i \tag{5.53}$$

or

$$\left[\mu(g^{ik} g^{jl} + g^{il} g^{jk}) + \lambda g^{ij} g^{kl}\right]\sigma_{kl,j} + \rho F^i = \rho a^i \tag{5.54}$$

where the results

$$g^{jm} g_{mn} = \delta^j_n \tag{5.55}$$

along with $\delta_j^i = 1$ for $i = j$ and $\delta_j^i = 0$ for $i \neq j$ have been used. Applying the general formalism for covariant differentiation given in Appendix A yields for the covariant derivative of σ_{kl}, appearing on the left hand-side of Eq. (5.54),

$$\sigma_{kl,j} = \frac{\partial \sigma_{kl}}{\partial x^j} - \Gamma_{jk}^n \sigma_{nl} - \Gamma_{jl}^n \sigma_{kn} \tag{5.56}$$

which may be expressed in terms of derivatives of the displacements, the u_i, by

$$\sigma_{kl,j} = \frac{1}{2}(u_{k,l} + u_{l,k})_{,j} = \frac{1}{2}u_{k,l,j} + \frac{1}{2}u_{l,k,j}$$

$$= \frac{1}{2}\frac{\partial}{\partial x^j}(u_{k,l} + u_{l,k}) - \frac{1}{2}\Gamma_{jk}^n(u_{n,l} + u_{l,n}) - \frac{1}{2}\Gamma_{jl}^n(u_{k,n} \mid u_{n,k})$$

$$= \frac{1}{2}\frac{\partial}{\partial x^j}\left(\frac{\partial u_k}{\partial x^l} + \frac{\partial u_l}{\partial x^k} - 2\Gamma_{lk}^\alpha u_\alpha\right) - \frac{1}{2}\Gamma_{jk}^n\left(\frac{\partial u_n}{\partial x^l} + \frac{\partial u_l}{\partial x^n} - 2\Gamma_{ln}^\beta u_\beta\right)$$

$$- \frac{1}{2}\Gamma_{jl}^n\left(\frac{\partial u_k}{\partial x^n} + \frac{\partial u_n}{\partial x^k} - 2\Gamma_{nk}^\alpha u_\alpha\right) \tag{5.57}$$

where the symmetry property of the Christoffal symbols, namely

$$\Gamma_{jk}^n = \Gamma_{kj}^n \tag{5.58}$$

has been used, as has the fact that in an inner product expression of the form $\Gamma_{jk}^n u_n$, n is a dummy index of summation and could be replaced by any letter. The acceleration, which is the intrinsic derivative of the velocity, may be written in terms of its contravariant components as

$$a^i = \frac{\delta v^i}{\delta t} = \frac{\partial v^i}{\partial t} + v_{,k}^i \frac{dx^k}{dt} \tag{5.59}$$

where

$$v_{,k}^i = \frac{\partial v^i}{\partial x^k} + \Gamma_{kn}^i v^n \tag{5.60}$$

If we now identify v^i as

$$v^i = \frac{du^i}{dt} \tag{5.61}$$

we have for a^i in terms of the displacements and the coordinates of the system in its unstressed state, the x^k,

$$a^i = \frac{\partial^2 u^i}{\partial t^2} + \left[\frac{\partial}{\partial x^k}\left(\frac{\partial u^i}{\partial t}\right) + \Gamma_{kn}^i \frac{\partial u^n}{\partial t}\right] \frac{dx^k}{dt} \tag{5.62}$$

The full equation of motion for infinitesimal elastic displacements in a general curvilinear coordinate system can now be written by substituting (5.62) and (5.57) into (5.54).

We now consider the reduction of these equations for the case of Cartesian coordinates in which (5.54) becomes, after using the various quantities of the form g^{kl} to raise the associated indices of $\sigma_{kl,j}$ and then noting that in Cartesian coordinates g^{ij} becomes δ^{ij},

$$\mu \left(\sigma^{ij}_{..,j} + \sigma^{ji}_{..,j} \right) + \lambda \delta^{ij} \sigma^{l}_{.l,j} + \rho F^i = \rho \frac{\partial^2 u^i}{\partial t^2} \tag{5.63}$$

where the dots indicate, as usual, the former locations of indices that have been raised (or lowered, as the case may be), and $\sigma_{kl,j}$ and a^i have become, respectively,

$$\sigma_{kl,j} = \frac{1}{2} \left(\frac{\partial^2 u_k}{\partial x^j \partial x^l} + \frac{\partial^2 u_l}{\partial x^j \partial x^k} \right) \tag{5.64}$$

and

$$a^i = \frac{\partial^2 u^i}{\partial t^2} \tag{5.65}$$

so that

$$\sigma^{ij}_{..,j} = \frac{1}{2} \left(\delta^{nj} \frac{\partial^2 u^i}{\partial x^j \partial x^n} + \delta^{mi} \frac{\partial^2 u^j}{\partial x^m \partial x^j} \right) \tag{5.66a}$$

$$\sigma^{ji}_{..,j} = \frac{1}{2} \left(\delta^{ni} \frac{\partial^2 u^j}{\partial x^n \partial x^j} + \delta^{mj} \frac{\partial^2 u^i}{\partial x^j \partial x^m} \right) \tag{5.66b}$$

$$\sigma^{l}_{.l,j} = \frac{1}{2} \left(\frac{\partial^2 u^l}{\partial x^j \partial x^l} + \delta^{ml} \frac{\partial^2 u_l}{\partial x^j \partial x^m} \right) = \frac{\partial^2 u^l}{\partial x^l \partial x^j} \tag{5.66c}$$

since in Cartesian coordinates, the covariant, contravariant, and physical components of vectors and tensors are all equal so that $u_l = u^l$ in that case. Substituting (5.66a)–(5.66c) into (5.63) yields

$$\mu \left(\delta^{mj} \frac{\partial^2 u^i}{\partial x^j \partial x^m} + \delta^{ni} \frac{\partial^2 u^j}{\partial x^j \partial x^n} \right) + \lambda \delta^{ij} \left(\frac{\partial^2 u^l}{\partial x^l \partial x^j} \right) + \rho F^i = \rho \frac{\partial^2 u^i}{\partial t^2} \tag{5.67}$$

for the equation of motion in Cartesian coordinates.

For the case of a stressed body held in equilibrium by forces applied everywhere to its surface, as in the scenario discussed at the beginning of the section, the condition for equilibrium can be expressed as

$$f^{ij} n_j = P^i \tag{5.68}$$

where the n_j are the covariant components of the unit vector normal to the surface at every point on it (often called the *unit normal*) and the P^i are the contravariant components of the applied surface forces per unit area of surface, sometimes called the *surface tractions*. The potential energy, $_pW$, "stored" in the body as a result of the elastic deformation of it is

$$_pW = \frac{1}{2}C^{ijkl}\sigma_{ij}\sigma_{kl} \tag{5.69}$$

Finite Deformations of an Elastic Medium

We now consider briefly the case of finite deformations. The strain tensor describing the deformed medium, which we denote by $\bar{\sigma}_{kl}$, may be written for this case as

$$\bar{\sigma}_{kl} = \frac{1}{2}(\bar{g}_{kl} - g_{kl}) \tag{5.70}$$

where g_{kl} is the metric tensor of the coordinate system embedded in the unstressed unsaturated body, and $\bar{g}_{kl}$ is the metric tensor of the coordinate system embedded in the deformed body that deforms with it. If the equations of the deformed coordinates are given in terms of the original coordinates, that is, $\bar{x}^i(x^k)$, then $\bar{g}_{kl}$ may be computed formally as

$$\bar{g}_{kl} = \frac{\partial \bar{x}^k}{\partial x^i}\frac{\partial \bar{x}^l}{\partial x^j}g_{ij} \tag{5.71}$$

where g_{ij} is computed from the unitary base vectors for the original system as discussed in Appendix A. The equations of motion in the deformed system have the same form as in the original system. For example, (5.52) becomes

$$\bar{f}^{ij}_{,j} + \rho\bar{F}^i = \rho\bar{a}^i \tag{5.72a}$$

where

$$\bar{f}^{ij} = \overline{C}^{ijkl}\bar{\sigma}_{kl} \tag{5.72b}$$

Also,

$$\bar{f}^{ij}\bar{n}_j = \overline{P}^i \tag{5.73}$$

and so on.

The development given has been for the case of an isothermal medium. Often, however, particularly when wave-motion solutions of the equations of motion are being considered, heat is generated locally in times short compared to those required for conduction to occur. In that case, the motion is adiabatic rather than isothermal. In this connection it is possible to develop expressions for the adiabatic bulk modulus, adiabatic Poisson's ratio, and adiabatic Young's modulus as an application of equilibrium thermodynamics formalism for use in the stress–strain relations and equations of motion in place of the isothermal expressions used previously.

We now consider the relation between stress and rate of strain and the equation of motion for a viscous compressible fluid.

5.2.2   Equation of Motion for a Viscous Fluid

The stress-rate of strain relation for a viscous, compressible fluid is similar in form to the stress–strain relation for an elastic solid and is, in general curvilinear coordinates,

$$t^{ij} = E^{ijkl}\dot{e}_{kl} \tag{5.74}$$

where t^{ij} is the viscous stress tensor, and represents the flow of momentum in the fluid system due to velocity gradients in the fluid, $\dot{e}_{kl}$ is the rate of strain tensor and is given by

$$\dot{e}_{kl} = \frac{1}{2}(v_{k,l} + v_{l,k}) \tag{5.75a}$$

where

$$v_k = g_{km}v^m \tag{5.75b}$$

and similarly for v_l. As an aside we note that if the position vector $\vec{r}(x^1, x^2, x^3)$ is measured relative to a set of fixed axes, then the vector velocity $\vec{v}$ may be written as the time rate of change of $\vec{r}$ as

$$\vec{v} = \frac{d\vec{r}}{dt} = \frac{\partial \vec{r}}{\partial x^k}\frac{dx^k}{dt} = \vec{a}_k\frac{dx^k}{dt} = \vec{a}_k v^k \tag{5.76}$$

so that for the contravariant components of velocity

$$v^k = \frac{dx^k}{dt} \tag{5.77}$$

where, as in Appendix A, $\vec{a}_k$ is the covariant unitary base vector for the fixed coordinate system and as usual,

$$g_{km} = \vec{a}_k \cdot \vec{a}_m \tag{5.78}$$

where g_{kl} is the metric tensor for the fixed coordinate system in which $\vec{v}$ is being measured and the $\vec{a}_i$ are the unitary base vectors for the system. The function $x^k(t)$ can be taken to follow the motion of the fluid or the motion of the observer if desired if the motion of the observer is different from that of the fluid velocity. For a viscous, compressible fluid

$$E^{ijkl} = \eta(g^{ik}g^{jl} + g^{il}g^{jk}) + \eta' g^{ij}g^{kl} \tag{5.79}$$

where η is the shear viscosity, or first coefficient of viscosity, the viscosity appearing in Poiseuille's law, for example, and η' is the second coefficient of viscosity and is used when the fluid is compressible; for an incompressible fluid, η' is taken to be zero. This form of E^{ijkl} is the same as C^{ijkl} for a homogeneous isotropic medium; that is, two parameters, determined by experiment, are required to characterize the medium, which is the expected result. The relation between rate of strain and stress can be written

$$\dot{e}_{kl} = G^{ij}_{kl} t_{ij} \tag{5.80}$$

where, as previously,

$$G^{ij}_{kl} = \frac{1}{4\eta}(\delta^i_k \delta^j_l + \delta^j_k \delta^i_l) - \frac{\eta'}{2\eta(3\eta' + 2\eta)} g_{kl} g^{ij} \tag{5.81}$$

and

$$t_{ij} = g_{im} g_{jn} t^{mn} \tag{5.82}$$

Substituting (5.75) and (5.79) into (5.74) and carrying out the indicated inner product summations yields

$$t^{ij} = 2\eta \dot{e}^{ij} + \eta' g^{ij} v^k_{,k} \tag{5.83}$$

so that with the addition of a pressure term the components of the full stress tensor in general curvilinear coordinates become

$$\tau^{ij} = t^{ij} - pg^{ij} = 2\eta \dot{e}^{ij} + \eta' g^{ij} v^k_{,k} - pg^{ij} \tag{5.84a}$$

or, in covariant components

$$\tau_{ij} = 2\eta \dot{e}_{ij} + \eta' g_{ij} v^k_{,k} - pg_{ij} \tag{5.84b}$$

The equation of motion is, as in (5.52),

$$\tau^{ij}_{,j} + \rho F^i = \rho a^i \tag{5.85}$$

or

$$g^{mj} \tau_{im,j} + \rho F_i = \rho a_i \tag{5.86}$$

where the F^i are the components of the sum of any body forces per unit mass, and the contravariant components of acceleration, the a^i, are given by

$$a^i = \frac{\delta v^i}{\delta t} = \frac{\partial v^i}{\partial t} + v^i_{,k} \frac{dx^k}{dt} \tag{5.87}$$

In (5.87), v^i is the velocity of the fluid relative to a fixed coordinate system and dx^k/dt is the velocity of the observer relative to the same fixed coordinate system. If dx^k/dt is taken to be equal to the kth component of the fluid velocity, then the observer of the fluid motion is moving with the fluid and (5.87) becomes

$$a^i = \frac{\partial v^i}{\partial t} + v^i_{,k} v^k \tag{5.88}$$

Similarly, the covariant components of fluid acceleration are written for the case in which the observer is moving with the fluid as

$$a_i = \frac{\partial v_i}{\partial t} + v_{i,k} v^k \tag{5.89}$$

It is common in the literature to see the expression for fluid acceleration written as $D\vec{v}/Dt$, the so-called material derivative, when the observer is moving with the fluid. This expression as it is usually written has no meaning in other than Cartesian coordinates and thus if generalization to other coordinate systems is required, (5.87) or another similar tensor expression that explicitly treats acceleration as the intrinsic derivative of velocity is actually the most appropriate and useful. Substituting (5.84b) into (5.86) yields for the Lagrangian form of the equation of motion of a viscous fluid

$$g^{mj} \left(2\eta \, \dot{e}_{im,j} + \eta' g_{im} v^k_{,k,j} - g_{im} \, p_{,j} \right) + \rho F_i = \rho a_i \tag{5.90}$$

5.3   VISCOELASTICITY VIA FRACTIONAL CALCULUS

In the developments given so far we have briefly discussed both elastic and viscous compressible media. Viscoelastic media have the combined properties of both viscosity and elasticity. An example is wet clay, which can flow, and which has elastic (and actually plastic) properties as well. Earth materials and the fluids moving through their pores behave viscoelastically in varying degrees under various circumstances of fluid saturation, pressure, temperature, etc. The mathematical description of viscoelasticity has taken many different forms over the years (see, for example, Flügge, 1967, or Christensen, 1982). A possible description, which will be developed here, is based on the observation that in the theory of elasticity, stress, or momentum flux (flow of momentum per unit area per unit time), is proportional to displacement(s) per unit length, or strain, while for a fluid, the stresses are proportional to the time rates of fluid displacement per unit length or rates of strain. Since elastic stresses are proportional to derivatives of the strains of order zero with respect to time, and fluid stresses are proportional to the derivatives of the strains with respect to time of order one, the suggestion has been made in the literature that viscoelastic substances be represented by functions of the fractional derivatives of the strain tensor of orders ranging from one to zero.

Fractional calculus is a well-established area of mathematics (Oldham and Spanier, 1974) that has experienced a revival in the past few years, though its origins actually date from the time of Newton and Leibnitz. There are a number of definitions of derivatives and integrals of fractional order, and all of them are given ultimately in terms of "ordinary" mathematical functions. We represent the relation between stress and rate of strain in an approximate assumed form, as follows. We write for the fourth-rank tensor, A_{ij}^{kl}, connecting components of viscoelastic stress, the χ_{ij}, with the fractional rates of strain, $d^\alpha e_{kl}/dt^\alpha$, where α is the order of the fractional derivative of e_{kl} with respect to time,

$$A_{ij}^{kl} \simeq \left[(1-\alpha)C_{ij}^{kl} + \alpha E_{ij}^{kl} \right] t_0^{\alpha - 1} \tag{5.91}$$

so that the stress-rate of strain relation for viscoelastic media is taken to be

$$\chi_{ij} = A_{ij}^{kl} \frac{d^\alpha e_{kl}}{dt^\alpha} = A_{ij}^{kl} \sigma_{kl}(\alpha) \tag{5.92}$$

where

$$e_{kl} = \frac{1}{2}(\xi_{k,l} + \xi_{l,k}) \tag{5.93}$$

The quantity $\sigma_{kl}(\alpha)$ is defined to be the (fractional) rate of strain tensor for a viscoelastic medium, and t_0 is the unit in which time is to be measured. Thus, the fractional powers of time introduced into the denominator of e_{kl} by the process of differentiation to fractional order α are multiplied by the factor $t_0^{\alpha-1}$, which is t_0 raised to the power $\alpha - 1$, so that time to an inverse power of one is always present. Thus, this formulation eliminates a problem in some power-law versions of rheological theories in which the constants are raised to powers such that the resulting equations are dimensionally incorrect. The ξ_k and ξ_l are the covariant components of the displacement of an element of viscoelastic material measured relative to a fixed coordinate system, and the $\xi_{k,l}$ are the covariant derivatives of the ξ_k and ξ_l, respectively. We now make the simplifying assumption that

$$\sigma_{kl}(\alpha) = \frac{1}{2} \left[\left(\frac{d^\alpha \xi_k}{dt^\alpha} \right)_{,l} + \left(\frac{d^\alpha \xi_l}{dt^\alpha} \right)_{,k} \right] \tag{5.94}$$

and see from this definition that for $\alpha = 1$, $\sigma_{kl}(\alpha)$ reduces to $\dot{e}_{kl}$, the rate of strain tensor for viscous media given in (5.75a), while for $\alpha = 0$, $\sigma_{kl}(\alpha)$ reduces to the strain tensor for elastic media, given previously in (5.17).

Fractional derivatives can be expressed in terms "ordinary" functions in various ways. A useful expression for the derivative of order α of a scalar function is

$$\frac{d^\alpha f(x)}{dx^\alpha} = D_x^\alpha f(x) = D_x^m D_x^{-\zeta} f(x)$$

$$= \frac{d^m}{dx^m} \left(\frac{1}{\Gamma(\zeta)} \int_0^x (x-\xi)^{\rho-1} f(\xi) d\xi \right) \tag{5.95}$$

where

$$\alpha = m - \zeta \tag{5.96}$$

where m is the least integer greater than α, so that $0 \leq \zeta \leq 1$, $\Gamma(\zeta)$ is the usual gamma function, and the operator $D_x^{-\zeta}$ represents fractional integration with respect to x of order ζ (Courant, 1936, pp. 339–340). Thus, the concept is that to carry out fractional differentiation of order α, we first integrate to order ζ and differentiate to order m such that (5.96) is satisfied. Integration to fractional order β is thus written

$$D_x^{-\beta} f(x) = \frac{1}{\Gamma(\beta)} \int_0^x (x - \xi)^{\beta - 1} f(\xi) \, d\xi \tag{5.97}$$

and, upon reversing the order of differentiation of $f(x)$ to orders m and ζ in (5.95), we have upon using successive integrations by parts and induction

$$D_x^{\alpha} f(x) = D_x^{-\zeta} D_x^m f(x) = \frac{1}{\Gamma(\zeta)} \int_0^x (x - \xi)^{\zeta - 1} f^{(m)}(\xi) \, d\xi \tag{5.98}$$

where $f^{(m)}(\xi)$ is the mth derivative of $f(\xi)$ with respect to ξ. In general

$$D_x^{\gamma} D_x^{\delta} f(x) = D_x^{\delta} D_x^{\gamma} f(x) \tag{5.99}$$

where γ and δ are real and represent differentiation (if positive) and integration (if negative) of $f(x)$ to arbitrary order. We now consider the fractional derivative of the displacement ξ_k with respect to time in more detail. Since ξ_k depends on position in the viscoelastic medium and time, and the coordinates of a given element of mass, $x^{\ell}(t)$, depend on time as well, we write

$$\xi_k \left[x^{\ell}(t) \right] = \xi_k \left[x^i(t), t \right] \tag{5.100}$$

where $1 \leq \ell \leq 4$, $1 \leq i \leq 3$, and

$$x^4(t) = t \tag{5.101}$$

We see from (5.100) that ξ_k is a function of a function of time [ξ_k is the outer function and $x^{\ell}(\ )$ is the inner function] and that a further fractional calculus result is needed to effect this differentiation. This result is provided by the creative work of Osler (1970a) and is, in essentially the notation of Osler

$$D_{g(z)}^{\alpha} [f(z)] = \sum_{n=0}^{\infty} \frac{1}{n!} \left(D_{g(z)}^{\alpha} \{ F(g, w) \, [h(z) - h(w)]^n \} \right)$$

$$\times D_{h(z)}^n \left(\frac{f(z)}{F(z, w)} \right) \Bigg|_{w=z} \tag{5.102}$$

where

$$D^\alpha_{g(z)} \left\{ F(g, w) \left[h(z) - h(w) \right]^n \right\} \Big|_{w=z}$$

$$= \sum_{r=0}^{n} \binom{n}{r} \left[-h(z) \right]^r D^\alpha_{g(z)} \left[F(z, w) h^{n-r}(z) \right] \Big|_{w=z} \quad (5.103)$$

The meaning of the symbols in (5.102) and (5.103) is as follows. The notation $D^\alpha_{g(z)} f(z)$ denotes the derivative of $f(z)$, with z a complex variable, of fractional order α, with respect to $g(z)$. For values of α less than zero, representing integration, this operator is the generalization to operations of fractional order of the Stieltjes integral, which indicates the integration of one function of a given independent variable with respect to another function of the independent variable rather than with respect to the independent variable itself. The function $f(z)$ in the present context means, again in the notation of Osler (loc. cit.)

$$f(z) = F_1 \left[h(z) \right] \quad (5.104)$$

where $F[]$ is the outer function and $h()$ is the inner function. The function $F(z, w)$ of the complex variables z and w can be chosen essentially arbitrarily for computational convenience or as an aid in generating particular results. An expression such as $D^\alpha_{g(z)} F(z, w) h(z)^{n-r} \Big|_{w=z}$ means that $D^\alpha_{g(z)}$ operates on the product $F(z, w) h(z)^{n-r}$ and that the result of the operation is evaluated for values of w such that $w = z$. The quantity $\binom{n}{r}$ is the usual binomial coefficient

$$\binom{n}{r} = \frac{\Gamma(n+1)}{\Gamma(r+1)\Gamma(n-r+1)} \quad (5.105)$$

where $\Gamma()$ is the usual gamma function. In the application of these results made here, we have $F_1[] = \xi_k[]$, $g(z) = z$, $z = t$, $F(z, w) = 1$, $h() = x^\ell()$, and thus,

$$f(z) \rightarrow \xi_k \left[x^\ell(t) \right] \quad (5.106)$$

Thus, (5.102) and (5.103) become for this application,

$$\frac{d^\alpha \xi_k}{dt^\alpha} = D^\alpha_t \left\{ \xi_k \left[x^\ell(t) \right] \right\}$$

$$= \sum_{n=0}^{\infty} \frac{1}{n!} D^\alpha_t \left\{ \left[x^\ell(t) - x^\ell(w) \right]^n \right\} \Big|_{w=t} D^n_{x^\ell(t)} \left\{ \xi_k \left[x^\ell(t) \right] \right\} \quad (5.107)$$

and

$$D_t^\alpha \left\{ \left[x^\ell(t) - x^\ell(w) \right]^n \right\} \Big|_{w=t} = \sum_{r=0}^{n} \binom{n}{r} (-1)^r \left[x^\ell(t) \right]^r D_t^\alpha \left\{ \left[x^\ell(t) \right]^{n-r} \right\}$$

(5.108)

We now write for compactness of notation

$$D_{x^\ell(t)}^n \left\{ \xi_k \left[x^\ell(t) \right] \right\} = \xi_k^{(n)}$$

(5.109)

We now consider the quantity $D_t^\alpha \left\{ \left[x^\ell(t) \right]^{n-r} \right\}$ appearing in (5.108). We first write for the fractional derivative of a product of two functions (a generalization of the usual Leibnitz rule for the nth derivative of the product of two functions) using essentially the notation of Osler (1970b)

$$D_z^\alpha \left[u(z)v(z) \right] = \sum_{p=0}^{\infty} \frac{\Gamma(\alpha + 1) D_z^{\alpha - p} \left[u(z) \right] D_z^p \left[v(z) \right]}{\Gamma(\alpha - p + 1) p!}$$

(5.110)

Setting $z = t$ and substituting the functions

$$u(t) = x^\ell(t)$$

(5.111a)

and

$$v(t) = \left[x^\ell(t) \right]^{n-r-1}$$

(5.111b)

into (5.110) yields

$$D_t^\alpha \left\{ \left[x^\ell(t) \right]^{n-r} \right\} = \sum_{p=0}^{\infty} \frac{\Gamma(\alpha + 1) D_t^{\alpha - p} \left[x^\ell(t) \right] D_t^p \left\{ \left[x^\ell(t) \right]^{n-r-1} \right\}}{\Gamma(\alpha - p + 1) p!}$$

(5.112)

Upon setting the parameter $-\beta$ appearing in (5.77) equal to $\alpha - p$, and noting that in this instance $\beta > 0$, and for $p > 0$, $p > \alpha > 0$ so that $\alpha - p < 0$ for all terms in the series beyond the first, we have for the quantity $D_t^{\alpha - p} \left[x^\ell(t) \right]$ appearing as part of the summand in (5.112)

$$D_t^{\alpha - p} \left[x^\ell(t) \right] = \frac{1}{\Gamma(p - \alpha)} \int_0^t (t - \xi)^{p - \alpha - 1} x^\delta(\xi) \, d\xi$$

(5.113)

which, since $p > \alpha > 0$, as just noted, represents integration to fractional order. For $p = 0$, we have from (5.95) with m set equal to 1,

$$D_t^\alpha \left[x^\ell(t) \right] = \frac{d}{dt} \frac{1}{\Gamma(1 - \alpha)} \int_0^t (t - \xi)^{-\alpha} x^\ell(\xi) \, d\xi$$

(5.114)

which represents the fractional differentiation of $x^\ell(t)$ to order α. For the quantity $D_t^p \left\{ [x^\ell(t)]^{n-r-1} \right\}$ we write (Schwatt, 1924, p. 12)

$$D_t^p \left\{ [x^\ell(t)]^{n-r-1} \right\} = \frac{d^p [x^\ell(t)]^{n-r-1}}{dt^p}$$

$$= \sum_{q=1}^{p} \frac{d^q \left\{ [x^\ell(t)]^{n-r-1} \right\}}{d [x^\ell(t)]^q} \sum_{a=1}^{q} \frac{(-1)^{q+a}}{q!} \binom{q}{a} [x^\ell(t)]^{q-a} \frac{d^p [x^\ell(t)]^a}{dt^p}$$

$$= \sum_{q=1}^{p} \frac{\Gamma(n-r)(x^\ell)^{n-r-1}}{\Gamma(n-r-q)} \sum_{a=1}^{q} \frac{(-1)^{q+a}}{q!} \binom{q}{a} [x^\ell(t)]^{-a} \frac{d^p \lfloor x^\ell(t) \rfloor^a}{dt^p} \quad (5.115)$$

Various approximations of these equations to obtain simplified expressions for $d^\alpha \xi_k \left[x^\ell(t) \right] /dt^\alpha$ of (5.94) are possible. For example, replacing the sum in (5.108) by the term containing the largest value, for a given value of n, of $\binom{n}{r}$, which, for even n is $\binom{n}{n/2}$ and that for odd n, which is $\binom{n}{(n+1)/2} = \binom{n}{(n-1)/2}$, and substituting the result into (5.87) yields

$$\frac{d^\alpha \xi_k \left[x^\ell(t) \right]}{dt^\alpha} \simeq \sum_{n=0}^{\infty} \frac{1}{2n!} \left[\binom{2n}{n} [x^\ell(t)]^n D_t^\alpha \left\{ [x^\ell(t)]^n \right\} \right] \xi_k^{(2n)}$$

$$- \sum_{n=0}^{\infty} \frac{1}{(2n+1)!} \left[\binom{2n+1}{n} [x^\ell(t)]^n D_t^\alpha [x^\ell(t)]^{n+1} \right] \xi_k^{(2n+1)} \quad (5.116)$$

where n in (5.108) has been replaced by $2n$ and $2n+1$ for the even and odd values of n in (5.107) so that instead of $n/2$ we now write $2n/2 = n$, and instead of $r = (n-1)/2$ for terms in which the values of n in (5.107) are odd, we write instead $r = [(2n+1) - 1]/2 = n$ Retaining the first two terms of the sum in (5.112) yields

$$D_t^\alpha \left\{ [x^\ell(t)]^{n-r} \right\} \simeq [x^\ell(t)]^{n-r-1} D_t^\alpha [x^\delta(t)]$$

$$+ \alpha(n-r-1) [x^\ell(t)]^{n-r-2} D_t^{\alpha-1} [x^\ell(t)] \quad (5.117)$$

Replacing $n - r$ in (5.117) by n and $n + 1$ as in (5.116) and substituting the results into (5.116) yields

$$\frac{d^\alpha \xi_k \left[x^\ell(t) \right]}{dt^\alpha} \simeq \sum_{n=0}^\infty \frac{1}{2n!} \xi_k^{(2n)} \left[x^\ell(t) \right]^n \binom{2n}{n} A(n)$$

$$- \sum_{n=0}^\infty \frac{1}{(2n+1)!} \xi_k^{(2n+1)} \left[x^\ell(t) \right]^n \binom{2n+1}{n} B(n)$$

$$\simeq \xi_k \left(\frac{1}{x^\ell(t)} D_t^\alpha \left[x^\ell(t) \right] - \frac{\alpha}{\left[x^\ell(t) \right]^2} D_t^{\alpha-1} \left[x^\ell(t) \right] \right) - \xi_k^{(1)} D_t^\alpha \left[x^\ell(t) \right] \quad \text{(5.118a)}$$

where

$$A(n) = \left[x^\ell(t) \right]^{n-1} D_t^\alpha \left[x^\ell(t) \right] + \alpha(n-1) \left[x^\ell(t) \right]^{n-2} D_t^{\alpha-1} \left[x^\ell(t) \right] \quad \text{(5.118b)}$$

$$B(n) = \left[x^\ell(t) \right]^n D_t^\alpha \left[x^\ell(t) \right] + \alpha n \left[x^\ell(t) \right]^{n-1} D_t^{\alpha-1} \left[x^\ell(t) \right] \quad \text{(5.118c)}$$

where the first term only of each sum in (5.118a) has been retained. Approximating further, we have

$$\left(\frac{d^\alpha \xi_k}{dt^\alpha} \right)_{,l} \simeq \left(\frac{1}{x^\ell(t)} D_t^\alpha \left[x^\ell(t) \right] - \frac{\alpha}{\left[x^\ell(t) \right]^2} D_t^{\alpha-1} \left[x^\ell(t) \right] \right) \xi_{k,l} \quad \text{(5.119)}$$

where, in order to obtain a linearized theory that reduces directly to the linear approximations for elasticity and viscous fluids given, the contribution from the second sum of (5.118) has been omitted. Thus, from (5.119) and (5.94) we have for $\sigma_{kl}(\alpha)$, the rate of strain tensor for viscoelastic media,

$$\sigma_{kl}(\alpha) \simeq \frac{1}{2} \left(\frac{1}{x^\ell(t)} D_t^\alpha \left[x^\ell(t) \right] - \frac{\alpha}{\left[x^\ell(t) \right]^2} D_t^{\alpha-1} \left[x^\ell(t) \right] \right) (\xi_{k,l} + \xi_{l,k})$$

$$\text{(5.120)}$$

where $D_t^\alpha [x^\ell(t)]$ and $D_t^{\alpha-1} \left[x^\ell(t) \right]$ are given by (5.114) and (5.113), respectively (this latter for $p = 1$), and the sum over ℓ is understood.

We now consider A_{ij}^{kl} further. Substituting (5.28) and (5.79) with the indices i and j lowered into (5.91) yields

$$A_{ij}^{kl} \simeq \left[(1-\alpha)\mu + \alpha\eta \right] t_0^{\alpha-1} \left(\delta_i^k \delta_j^l + \delta_i^l \delta_j^k \right) + \left[(1-\alpha)\lambda + \alpha\eta' \right] t_0^{\alpha-1} g_{ij} g^{kl}$$

$$\text{(5.121)}$$

where the symbol $\simeq$ is employed to point up the fact that in an actual experimental situation involving a viscoelastic medium the values of μ, η, λ, and η' will be somewhat different in value and interpretation from those of their limiting cases ($\alpha = 0$ or 1). Also, as mentioned, a theory based on a linear relation between stress and strain is approximate at best, though in many cases the approximation is a good description of reality. The equation of motion for a viscoelastic medium is formally given in this representation by

$$\chi^{ij}_{,j} + \rho F^i = \rho a^i \qquad (5.122)$$

where F^i is the sum of the body forces per unit mass and a^i is the acceleration of the medium given as in (5.64) by

$$a^i = \frac{\partial^2 \xi^i}{\partial t^2} + \left[\frac{\partial}{\partial x^m} \left(\frac{\partial \xi^i}{\partial t} \right) + \Gamma^i_{mn} \frac{\partial \xi^n}{\partial t} \right] \frac{dx^m}{dt} \qquad (5.123)$$

Raising the indices i and j of A^{kl}_{ij} as given by (5.91) as was done for C^{kl}_{ij} in the development leading to (5.53) and substituting the result into (5.121) yields

$$\rho a^i = \{ [(1 - \alpha)\mu + \alpha\eta]\, t_0^{\alpha - 1} \left(g^{ik} g^{jl} + g^{il} g^{jk} \right)$$

$$+ [(1 - \alpha)\lambda + \alpha\eta']\, t_0^{\alpha - 1} g^{ij} g^{kl} \} \, \sigma(\alpha)_{kl,j} + \rho F^i \qquad (5.124)$$

where, as in (5.120) and following (5.57), we approximate the covariant derivative of the fractional rate of strain tensor, $\sigma_{kl,j}(\alpha)$, to yield

$$\sigma_{kl,j}(\alpha) \simeq \frac{1}{4} \left(\frac{1}{x^\delta(t)} D_t^\alpha \left[x^\delta(t) \right] - \frac{\alpha}{[x^\delta(t)]^2} D_t^{\alpha - 1} \left[x^\delta(t) \right] \right)$$

$$\times \left[\frac{\partial}{\partial x^j} \left(\frac{\partial \xi_k}{\partial x^l} + \frac{\partial \xi_l}{\partial x^k} - 2\Gamma^m_{lk}\xi_m \right) - \Gamma^m_{jk} \left(\frac{\partial \xi_m}{\partial x^l} + \frac{\partial \xi_l}{\partial x^m} - 2\Gamma^\beta_{lm}\xi_\beta \right) \right.$$

$$\left. - \Gamma^m_{jl} \left(\frac{\partial \xi_k}{\partial x^m} + \frac{\partial \xi_m}{\partial x^k} - 2\Gamma^\beta_{mk}\xi_\beta \right) \right] \qquad (5.125)$$

Up to now, we have considered the equations of motion as if there were no porous medium present. We now consider the forms of the equations of motion, continuity, and energy, all of which are required, along with the equation(s) of state for a complete description of the motion of the fluid(s) in a porous medium, in more detail.

5.4 EQUATIONS OF CONTINUITY

We consider a curvilinear coordinate system having axes x^1, x^2, and x^3 fixed in space. The space is assumed to be filled with a homogeneous and isotropic porous medium having constant grain density, but not necessarily with a constant bulk density or porosity. We next assume an elementary volume, δV, having its sides parallel to the coordinate axes and fixed in space relative to them. At time $t = t_0$, we further assume the existence of an elementary volume $\delta^{(0)}V_0$ coincident with δV but fixed in the porous medium as denoted by the subscript zero which does not connote a tensor index in this case. We now recall that the Lagrangian description of fluid flow involves following the motion of the material initially at a given location, in this case the material contained in δV at $t = t_0$, through space; that is, for the porous medium this means following the motion of the material initially contained in $\delta^{(0)}V_0$ through space. Each elementary amount of fluid and contaminant of species i contained in δV at $t = t_0$ is thought of as having the same initial elementary volume $\delta^{(i)}V_0$ and of being followed through space as time passes. As time passes, each such elementary volume of material of species i can occupy changing positions relative to every other initial elementary volume, and moreover, each such elementary volume contains the same material (constant mass) as it moves through the system, even though the shape, size, and bulk density of each elementary volume in general alters as it moves and the masses of the individual elementary volumes of different species of material interpenetrate each other. If the coordinates of the center of the initial elementary volume, $\delta^{(i)}V_0$, are $^{(i)}x_0^1$, $^{(i)}x_0^2$, and $^{(i)}x_0^3$, for all species and the coordinates of the center of the deformed parallelepiped of moving material of species i are $^{(i)}x^1$, $^{(i)}x^2$, and $^{(i)}x^3$, then the volume of the deformed parallelepiped, $\delta^{(i)}V$, is (Lamb, 1945, pp. 12–14)

$$\delta^{(i)}V = \frac{\partial(^{(i)}x^1, ^{(i)}x^2, ^{(i)}x^3)}{\partial(^{(i)}x_0^1, ^{(i)}x_0^2, ^{(i)}x_0^3)} \, \delta^{(i)}V = {}^{(i)}D \, \delta^{(i)}V_0 \qquad (5.126)$$

where

$$^{(i)}D = \det \begin{vmatrix} \frac{\partial^{(i)}x^1}{\partial^{(i)}x_0^1} & \frac{\partial^{(i)}x^1}{\partial^{(i)}x_0^2} & \frac{\partial^{(i)}x^1}{\partial^{(i)}x_0^3} \\[2mm] \frac{\partial^{(i)}x^2}{\partial^{(i)}x_0^1} & \frac{\partial^{(i)}x^2}{\partial^{(i)}x_0^2} & \frac{\partial^{(i)}x^2}{\partial^{(i)}x_0^3} \\[2mm] \frac{\partial^{(i)}x^3}{\partial^{(i)}x_0^1} & \frac{\partial^{(i)}x^3}{\partial^{(i)}x_0^2} & \frac{\partial^{(i)}x^3}{\partial^{(i)}x_0^3} \end{vmatrix} \qquad (5.127)$$

is the usual Jacobian determinant written in this case for the movement of species i. We reiterate that in the flow situation envisioned here, the $^{(i)}x_0^j$ and the $\delta^{(i)}V_0$ are the same for each of the i species since the initial element of volume is the same bulk volume of each species.

The equation of continuity for the ith species in the Lagrangian formulation in terms of the bulk mass density, $^{(i)}\rho_b$, equals the mass of species i contained in the elementary volume, $\delta^{(i)}V_0$ divided by $\delta^{(i)}V_0$, while $^{(i)}\rho_{0b}$ is defined in a similar manner for the initial value of $\delta^{(i)}V$, $\delta^{(i)}V_0$, so that at time $t = t_0$, $^{(i)}\rho_0 = {}^{(i)}\rho_{0b}$. Thus

$$^{(i)}\rho_b \, {}^{(i)}D = {}^{(i)}\rho_{0b} \qquad (5.128)$$

Later spatial locations of the centers of the moving and deforming parallelepipeds of material of species i are given by

$$^{(i)}x^j = {}^{(i)}x^j \left({}^{(i)}x_0^1, \, {}^{(i)}x_0^2, \, {}^{(i)}x_0^3, t \right) \tag{5.129}$$

In the Lagrangian formulation we have for the velocity and acceleration respectively, of the moving parallelepiped of species i relative to the original fixed coordinate system

$$^{(i)}v^j = \frac{d\,^{(i)}x^j}{dt} \tag{5.130}$$

$$^{(i)}a^j = \frac{\delta\,^{(i)}v^j}{\delta t} - \frac{d\,^{(i)}v^j}{dt} \mid \Gamma_{kn}^j \, ^{(i)}v^n \frac{d\,^{(i)}x^k}{dt} = \frac{\partial\,^{(i)}v^j}{\partial t} + {}^{(i)}v_{,k}^j \frac{d^{(i)}x^k}{dt} \tag{5.131}$$

where $d^{(i)}x^k/dt$ is the kth component of the velocity of the observer of the motion of the ith species of fluid. If the observer of the ith component of fluid motion is assumed to move with it, $d^{(i)}x^k/dt$ becomes, as previously,

$$\frac{d\,^{(i)}x^k}{dt} = {}^{(i)}v^k \tag{5.132}$$

relative to the fixed system. Thus, the acceleration of the ith component of fluid measured relative to the fixed system as "seen" by an observer moving with velocity v^k relative to the fixed system is

$$^{(i)}a^j = \frac{d\,^{(i)}v^j}{dt} + \Gamma_{kn}^j \, ^{(i)}v^n \, ^{(i)}v^k = \frac{\partial\,^{(i)}v^j}{\partial t} + {}^{(i)}v_{,k}^j \, ^{(i)}v^k \tag{5.133}$$

The lth component of the mass flux of species i, $^{(i)}j^l$, leaving δV can be written as

$$^{(i)}j^l = {}^{(i)}\rho_b \, ^{(i)}v^l = {}^{(i)}\phi \, ^{(i)}v^l \, ^{(i)}\rho = {}^{(i)}v^l \, ^{(i)}\theta \, ^{(i)}\rho \, ^{(0)}\epsilon \tag{5.134}$$

where $^{(i)}\phi$ is the fraction of δV occupied by material of species i. Thus, the elementary volume initially occupied by material of species i equals $^{(i)}\phi \delta V$ and the mass density of the pure material of species i is $^{(i)}\rho = {}^{(i)}\rho_b/{}^{(i)}\phi$. Note that in this formulation $^{(i)}\phi$ is the volumetric "saturation" of the ith species– a modest generalization of the usual concept of volumetric saturation here applied to all species instead of only liquid water in the porous medium. We take $^{(0)}\epsilon$ to be the porosity of the porous medium itself. Note in passing that $^{(i)}\epsilon = 1 - {}^{(i)}\phi$ and that for $n + 1$ interpenetrating continuae,

$$\sum_{i=0}^{n} {}^{(i)}\epsilon = 0 \tag{5.135}$$

Note also that $^{(0)}k\,^{(i)}\theta = \,^{(i)}\phi$ for $i > 1$. The composite (total) density, $_c\rho$, of the n species of material initially contained in the element of porous medium having volume δV is given in this formulation by

$$_c\rho = \sum_{i=1}^{n} \,^{(i)}\rho_b \tag{5.136}$$

and

$$\rho = \,^{(0)}\rho_b + \,_c\rho \tag{5.137}$$

where $^{(0)}\rho_b$ is the bulk density of the porous medium and ρ is the total composite density of an elementary volume of porous medium and the fluids it contains.

If the separations between the centers of the n image parallelepipeds of δV are such that they all overlap to some extent, then a common center of mass may be defined and $_c\rho$ computed at successive times will be the total composite density of the species moving relative to the porous medium. The equation of continuity for the ith species in the Lagrange formulation but in terms of Eulerian independent variables may be written in the form, alternative to (5.127)

$$\frac{\delta\,^{(i)}\rho_b}{\delta t} = -\,^{(i)}\rho_b\,^{(i)}v^j_{,j} \tag{5.138}$$

where the intrinsic derivative of a scalar, $\delta\,^{(i)}\rho_b/\delta t$ in this case, reduces to the ordinary total derivative. The divergence of $^{(i)}v^j$, namely $^{(i)}v^j_{,j}$, is computed as follows. The notation $^{(i)}v^j_{,k}$ indicates, as in (5.131) and (5.134), the covariant derivative of $^{(i)}v^j$ with respect to x^k, the generalization of the gradient operation from scalar quantities to vectors and tensors. In general

$$^{(i)}v^j_{,k} = \frac{\partial\,^{(i)}v^j}{\partial x^k} + \Gamma^j_{kn}\,^{(i)}v^n \tag{5.139}$$

When the index k is set equal to j, contracted, so called, the result is the divergence of $^{(i)}v^j$ (the divergence is by definition the contraction of the covariant derivative), and we have for the quantity appearing on the right-hand side of (5.138)

$$^{(i)}v^j_{,j} = \frac{\partial\,^{(i)}v^j}{\partial x^j} + \Gamma^j_{jn}\,^{(i)}v^n \tag{5.140}$$

The connection between (5.128), the equation of continuity in terms of Lagrangian independent variables, and (5.138), the equation of continuity written for an observer following the motion, and so Lagrangian in form, in terms of the x^k, is given by

$$^{(i)}v^j_{,j} = \frac{1}{^{(i)}D} \frac{\partial\, ^{(i)}D}{\partial t} \tag{5.141}$$

(Courant, 1936, p. 213, Kaplan, 1952, p.300), where $^{(i)}v^j_{,j}$ is written in terms of Eulerian independent variables, and $^{(i)}D$ is given by (5.127). To derive (5.141), eq.(5.128) is differentiated with respect to time, and it is noted that $\delta\, ^{(i)}\!\rho_{0b}/\delta t = 0$, where $^{(i)}\!\rho_{0b}$ is the initial bulk density of the material of species i, and the result is combined with (5.138).

The Eulerian form of the equation of continuity, in which events at a given location in the flow field are examined, is

$$\frac{\partial\, ^{(i)}\!\rho_b}{\partial t} = - \left[^{(i)}\!\rho_b\, ^{(i)}v^l \right]_{,l} = -\,^{(i)}j^l_{,l} \tag{5.142}$$

where $^{(i)}j^l$ is given by (5.134). We note that each distinct choice of location to be observed corresponds to a different implicit choice of the initial coordinates and hence of an elementary volume of fluid initially located at them. We now turn to consideration of the equation(s) of motion of multiple species in a porous medium.

5.5 MULTISPECIES EQUATIONS OF MOTION

The equation of motion of the ith component of the fluid in a porous medium written from the point of view of an observer following the fluid motion, the Lagrangian point of view, is

$$^{(i)}\tau^{lj}_{,j} + {}^{(i)}\!\rho_b\, ^{(i)}F^l = {}^{(i)}\!\rho_b\, ^{(i)}a^l \tag{5.143}$$

where $^{(i)}\tau^{lj}$ is the stress tensor for species i, to be discussed in more detail shortly, $^{(i)}F^l$ is the body force per unit mass of component i, and $^{(i)}a^l$ is given by (5.134). To go from Eulerian to Lagrangian coordinates in (5.143), we write

$$\frac{\partial\, ^{(i)}_0 x^m}{\partial\, ^{(i)}x^l}\, ^{(i)}\tau^{lj}_{,j} + \frac{\partial\, ^{(i)}x_0{}^m}{\partial\, ^{(i)}x^l}\, ^{(i)}\!\rho_b\, ^{(i)}F^l = \frac{\partial\, ^{(i)}x_0^m}{\partial^{(i)}x^l}\, ^{(i)}\!\rho_b\, ^{(i)}a^l \tag{5.144a}$$

which is the usual tensor transformation law for contravariant components of vectors (which are tensors of rank one) and tensors, to yield

$$^{(i)}\tau^{mj}_{0,j} + {}^{(i)}\!\rho_{0b}\, ^{(i)}F_0^m = {}^{(i)}\!\rho_{0b}\, ^{(i)}a_0^m \tag{5.144b}$$

and the $^{(i)}x^l$ are identified with the locations in the flow field of interest, namely the x^l. Since the choices of the $^{(i)}x_0^j$ and time determine the corresponding $^{(i)}x^l$, a corresponding observation point in the fluid flow field, x^l, cannot be readily chosen arbitrarily. Note in passing that the elementary components of fluid filling an elementary volume passing a given location in a flow field at a given instant of time may well have come from a number of distinct initial positions. In multispecies flow in a porous medium, the fluid passing a given location may be made up of parts of the fluid species initially present in a number of spatially distinct elementary initial fluid volumes, possibly widely separated at time $t = 0$. These may interpenetrate and partially disperse again and partially interpenetrate again at other spatial locations as time passes.

In the interest of completeness, we note that rearrangement of (5.143) yields for the Eulerian form of the equation of motion of the ith species

$$\frac{\partial \left(^{(i)}\rho_b \, ^{(i)}v^l \right)}{\partial t} = -^{(i)}v^l \left[^{(i)}\rho_b \, ^{(i)}v^m \right]_{,m} - \, ^{(i)}\rho_b \, ^{(i)}v^l_{,k} \, ^{(i)}v^k +^{(i)}\tau^{lj}_{,j} +^{(i)}\rho_b \, ^{(i)}F^l$$

$$= -^{(i)}v^l \, ^{(i)}j^m_{,m} -^{(i)}\rho_b \, ^{(i)}v^l_{,m} \, ^{(i)}v^m +^{(i)}\tau^{lj}_{,j} + \, ^{(i)}\rho_b \, ^{(i)}F^l \quad (5.145)$$

where the identifications

$$\rho_b \, a^l = \rho_b \frac{\delta v^l}{\delta t} = \rho_b \frac{\partial v^l}{\partial t} +\rho_b(v^l)_{,m} v^m \qquad (5.146)$$

$$\rho_b \frac{\delta v^l}{\delta t} = \rho_b \frac{\partial v^l}{\partial t} - v^l \left(\rho_b v^k\right)_{,k} + v^l \left(\rho_b v^k\right)_{,k} + \rho_b v^l_{,m} v^m$$

$$= \rho_b \frac{\partial v^l}{\partial t} + v^l \frac{\partial \rho}{\partial t} + v^l \left(\rho_b v^k\right)_{,k} + \rho_b v^l_{,m} v^m$$

$$= \frac{\partial(\rho_b v^l)}{\partial t} + v^l \left(\rho_b v^k\right)_{,k} + \rho_b v^l_{,m} v^m \qquad (5.147)$$

have been made for each species, (5.147) has been used in writing (5.145), and (5.142) has been used in writing (5.147).

The stress tensor, $^{(i)}\tau^{ij}$, for the ith component of the fluid system has been examined by Bearman and Kirkwood (1967) using statistical mechanics and an expression developed for it that is microscopic rather than macroscopic in nature. The statement has been made in the literature that $^{(i)}\tau^{ij}$ cannot be defined directly macroscopically (Fitts, 1962, p. 16). The same physically based point could be raised concerning the microscopic shear viscosity, $^{(i)}\eta$, and the microscopic bulk viscosity, $^{(i)}\eta'$, for each fluid species. However, Bearman and Kirkwood (loc. cit., p. 169) give microscopic expressions for the shear viscosity, $^{(i)}\eta$, and the bulk viscosity, $^{(i)}\eta'$, for each fluid species and then sum these to obtain the total shear and bulk viscosities, respectively, for the composite fluid system. We therefore assume with Bearman and Kirkwood that these quantities exist and write the stress tensor, $^{(i)}\chi^{ij}$, for the ith species of a viscoelastic fluid

$$^{(i)}\chi^{ij} = {}^{(i)}A^{ijkl}\frac{d^\alpha\,{}^{(i)}e_{kl}}{dt^\alpha} = {}^{(i)}A^{ijkl}\,{}^{(i)}\sigma_{kl}(\alpha) \tag{5.148}$$

where

$$^{(i)}A^{ijkl} \simeq \left[(1-\alpha)\,{}^{(i)}C^{ijkl} + \alpha\,{}^{(i)}E^{ijkl}\right]t_0^{\alpha-1} \tag{5.149}$$

and α is the order of the fractional derivative of the strain tensor for the ith species, $^{(i)}e_{kl}$, which is given by

$$^{(i)}e_{kl} = \frac{1}{2}\left[{}^{(i)}\xi_{k,l} + {}^{(i)}\xi_{l,k}\right] \tag{5.150}$$

The quantity $^{(i)}\sigma_{kl}(\alpha)$ is defined to be the (fractional) rate of strain tensor for a viscoelastic medium, and is approximated by

$$^{(i)}\sigma_{kl}(\alpha) = \frac{1}{2}\left[\left(\frac{d^\alpha\,{}^{(i)}\xi_k}{dt^\alpha}\right)_{,l} + \left(\frac{d^\alpha\,{}^{(i)}\xi_l}{dt^\alpha}\right)_{,k}\right] \tag{5.151}$$

The parameter t_0 is the unit in which time is to be measured so that the fractional powers of time introduced into the denominator of (5.148) by the process of differentiation to fractional order are multiplied by the factor $t_0^{\alpha-1}$ so that time to an inverse power of one is always present as noted in connection with (5.92). The tensor of elasticity for the ith species, $^{(i)}C^{ijkl}$, is given by

$$^{(i)}C^{ijkl} = {}^{(i)}\mu(g^{ik}g^{jl} + g^{il}g^{jk}) + {}^{(i)}\lambda g^{ij}g^{kl} \tag{5.152}$$

where $^{(i)}\mu$ and $^{(i)}\lambda$ are the first and second coefficients of elasticity for the ith species respectively. The viscosity tensor for the ith species, $^{(i)}E^{ijkl}$, is given by

$$^{(i)}E^{ijkl} = {}^{(i)}\eta(g^{ik}g^{jl} + g^{il}g^{jk}) + {}^{(i)}\eta'g^{ij}g^{kl} \tag{5.153}$$

where $^{(i)}\eta$ and $^{(i)}\eta'$ are the first and second coefficients of viscosity for the ith species respectively. Finally, the quantity $^{(i)}\xi_k$ is the displacement of the ith species and $^{(i)}\xi_{k,l}$ is its covariant derivative. Writing (5.148) for the special case for $\alpha = 1$ yields for the viscous stress tensor, $^{(i)}\tau_{ij}$ for the ith species

$$^{(i)}\tau_{ij} = {}^{(i)}\eta\left({}^{(i)}v_{i,j} + {}^{(i)}v_{j,i}\right) + {}^{(i)}\eta'\,g_{ij}\,{}^{(i)}v^k_{,k} - p\,g_{ij} \tag{5.154}$$

where $^{(i)}v_{i,j}$ is the covariant derivative of the covariant components of the velocity of the ith species of fluid, and similarly for $^{(i)}v_{j,i}$, while $^{(i)}\eta$ and $^{(i)}\eta'$ have been previously defined. Similarly for $^{(i)}\tau^{ij}$ we have

$$^{(i)}\tau^{ij} = {}^{(i)}\eta\left({}^{(i)}v_{\alpha,\beta} + {}^{(i)}v_{\beta,\alpha}\right)g^{\alpha i}g^{\beta j} + {}^{(i)}\eta'g^{ij}\,{}^{(i)}v^k_{,k} - p\,g^{ij} \tag{5.155}$$

in obvious notation. Substituting (5.155) into (5.143) yields the requisite expanded Lagrangian form of the equation of motion for the ith species in a porous medium. The velocity terms are given by the appropriate flux law, such as Darcy's law, Fick's law, etc. with the gradients of head, concentration, etc. acting as scalar velocity potentials. Note that these choices for scalar velocity potentials assume that coupled effects in the sense of nonequilibrium thermodynamics can be neglected and that direct effects only are important. Note further that for the case of a vector quantity, a scalar velocity potential function is only part of the total potential on which it can depend and that it can depend on the curl of a vector potential as well. Thus, for the case of the velocity of the ith species of fluid in a porous medium, we have, on projecting the curl of the vector potential for the ith species, $^{(i)}A_{p,q} - {}^{(i)}A_{q,p}$, which is an antisymmetric, second-rank tensor, on the direction of flow,

$$^{(i)}v^p = -K^{pl}\,^{(i)}\phi_{,l} + \left({}^{(i)}A_{m,q} - {}^{(i)}A_{q,m} \right)g^{pm}n^q \qquad (5.156)$$

where $^{(i)}\phi$ is a scalar potential function for species i, K^{pl} is a phenomenological coefficient, $^{(i)}A$ is the vector potential for species i on which $^{(i)}v^p$ could also depend, and the n^q are the components of a unit normal in the direction of v^q. The use of a vector potential function has not been explored in the available descriptions of the flow of fluids in porous media. Such exploration of the use of the vector potential will not be carried out here, but is a potentially important subject for future work.

The equation of state of a compressible fluid is given by

$$\frac{\partial\,^{(i)}p}{\partial t} = -\frac{1}{^{(i)}\kappa}\vec{\nabla}\cdot\,^{(i)}\vec{v} \qquad (5.157a)$$

or

$$\frac{\partial\,^{(i)}p}{\partial t} = -\frac{1}{^{(i)}\kappa}\,^{(i)}v^q_{,q} \qquad (5.157b)$$

where the symbols are as previously defined. For a mixture of fluids the compressibility of the ith species must be derived from a microscopic (molecular) model of the interpenetrating fluids using statistical mechanics as discussed for the viscosity and the stress tensor.

In order to have a completely determined equation set describing the motion of each species, the equation of energy flow is required as well and is now discussed.

5.6   ENERGY EQUATIONS

The total energy per unit volume of the moving continuous medium of species i may be expressed for an element of volume in space through which the medium is passing as

$$^{(i)}\rho_b \, ^{(i)}E = \, ^{(i)}\rho_b \left(\frac{1}{2} \, ^{(i)}\vec{v} \cdot \, ^{(i)}\vec{v} + \, ^{(i)}U \right) \tag{5.158a}$$

or

$$^{(i)}\rho_b \, ^{(i)}E = \, ^{(i)}\rho_b \left(\frac{1}{2} \, ^{(i)}v^m \, ^{(i)}v_m + \, ^{(i)}U \right) \tag{5.158b}$$

where $^{(i)}\rho_b$ and $^{(i)}\vec{v}$ are as defined previously, $^{(i)}E$ is the total energy per unit mass of species i, and $^{(i)}U$ is the internal energy per unit mass, in the thermodynamic sense, of species i. Thus, $^{(i)}\rho_b \, ^{(i)}U$ is the internal energy per unit volume of species i and $^{(i)}\rho_b \, ^{(i)}E$ is the total energy per unit volume of species i. Considering now a volume fixed in space through which material is flowing, we have for the energy balance, in vector notation (compare Lindsay, 1960, p. 305)

$$\frac{\partial \left(^{(i)}E \, ^{(i)}\rho_b \right)}{\partial t} = -\vec{\nabla} \cdot \left[^{(i)}\rho_b \, ^{(i)}\vec{v} \left(^{(i)}U + \frac{1}{2} \, ^{(m)}\vec{v} \cdot \, ^{(m)}\vec{v} \right) \right] - \vec{\nabla} \cdot \vec{q}$$

$$+ \, ^{(i)}\rho_b \, ^{(i)}\vec{v} \cdot \vec{g} - \vec{\nabla} \cdot \left(^{(i)}\overleftrightarrow{\chi} \cdot \, ^{(i)}\vec{v} \right) \tag{5.159}$$

where $^{(i)}\rho_b \, ^{(i)}\vec{v} \left[^{(i)}U + \frac{1}{2} \, ^{(i)}\vec{v} \cdot \, ^{(i)}\vec{v} \right]$ is the rate of flow of energy per unit area of volume element into and/or out of the volume element, $\vec{q}$ is the net heat flux, in units of energy per unit area per unit time, if any, flowing into and out of the volume element, $\vec{g}$ is the vector (direction and magnitude) acceleration due to gravity, $^{(i)}\overleftrightarrow{\chi}$ is the vector notation expression for the stress tensor for the possibly viscoelastic species i as given in (5.148) above, and $^{(i)}\overleftrightarrow{\chi} \cdot \, ^{(i)}\vec{v}$ is the rate per unit area of the elementary volume at which work is done on it by the surface stresses of the moving medium. In tensor notation (5.159) becomes

$$\frac{\partial}{\partial t} \left\{ ^{(i)}\rho_b \left[\left(\frac{1}{2} \, ^{(i)}v^m \, ^{(i)}v_m \right) + \, ^{(i)}U \right] \right\}$$

$$= - \left[^{(i)}\rho_b \, ^{(i)}v^j \left(^{(i)}U + \frac{1}{2} \, ^{(i)}v^m \, ^{(i)}v_m \right) \right]_{,j} - q^j_{,j}$$

$$+ \, ^{(i)}\rho_b \, ^{(i)}v^m \, g_m - \left(^{(i)}\chi^{mj} \, ^{(i)}v_m \right)_{,j} \tag{5.160}$$

The equation describing the total energy flow in the system of interest due to the motion of species i in which the derivative of $^{(i)}E$ with respect to time follows the motion of the fluid is

$$^{(i)}\rho_b \left(\frac{\partial\, ^{(i)}E}{\partial t} + {}^{(i)}E_{,j}\, ^{(i)}v^j \right) + {}^{(i)}E \left(\frac{\partial\, ^{(i)}\rho_b}{\partial t} + \left(^{(i)}\rho_b\, ^{(i)}v^j \right)_{,j} \right)$$

$$= -q^j_{,j} + {}^{(i)}\rho_b\, ^{(i)}v^m g_m - \left(^{(i)}\chi^{mj}\, ^{(i)}v_m \right)_{,j} \qquad (5.161)$$

We now derive the equation describing the flow of internal energy in the multicomponent system. We write for the flow of mechanical energy per unit volume

$$^{(i)}\rho_b \left(\frac{\partial}{\partial t} \left(\frac{1}{2}\, ^{(i)}v^m\, ^{(i)}v_m \right) + {}^{(i)}v^j \left[\frac{1}{2}\, ^{(i)}v^m\, ^{(i)}v_m \right]_{,j} \right)$$

$$= {}^{(i)}v_m\, ^{(i)}\chi^{mj}_{,j} + {}^{(i)}\rho_b\, ^{(i)}v^m g_m \qquad (5.162)$$

Subtracting (5.162) from (5.161) yields

$$^{(i)}\rho_b \left(\frac{\partial\, ^{(i)}U}{\partial t} + {}^{(i)}U_{,j}\, ^{(i)}v^j \right) + {}^{(i)}E \left(\frac{\partial\, ^{(i)}\rho_b}{\partial t} + \left[^{(i)}\rho_b\, ^{(i)}v \right]_{,j} \right)$$

$$= - \left[^{(i)}\chi^{mj}\, ^{(i)}v_m \right]_{,j} - q^j_{,j} - {}^{(i)}\chi^{mj}_{,j}\, ^{(i)}v_m \qquad (5.163)$$

which is the equation describing the flow of internal energy per unit mass of species i due to its motion with the intrinsic derivative of $^{(i)}U$ with respect to time following the motion of the medium.

5.7   MULTISPECIES THERMODYAMICS

In Chapter 3, applications of equilibrium thermodynamics to surface tension, matric potential, etc. were discussed. It was there seen that the work term in the differential form of the combined first and second laws is the primary designator required for applications to a given system. Thus, since stress is an intensive, and strain an extensive, quantity, we have for the work term for a possibly viscoelastic species i

$$d\, ^{(i)}W = {}^{(i)}\chi^{mj}\, d\, ^{(i)}\sigma_{mj}(\alpha) \qquad (5.164)$$

The total differential of the internal energy per unit volume of species i for the case of constant mol numbers is

$$d^{(i)}U = T d^{(i)}S + {}^{(i)}\chi^{mj} d^{(i)}\sigma_{mj}(\alpha)$$

$$+ \frac{\partial^{(i)}U}{\partial^{(i)}x^k(t)} d^{(i)}x(t) + \frac{\partial^{(i)}U}{\partial t} dt \qquad (5.165)$$

where ${}^{(i)}S$ is the entropy per unit mass of species i, T is temperature, the term containing ${}^{(i)}x(t)$ denotes the explicit dependence of ${}^{(i)}U$ on spatial location in the medium and ${}^{(i)}x(t)$ denotes the dependence of the location of an elementary mass of medium of species i on time and α, the order of the fractional derivative, depends on both the species i and its relative concentration. These dependencies are at first sight somewhat at variance with the concept of equilibrium thermodynamics as describing equilibrium only. It has just been seen, however, and will be seen in the next chapter as well, that the concepts of thermodynamics do indeed apply in time-dependent flow situations. In fact, equilibrium statistical mechanical calculations of globally time-dependent thermodynamic quantities are carried out assuming local—on the level of a volume element—equilibrium. This process is called coarse-graining and is sometimes appealed to in nonequilibrium thermodynamics calculations as well. The last term on the right-hand side of (5.164) represents possible explicit time dependence of the partial differential equations for the flow of total energy, (5.161), and the flow of internal energy, (5.163). We have from (5.165)

$$\left(\frac{\partial^{(i)}U}{\partial^{(i)}\sigma_{mj}(\alpha)} \right)_{(i)S} = {}^{(i)}\chi^{mj} \qquad (5.166)$$

so that ${}^{(i)}U$ may be viewed as a thermodynamic potential at constant entropy for the generalized force ${}^{(i)}\chi^{mj}$ in this case.

Carrying out a Legendre transform on T in (5.165) yields expressions for both the differential and integrated forms of the Helmholtz free energy for species i, viz.,

$$^{(i)}F = {}^{(i)}U - T \cdot {}^{(i)}S \qquad (5.167)$$

and

$$d^{(i)}F = - {}^{(i)}S \, dT + {}^{(i)}\chi^{mj} d^{(i)}\sigma_{mj}(\alpha)$$

$$+ \frac{\partial^{(i)}F}{\partial x^m(t)} d^{(i)}x^m + \frac{\partial^{(i)}F}{\partial t} dt \qquad (5.168)$$

where ${}^{(i)}F$ is the Helmholtz free energy per unit mass of species i and the other symbols are as previously defined. From (5.165) we have

$$\left(\frac{\partial^{(i)}F}{\partial^{(i)}\sigma_{mj}(\alpha)} \right)_T = {}^{(i)}\chi^{mj} \qquad (5.169)$$

Thus, $^{(i)}F$ may be considered to be the thermodynamic potential function for $^{(i)}\chi^{mj}$ at constant temperature. The integrated form of $^{(i)}F$ for the viscoelastic energy stored in species i per unit mass at constant temperature is

$$^{(i)}F = \frac{1}{2}\,^{(i)}\chi^{mj}\,^{(i)}\sigma_{mj}(\alpha) \qquad (5.170)$$

Other thermodynamic functions may be similarly defined. When the medium being described is the water in the pores of the porous medium, additions to the work term to account for capillary effects and gravitational effects on solution concentration are required. Alternatively, capillary effects (matric potential) may be included as part of the stress tensor via Darcy's law.

We now write down Maxwell relations derivable from (5.165) as follows. We have

$$\left(\frac{\partial T}{\partial\,^{(i)}\sigma_{mj}(\alpha)}\right)_{^{(i)}S,\,^{(i)}x^k,t} = \left(\frac{\partial\,^{(i)}\chi^{mj}}{\partial\,^{(i)}S}\right)_{^{(i)}\sigma_{mj},\,^{(i)}x^k,t} \qquad (5.171)$$

$$\left(\frac{\partial\,^{(i)}\chi^{mj}}{\partial\,^{(i)}x^k}\right)_{^{(i)}\sigma_{mj}(\alpha),\,^{(i)}S,t} = \left[\frac{\partial\left(\frac{\partial\,^{(i)}U}{\partial\,^{(i)}x^k}\right)}{\partial\,^{(i)}\sigma_{mj}(\alpha)}\right]_{^{(i)}x^k,\,^{(i)}S,t} \qquad (5.172)$$

$$\left(\frac{\partial T}{\partial\,^{(i)}x^k}\right)_{^{(i)}S,\,^{(i)}\sigma_{mj}(\alpha),t} = \left[\frac{\partial\left(\frac{\partial\,^{(i)}U}{\partial\,^{(i)}x^k}\right)}{\partial\,^{(i)}S}\right]_{^{(i)}\sigma_{mj}(\alpha),t,\,^{(i)}x^k} \qquad (5.173)$$

$$\left(\frac{\partial T}{\partial t}\right)_{^{(i)}S,\,^{(i)}\sigma_{mj}(\alpha),\,^{(i)}x^k} = \left[\frac{\partial\left(\frac{\partial\,^{(i)}U}{\partial t}\right)}{\partial\,^{(i)}S}\right]_{^{(i)}\sigma_{mj}(\alpha),\,^{(i)}x^k,t} \qquad (5.174)$$

$$\left(\frac{\partial\,^{(i)}\chi^{mj}}{\partial t}\right)_{^{(i)}S,\,^{(i)}\sigma_{mj}(\alpha),\,^{(i)}x^k} = \left[\frac{\partial\left(\frac{\partial\,^{(i)}U}{\partial t}\right)}{\partial\,^{(i)}\sigma_{mj}(\alpha)}\right]_{^{(i)}S,\,^{(i)}x^k,t} \qquad (5.175)$$

$$\frac{\partial}{\partial t}\left(\frac{\partial\,^{(i)}U}{\partial\,^{(i)}x^k}\right)\bigg|_{^{(i)}S,\,^{(i)}\sigma_{mj}(\alpha),\,^{(i)}x^k} = \frac{\partial}{\partial\,^{(i)}x^k}\left(\frac{\partial\,^{(i)}U}{\partial t}\right)\bigg|_{^{(i)}S,\,^{(i)}\sigma_{mj}(\alpha),t} \qquad (5.176)$$

A similar set of Maxwell relations may be written down from (5.160).

We now consider the viscoelastic quantities thermal stress and thermal strain as follows. We write

$$^{(i)}b^{mj} = \frac{\partial\,^{(i)}\chi^{mj}}{\partial T} \qquad (5.177)$$

$$^{(i)}a^{kl} = \frac{\partial \,^{(i)}\sigma_{kl}}{\partial T} \tag{5.178}$$

for the coefficients of thermal stress and thermal strain, respectively. Note that the $^{(i)}a_{kl}$ would reduce to a single coefficient of thermal expansion for a homogeneous isotropic medium. Combining (5.177) and (5.178) with (5.148) yields

$$^{(i)}a_{kl} = \,^{(i)}B_{mjkl}\,^{(i)}b^{mj} \tag{5.179}$$

which would reduce, for the case of a homogeneous isotropic system, to the usual thermodynamic relation

$$^{(i)}a = \,^{(i)}\kappa \left(\frac{\partial P}{\partial T} \right)_V \tag{5.180}$$

where $^{(i)}a$ is the usual coefficient of thermal expansion for species i, $^{(i)}\kappa$ is the isothermal compressibility of species i, and P, T, and V, are pressure, temperature, and volume, respectively. Finally, we notice the result, adapted to the present case, that

$$^{(i)}C_{\chi^{mj}} = \,^{(i)}C_{\sigma_{mj}(\alpha)} + \,^{(i)}v\, T\,^{(i)}\chi^{nj}\,^{(i)}\sigma_{nj}(\alpha) \tag{5.181}$$

where $^{(i)}v$ is the volume per mol of species i, and

$$^{(i)}C_{\chi^{mj}} = T \left(\frac{\partial \,^{(i)}S}{\partial T} \right)_{\chi^{mj}} \tag{5.182}$$

where $^{(i)}C_{\chi^{mj}}$ is the molar specific heat of species i at constant stress, while

$$^{(i)}C_{\sigma_{mj}} = T \left(\frac{\partial \,^{(i)}S}{\partial T} \right)_{\sigma_{mj}} \tag{5.183}$$

is the molar specific heat at constant strain.

SELECTED REFERENCES

Bearman, J., and J. G. Kirkwood, "The Statistical Mechanics of Transport Processes. XI. Equations of Transport in Multi-Component Systems." In *Selected Topics in Statistical Mechanics* (I. Oppenheim, ed.) Gordon and Breach, New York, 1967.

Christensen, R. M., *Theory of Viscoelasticity*, Academic Press, New York, 1982.

Courant, R., *Differential and Integral Calculus*, Vol. II, Interscience Publishers, Inc., New York, 1936.

Fitts, D. D., *Nonequilibrium Thermodynamics*, McGraw-Hill Book Co., Inc., New York, 1962.

Flügge, W., *Viscoelasticity*, Blaisdell Publishing Co., Waltham, Massachusetts, 1967.

Green, A. E., and W. Zerna, *Theoretical Elasticity*, Oxford University Press, New York, 1974.

Kaplan, W., *Advanced Calculus*, Addison–Wesley Publishing Co., Inc., Reading, Massachusetts, 1952.

Lamb, H., *Hydrodynamics*, Dover Publications, New York, 1945.

Lindsay, R. B., *Mechanical Radiation*, McGraw–Hill Book Co., New York, 1960.

Oldham, K. B., and J. Spanier, *The Fractional Calculus*, Academic Press, New York, 1974.

Osler, T. J., "The Fractional Derivative of a Composite Function," *SIAM J. Math. Anal.*, Vol. 2, pp. 288–293, 1970a.

Osler, T. J., "Leibnitz Rule for Fractional Derivatives Generalized and an Application to Infinite Series," *SIAM J. of Applied Mathematics*, Vol. 18, pp. 658–674, 1970b.

Schwatt, I. J., *An Introduction to the Operations with Series*, Chelsea Publishing Co., New York, 1924.

PROBLEMS

5.1 Given that $\sigma_{kl,j} = \frac{1}{2}\left(\partial^2 u_k/\partial x^j\,\partial x^l + \partial^2 u_l/\partial x^j\,\partial x^k\right)$, show that in Cartesian coordinates

$$\sigma^k_{l,j} = \frac{1}{2}\left(\frac{\partial^2 u_k}{\partial x^j\,\partial x^l} + g^{mk}\frac{\partial^2 u_l}{\partial x^j\,\partial x^m}\right)$$

by raising the index k and noting that in Cartesian coordinates $g^{mk} = \delta^{mk}$.

5.2 Given the expression for $\sigma_{kl,j}$ in Problem 5.1, show that

$$\sigma^{kl}_{,j} = \frac{1}{2}\left(\delta^{nl}\frac{\partial^2 u^k}{\partial x^j\,\partial x^n} + \delta^{mk}\frac{\partial^2 u^l}{\partial x^j\,\partial x^m}\right)$$

and thus that

$$\sigma^{ij}_{,j} = \frac{1}{2}\left(\delta^{nj}\frac{\partial^2 u^i}{\partial x^j\,\partial x^n} + \delta^{mi}\frac{\partial^2 u^j}{\partial x^j\,\partial x^m}\right)$$

5.3 Given the results of Problem 5.2, write down the expression for $\sigma^{ji}_{,j}$, and, noting that m and n are dummy indices that could be replaced by any letter, show that $\sigma^{ij}_{,j} + \sigma^{ji}_{,j}$ can be written as

$$\sigma^{ij}_{,j} + \sigma^{ji}_{,j} = \delta^{mj}\frac{\partial^2 u^i}{\partial x^j\,\partial x^m} + \delta^{ni}\frac{\partial^2 u^j}{\partial x^j\,\partial x^n}$$

6

NONEQUILIBRIUM
THERMODYNAMICS

6.1 INTRODUCTION

The theory of nonequilibrium thermodynamics is ideally suited to the description of the various coupled fluxes occurring in unsaturated flow. For the purposes of this discussion, these will be taken to be (1) movement of liquid water, (2) flow of heat, (3) movement of electric charge, (4) movement of vapor, and (5) movement of chemical species in the water. The movement of any one of these can affect the movement of any other of them. The explicit recognition of this fact and the formalism expressing the coupling of the various flows via phenomenological coupling coefficients is the subject of nonequilibrium thermodynamics. This is the only formalism that combines the various coupled flows into one equation set and shows the relationship of one flux to another. The various flux laws including their associated phenomenological coefficients, such as Darcy's law, which has hydraulic conductivity for a phenomenological transport coefficient, and Fourier's law (heat diffusion), which has the diffusivity as a phenomenological transport coefficient, were discovered one at a time in a piecemeal fashion and do not in themselves include any information on the interaction, for example, of flows of heat and matter. All such laws are special cases of the more comprehensive formalism of nonequilibrium thermodynamics, which predicts the existence of and embodies formalism for calculating the coupled flows that can arise. An important review of nonequilibrium thermodynamics and its application to groundwater flow has been carried out by Carnahan (1976), and some of the material in this section is drawn from that work. In nonequilibrium thrmodynamics, each of the fluxes, quantity per unit area per unit time, which we denote by $_iJ$, is written in terms of all the generalized forces acting on the system, denoted by $_kF$, and the phenomenological coefficients, L_{ik}, connecting them. In compact notation for the case of m fluxes,

$$_iJ = \sum_{k=l}^{m} L_{ik}\ _kF \tag{6.1}$$

where the Einstein summation convention is not employed since the L_{ik} are not necessarily tensor quantities. In this formulation, it is assumed that the fluxes and forces are chosen in such a way that no linear homogeneous relations exist within both the set of fluxes and the set of forces. This means that all the $_iF$ and concomitantly the $_iJ$ are distinct. The direct effects are represented by conjugate forces and fluxes that are related by the coefficients L_{ii}, while the forces and fluxes of the coupled effects are connected by the L_{ik}, where $i \neq k$. Writing (6.1) in matrix form, (6.2) shows the coefficients for the direct effects on the main diagonal of the coefficient matrix and coupled effects as off-diagonal elements of the coefficient matrix:

$$\begin{bmatrix} _1J \\ _2J \\ _3J \\ \vdots \\ _mJ \end{bmatrix} = \begin{bmatrix} L_{11} & L_{12} & L_{13} & \cdots & L_{1m} \\ L_{21} & L_{22} & L_{23} & \cdots & L_{2m} \\ L_{31} & L_{32} & L_{33} & \cdots & L_{3m} \\ \vdots & \vdots & \vdots & \vdots & \vdots \\ L_{m1} & L_{m2} & L_{m3} & \cdots & L_{mm} \end{bmatrix} \begin{bmatrix} _1F \\ _2F \\ _3F \\ \vdots \\ _mF \end{bmatrix} \tag{6.2}$$

If the forces and fluxes are properly chosen, then $L_{ik} = L_{ik}$, which are the Onsager reciprocal relations. The generalized forces and fluxes chosen to make L_{ik} symmetric are called the *Onsager fluxes and forces*. In general, the L_{ik} can have any tensorial rank between zero and four (Tensors of rank zero are scalars, tensors of rank one are vectors, such as gradients of head and specific discharge, or volume flux, or Darcy velocity in the tensor form of Darcy's law.) Tensors of rank two relate processes in anisotropic media, such as polarization in crystals in the presence of an applied electric field, depending on the tensorial rank of the generalized forces being connected to a given flux. Also, in addition to the symmetry between coefficients, the coefficients themselves possess a symmetry that arises from the symmetries of the spatial coordinates in a given problem and depends on the tensorial rank of the coefficient. The fact that the fluxes and their conjugate forces can be chosen such that $L_{ik} = L_{ki}$ is very probably the single most important part of nonequilibrium thermodynamics both in terms of the theory itself and for its applications. A considerable body of experimental work has developed in recent years to measure/verify the reciprocal relations.

The dissipation function, Φ, which is the rate of production of entropy per unit volume, ξ, in the system of interest, as the result of an irreversible process, multiplied by absolute temperature, T, may be written as

$$\Phi = \xi T = \sum_{i=l}^{m} {}_iJ\,{}_iF \tag{6.3}$$

The dissipation function represents the rate of dissipation of free energy by irreversible processes or the degradation of the ability of the system to do work. An increase in entropy in a system represents a corresponding loss of information or loss of order in the system as well. The choice of the form of the forces and fluxes to make $L_{ik} = L_{ki}$ must satisfy the additional requirements that: (1) The product of a generalized force with its conjugate flux must have the dimensions of the dissipation function (energy per unit time per unit volume) of $Mt^{-3}L$, where M is mass, t is time, and L is length, and (2) for any given system, the dissipation function must be invariant, that is, have the same value under any change of fluxes and forces from one coordinate system to another. Since the rate of production of entropy is positive, that is, entropy increases when an irreversible process occurs, the dissipation function Φ is always positive. Substituting (6.1) into (6.3) yields

$$\Phi = \sum_{i=l}^{m} \sum_{k=l}^{m} L_{ik}\,{}_kF\,{}_iF \tag{6.4}$$

Separating the double sum on the right of (6.4) into two terms, the first of which $(i = k)$ represents the contribution of the direct effects to Φ and the second of which $(i \neq k)$ represents the contribution of the coupled effects to Φ, yields

$$\Phi = \sum_{i=k}^{m} L_{ii}\,{}_iF^2 + \sum_{i=l}^{m} \sum_{k=l}^{m} L_{ik}\,{}_kF\,{}_iF \tag{6.5}$$

which shows that Φ is a positive definite quadratic form and the matrix of coefficients given in (6.2) is also positive definite. This means that every principal minor in the L matrix is positive, and the result

$$L_{ii}L_{kk} - L_{ik}L_{ki} > 0 \tag{6.6}$$

can be derived by algebraic manipulation of the elements of L. Since any of the ${}_iF$ or equivalently the ${}_kF$, may be set equal to zero in (6.2), we see from (6.5) that

$$L_{ii} > 0 \tag{6.7}$$

$$L_{kk} > 0 \tag{6.8}$$

Using (6.7), (6.8), and (6.3) in conjunction with (6.6) yields the condition

$$L_{ii}L_{kk} > L_{ik}^2 \qquad (6.9)$$

Note that since L_{ik} enters (6.9) as a squared term, (6.8) includes the possibility of negative values of the coupling coefficients. Physically, this possibility means that the coupled flow contribution to the transport of a given entity, water, for example, may act in a direction opposite to that of the direct effect characterized by L_{ii}. In such a case, the ratio $-|L_{ik}|/L_{ii}$, where the bars denote the absolute value of L_{ik}, is sometimes called the *reflection coefficient*.

6.2   GENERALIZED FLUXES IN UNSATURATED FLOW

In order to apply more detailed considerations of the fluxes and their conjugate forces to a porous medium, it will be assumed that an appropriate volume averaging process has been carried out so that the most elementary volume considered, to which derivatives apply, etc., is small macroscopically but large microscopically, so that the notion of a pore size distribution is viable, but the specific interactions of the fluid with the pore/capillary walls, such as viscous dissipation via flow, are averaged by the act of measuring macroscopic transport coefficients and so do not appear explicitly. We now write expressions for the forces and fluxes that would be expected in an unsaturated flow system (Carnahan, 1976).

If $_1\vec{J}$ is taken to be the vector flux of liquid water, that is, volume of water per unit area per unit time, then $_1\vec{F} = -(\vec{\nabla}^{(1)}\Psi + \rho g \vec{\nabla} z)$, where ρ is the density of liquid water, g is the acceleration due to gravity, $^{(1)}\Psi$ is matric potential due to water in the porous medium, $\vec{\nabla}$ is the gradient operator, and z is a vertical position coordinate, is a possible direct effect conjugate force.

If $_2\vec{J}$ is taken to be the flux of heat in the system, heat energy per unit area per unit time, then $_2F = -\vec{\nabla}\ln[T(\vec{r})]$ is a possible direct effect conjugate force. In this expression the temperature T is a function of position as indicated by the general position vector $\vec{r}$.

If $_3\vec{J}$ is the vector flux of charge, then $-\vec{\nabla}\varphi(\vec{r}) = {_3}\vec{F}$ is a possible direct effect conjugate force, where $\varphi(\vec{r})$ is the electric potential, which is a function of position, as indicated by its dependence on the position vector $\vec{r}$.

If $_4\vec{J}$ is taken to be the vector flux of water vapor in the system, then $_4\vec{F} = -\vec{\nabla}\mu_v(\vec{r})$ is a possible conjugate force. In this expression, $\mu_v(\vec{r})$ is the chemical potential of the vapor given as a function of position.

If $_l\vec{J}$ is the vector flux of the lth solute in the water, and if the description of the movement of the water itself is contained in $_1\vec{J}$, then $_l\vec{F} = \vec{\nabla}\mu_l^c(\vec{r})$, where $\mu_l^c(\vec{r})$ is the chemical potential of the lth species of solute ($5 < l < m$) and $m - l$ is the total number of solutes. It can be shown that while the movement of solutes couples with the movements of water, heat, charge, vapor, and each other, the chemical reactions that may take place among the $m-5$ chemical species do not couple with the other flows and coordinate transformations can be found in which the possible chemical reactions do not couple with each other in the sense of nonequilibrium thermodynamics. Combining the terms suggested, we write the dissipation function as

$$\Phi = {}_1\vec{J}\cdot{}_1\vec{F} + {}_2\vec{J}\cdot{}_2\vec{F} + {}_3\vec{J}\cdot{}_3\vec{F} + {}_4\vec{J}\cdot{}_4\vec{F} + \sum_{l=5}^{m} {}_l\vec{J}\cdot{}_l\vec{F}$$

$$= -{}_1\vec{J}\cdot\vec{\nabla}\ln[T(\vec{r})] - {}_2\vec{J}\cdot(\vec{\nabla}\Psi + \rho g\vec{\nabla}z) - {}_3\vec{J}\cdot\vec{\nabla}\varphi(\vec{r})$$

$$- {}_4\vec{J}\cdot\vec{\nabla}\mu_v(\vec{r}) - \sum_{l=5}^{m} {}_l\vec{J}\cdot\vec{\nabla}\mu_l^c(\vec{r}) \tag{6.10}$$

We now examine the terms on the right-hand side of (6.10) to verify that they satisfy the criterion mentioned that the product of the generalized force and its conjugate flux have the dimensions of the rate of production of entropy per unit volume per unit time.

The dimensions of $_1\vec{F}$ are force per unit volume, or $ML/t^2L^3 = M/t^2L^2$. Thus $_1\vec{J}$ must have the dimensions of L/t to be admissible. Defining $_1\vec{J}$ as the volumetric flux or the volume of water per unit area per unit time yields the dimensions L/t.

The dimensions of $\vec{\nabla}\ln[T(\vec{r})] = -\vec{\nabla}T(\vec{r})$ are L^{-1} since the dimensions of temperature cancel and the gradient operator introduces a dimension of L^{-1} so that $_2\vec{J}$ must have dimensions of Mt^{-3} to be admissible. Since $_2\vec{J}$ is the flux of heat, we may express it as energy, in joules, for example, per unit area per unit time or, $(ML^2/t^2L^2t = M/t^3)$, which is correct.

The generalized force $_3\vec{F}$ has dimensions of voltage per unit length. Since a volt can be expressed as a joule/coulomb, we see that $_3\vec{F}$ has dimensions of ML/t^2q, where q is charge. This yields for acceptable dimensions for $_3\vec{J}$, q/L^2t or charge per unit area per unit time, which is correct.

If $_4\vec{J}$ is taken to be the flux of vapor, then $_4\vec{F}$ is $-\vec{\nabla}\mu_v(\vec{r})$, where $\mu_v(\vec{r})$ is the chemical potential of the vapor as a function of position $\vec{r}$. If chemical potential is taken to be the Gibbs free energy per unit mass (precisely speaking, chemical potential is defined as the Gibbs free energy per mol), the dimensions of $_4\vec{F}$ become L/t^2 or force per unit mass. Thus $_4\vec{J}$ must have dimensions of M/L^2t, which indeed corresponds to a mass flux.

The vector fluxes of the diffusing chemical species, $_1\vec{J}$, have the same dimensions as $_4\vec{J}$ and so are also acceptable. We now write the set of phenomenological equations corresponding to (6.10)

$$_1\vec{J} = \overleftrightarrow{L}_{11} \cdot {}_1\vec{F} + \overleftrightarrow{L}_{12} \cdot {}_2\vec{F} + \overleftrightarrow{L}_{13} \cdot {}_3\vec{F} + \overleftrightarrow{L}_{14} \cdot {}_4\vec{F} + \sum_{l=5}^{m} \overleftrightarrow{L}_{1l} \cdot {}_l\vec{F} \tag{6.11a}$$

$$_2\vec{J} = \overleftrightarrow{L}_{21} \cdot {}_1\vec{F} + \overleftrightarrow{L}_{22} \cdot {}_2\vec{F} + \overleftrightarrow{L}_{23} \cdot {}_3\vec{F} + \overleftrightarrow{L}_{24} \cdot {}_4\vec{F} + \sum_{l=5}^{m} \overleftrightarrow{L}_{2l} \cdot {}_l\vec{F} \tag{6.11b}$$

$$_3\vec{J} = \overleftrightarrow{L}_{31} \cdot {}_1\vec{F} + \overleftrightarrow{L}_{32} \cdot {}_2\vec{F} + \overleftrightarrow{L}_{33} \cdot {}_3\vec{F} + \overleftrightarrow{L}_{34} \cdot {}_4\vec{F} + \sum_{l=5}^{m} \overleftrightarrow{L}_{3l} \cdot {}_l\vec{F} \tag{6.11c}$$

$$\vdots$$

$$_m\vec{J} = \overleftrightarrow{L}_{ml} \cdot {}_1\vec{F} + \overleftrightarrow{L}_{m2} \cdot {}_2\vec{F} + \overleftrightarrow{L}_{m3} \cdot {}_3\vec{F} + \overleftrightarrow{L}_{m4} \cdot {}_4\vec{F} + \sum_{l=5}^{m} \overleftrightarrow{L}_{ml} \cdot {}_l\vec{F} \tag{6.11d}$$

6.3 PHENOMENOLOGICAL COEFFICIENTS

We shall now examine the meaning of each of the phenomenological coefficients in turn. In their most general form, in this case, each of the L_{ik} is a second-rank tensor written in dyadic vector notation as $\overleftrightarrow{L}_{ik}$, and the inner product formed between it and the corresponding force yields a vector, the direction of which is likely to be in a direction different from the force and which has been anticipated in the notation of (6.11).

The coefficient $\overleftrightarrow{L}_{11}$ is a direct coefficient for fluid flow in the generalized form of Darcy's law, which is Eq. (6.11b), in which $_1\vec{J}$, the Darcy velocity, is the volume flux of moisture, volume per unit area per unit time. Writing Darcy's law in its usual form, $_1\vec{J} = -^{(1)}\overleftrightarrow{K} \cdot \vec{\nabla}h$, where $^{(1)}\overleftrightarrow{K}$ is the usual second-rank hydraulic conductivity tensor, and H the hydraulic head, $^{(1)}\Psi/\rho g + z$, we have on comparison with (6.10) in which $_1\vec{J} = -\vec{\nabla}\left(^{(1)}\Psi + \rho g z\right)$, $\overleftrightarrow{L}_{11} = {}^{(1)}\overleftrightarrow{K}/\rho g$ for constant ρ, which is a questionable assumption since ρ is temperature dependent and gradients of temperature in the system are important in the unsaturated flow of moisture. The tensorial (second-rank symmetric) nature of $\overleftrightarrow{L}_{11}$ has been explicitly displayed. The coefficient $\overleftrightarrow{L}_{13} = \overleftrightarrow{L}_{31}$ represents coupling between the flows of water and charge. The electro-osmosis effect, which is the movement of mass due to a gradient of electrical potential, is characterized by $\overleftrightarrow{L}_{13}$, and the streaming current, which is the movement of charge due to a flux of mass, is characterized by $\overrightarrow{L}_{31}$. These effects, discussed briefly in Chapter 3, depend on the presence of charges distributed throughout the porous matrix and may be quite significant in the vicinity of unsaturated clay layers.

The coefficients $\overrightarrow{L}_{14}=\overrightarrow{L}_{41}$ represent the coupling of flows of liquid (water) and vapor. Since vapor transport becomes one of the dominant mechanisms for transport as the saturation drops, this coupling is expected to be quite significant.

$_2\vec{J}$ represents a flux of heat and L_{22} represents the direct effect of a temperature gradient in the system on this flux. Thus, we may write Fourier's law for heat conduction as:

$$_2\vec{J} = - {}^{(2)}\overrightarrow{k} \cdot \vec{\nabla}T \tag{6.12}$$

where $^{(2)}\overrightarrow{k}$ is the second rank thermal conductivity tensor for the medium. Comparing this with the first term of (6.11a) yields

$$\overrightarrow{L}_{22} \cdot {}_2\vec{F} = \overrightarrow{L}_{22} \cdot \vec{\nabla}[\ln(T)] = \overrightarrow{L}_{22} \cdot \frac{\vec{\nabla}T}{T} \tag{6.13}$$

where the tensorial nature of $\overrightarrow{L}_{22}$ has been explicitly noted, and we see that $\overrightarrow{L}_{22}= T{}^{(2)}\overrightarrow{k}$. The coefficient $\overrightarrow{L}_{21}=\overrightarrow{L}_{12}$ represents the coupling of the volume flux of water, $_1\vec{J}$, the Darcy velocity, with the heat flux $_2\vec{J}$. The coefficient $\overrightarrow{L}_{21}$ represents thermoinfiltration and $\overrightarrow{L}_{12}$ represents thermoosmosis. The quantities $\overrightarrow{L}_{21}$ and $\overrightarrow{L}_{12}$ are positive and refer to the transport of molecules of water absorbed in thin layers on the surfaces of the porous matrix. The coefficient $\overrightarrow{L}_{23}=\overrightarrow{L}_{32}$ represents the coupling of the flux of heat with the flux of charge $_3\vec{J}$, that is, thermoelectric effects. The coefficient $\overrightarrow{L}_{23}$ represents the establishing of a potential difference due to the flow of heat, and $\overrightarrow{L}_{32}$ represents the establishing of a gradient of temperature due to the flow of electric charge. These effects are expected to be small in earth materials. The coefficients $\overrightarrow{L}_{24}=\overrightarrow{L}_{42}$ represent the coupling of heat and vapor fluxes. In cases of low saturation, $^{(1)}\theta < 0.2$, where $^{(1)}\theta$ is the fraction of the pore space containing liquid water, vapor transport is one of the principal mechanisms for coupled heat and moisture movement.

The coefficient $\overrightarrow{L}_{33}$ characterizes the direct effect for the flow of charge. Writing $_3\vec{J}$ for the flow of charge per unit area per unit time, which is the same as the current density (current per unit area) as $_3\vec{J} = - {}^{(3)}\overrightarrow{k} \cdot \vec{\nabla}\varphi(\vec{r})$, where $^{(3)}\overrightarrow{k}$ is the electrical conductivity tensor, we have by comparison with (6.11c), $\overrightarrow{L}_{33}= {}^{(3)}\overrightarrow{k}$.

The coefficients $\overset{\leftrightarrow}{L}_{34}=\overset{\leftrightarrow}{L}_{43}$ represent the coupling of flows of vapor and charge. There is no experimental information on this effect in soils. However, it is possible to suggest that the effect is small because the water vapor in the soil is not expected to be ionized.

The coefficient $\overset{\leftrightarrow}{L}_{44}$ characterizes the direct effect of the flow of vapor in the porous medium. Writing $_4\vec{J} = - {}^{(4)}\overset{\leftrightarrow}{K} \cdot \vec{\nabla}\mu_v$, where ${}^{(4)}\overset{\leftrightarrow}{K}$ is the tensor conductivity for water vapor in the soil, we have that $\overset{\leftrightarrow}{L}_{44}= {}^{(4)}\overset{\leftrightarrow}{K}$. Note that in this formulation gradients of chemical potential are used instead of gradients of density as being the more thermodynamically correct driving forces. The two are related via

$$\vec{\nabla}\mu = \frac{\partial\mu}{\partial\rho}\vec{\nabla}\rho \tag{6.14}$$

The coefficients $\overset{\leftrightarrow}{L}_{kl}$ for $l < k < m$ represent coupling of the mobile chemical species in the water with the other fluxes in the system and with each other, which is known as dispersion–dispersion coupling. A formalism of the dispersivity tensor in a porous medium that includes the diffusion of dissolved species in water and that, for the first time, fits into the scheme of nonequilibrium thermodynamics, has been given by Carnahan (1976).

The coupling between the flux of heat and the dissolved chemical species is represented by cross-coefficients of the form $\overset{\leftrightarrow}{L}_{2l}$ where $5 < k < m$. The coefficients $\overset{\leftrightarrow}{L}_{2l}$ represent the diffusion of heat due to a flux of species dissolved in water, the Dufour effect, and the $\overset{\leftrightarrow}{L}_{k2}$ represent the movement of dissolved species due to the flux of heat arising from gradients of temperature.

6.4   COUPLING OF MOISTURE, HEAT, & IMPURITIES

In the development given in Chapter 5, it was implicitly assumed that the off-diagonal coupling coefficients of Eq. (6.2) were equal to zero, or sufficiently small compared to the coefficients on the diagonal representing the direct effects. We have seen that each flux depends on all the generalized forces present in the system so that, for example, for the case of water containing one chemical impurity in a porous medium in the presence of heat flows, we have

$$_m\vec{J} = {}_m\overset{\leftrightarrow}{L}_1 \cdot {}_1\vec{F} + {}_m\overset{\leftrightarrow}{L}_2 \cdot {}_2\vec{F} + {}_m\overset{\leftrightarrow}{L}_5 \cdot {}_5F \tag{6.15}$$

where $m = 1$ for a flux of water, $m = 2$ for a flux of heat, and $m = 3$ for an impurity flux. The quantities appearing in (6.15) are as discussed.

The equation describing the flow of energy in the system is, as given in Eq. (5.159),

$$\frac{\partial \left({}^{(i)}E\,{}^{(i)}\rho_b \right)}{\partial t} = \vec{\nabla} \cdot \left({}^{(i)}\rho_b\,{}^{(i)}\vec{v}\,{}^{(i)}E \right) - \vec{\nabla} \cdot {}_2\vec{J} + {}^{(i)}\rho_b\,{}^{(i)}\vec{v} \cdot \vec{g} - \vec{\nabla} \cdot \left({}^{(i)}\overset{\leftrightarrow}{\chi} \cdot {}^{(i)}\vec{v} \right)$$

(6.16)

where ${}^{(i)}E$ is given in (5.158a), and in this case ${}^{(1)}\vec{v} = {}_1\vec{J}\,{}^{(1)}\rho/{}^{(1)}\rho_b$ from Eqs. (1.22) and (1.24) and ${}^{(5)}\vec{v} = {}_5\vec{J}\,{}^{(5)}\rho/{}^{(5)}\rho_b$ from Eq. (1.24) and ${}_2\vec{J}$ and ${}_5\vec{J}$ are given by (6.15) with $m = 2$ and 5, respectively. The ijth component of the viscoelastic stress tensor for the ith species, ${}^{(i)}\chi^{ij}$, is given by

$$ {}^{(i)}\chi^{ij} = {}^{(i)}A^{ijkl}\,{}^{(i)}\sigma_{kl}(\alpha) $$

(6.17)

where the indices i and j in (5.91) have been raised and ${}^{(i)}E^{ijkl}$ of (5.79) is written for a general species i by writing ${}^{(i)}\eta$ and ${}^{(i)}\eta'$ in place of η and η', respectively, α is set equal to one, and ${}^{(i)}\sigma(1)$ of (5.94) is written for a general species i as

$$ {}^{(i)}\sigma_{kl}(1) = \frac{1}{2}\left[\left(\frac{d\,{}^{(i)}\xi_k}{dt} \right)_{,l} + \left(\frac{d\,{}^{(i)}\xi_l}{dt} \right)_{,k} \right] $$

(6.18)

The Lagrangian form of the equation of motion is given by (5.143) with the expressions for ${}^{(1)}\vec{v}$ and ${}^{(5)}\vec{v}$ as given previously. In order to have a complete equation set, equation of state information, namely pressure as a function of density and temperature, as well as the dependence of ${}^{(i)}U$, the internal energy of species i, on bulk density and temperature are required as well. Note that the expressions for the velocities given are to be substituted into the expression for ${}^{(i)}E$ of (5.158a) as well.

To summarize, the complete set of equations describing coupled multispecies flow in porous media consists of (1) the Lagrangian form of the equation of motion based on conservation of momentum, (2) the equations of continuity for the individual species, (3) the equation of energy flow, (4) the equations of state, the dependence of pressure on density and temperature, or on compressibility and temperature, for each species, (5) the flux law for each species containing a gradient (generalized force) term for each type of flow being considered expressed in terms of the velocity of that species. This last expression would contain only the direct effect generalized force in situations in which the coupled effects are small compared to the direct effects, as was assumed to be the case in the developments of Chapters 4 and 5.

SELECTED REFERENCES

Carnahan, C. L., "Non-Equilibrium Thermodynamics of Groundwater Flow Systems: Symmetry Properties of Phenomenological Coefficients and Considerations of Hydrodynamic Dispersion," *J. of Hydrology*, Vol. 31, 1976, pp. 125–150.

de Groot, S. R., and P. Mazur, *Non-Equilibrium Thermodynamics*, Dover Publications, Inc., New York, 1984.

Haase, R., *Thermodynamics of Irreversible Processes*, Addison–Wesley Publishing Co., Reading, Massachusetts, 1969.

PROBLEMS

6.1 Using the relation

$$\vec{v} = \frac{\sum_{i=0}^{n} {}^{(i)}\rho_b \, {}^{(i)}v}{\sum_{i=0}^{n} {}^{(i)}\rho_b}$$

where the sum is over species of interpenetrating continuae, the porous medium and the fluids contained within it show, by summing over the equation of continuity for a single species, that

$$\frac{\partial \rho}{\partial t} = -\vec{\nabla} \cdot (\rho \vec{v})$$

where ρ is the total composite density of the interpenetrating continuae at any given location.

6.2 Given the relation defining the isothermal compressibility, κ,

$$\kappa = -\frac{1}{V}\frac{dV}{dp} = \frac{1}{\rho}\frac{d\rho}{dp}$$

where ρ is the density of the material contained in volume V at pressure p, show that the isothermal compressibility of an ideal gas is $1/p$ using each form of the definition given.

6.3 Given the relation defining the isothermal bulk modulus, B, the reciprocal of the isothermal compressibility,

$$B = -V\frac{dp}{dV} = \rho\frac{dp}{d\rho}$$

show directly from each form of the definition that B for an ideal gas equals p.

7

SELECTED MEASUREMENT TECHNIQUES

7.1   INTRODUCTION

In this chapter we discuss in a general way, based in part on what has gone before, the principles underlying selected experimental methods and some general aspects of the operation of the related instrumentation. No attempt has been made to discuss specific equipment available from specific manufacturers, since such a discussion would necessarily be both invidious and soon out of date.

7.2   MEASUREMENT OF MATRIC POTENTIAL

7.2.1   The Tensiometer

The tensiometer measures matric potential. Separate measurements of the degree of soil saturation at known temperatures as will be described are required to obtain the family of moisture characteristic curves for a given type of soil. The principle of operation of the tensiometer is as follows. Consider a tube with a ceramic tip filled with water, or better still, soil solution, though this ideal is seldom realized in practice, with the damp ceramic tip initially exposed to air and the upper end of the water column not exposed to the atmosphere as shown in Figure 7.1.

The ceramic of which the tip is made is typically a "twenty bar" ceramic by which is meant that a gas pressure of nominally twenty bars must be applied across a sample of ceramic saturated with water, but exposed to air, to remove the water from it. (Note that a pressure of 10^6 dyn/cm^2

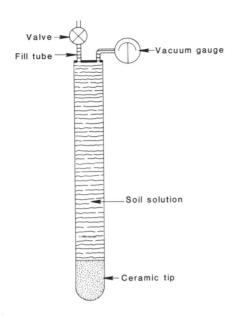

Figure 7.1: Schematic diagram of a filled tensiometer in air. The water is held in the column against gravity by the capillary action of water menisci in the ceramic tip. The vacuum gauge is set to read zero.

equals 1 bar.) Similar meanings are attached to the terms 1-bar, 5-bar, etc. ceramics. The zero reading on the vacuum gauge is generally set to zero before the damp ceramic tip is inserted in soil to begin matric potential measurements. The actual pressure distribution in a vertical column of water terminating in a horizontal capillary tube is as shown in Figure 7.2. The pressure required to force water out of the circular capillary opening of radius a is $p_a + 2\gamma \cos(\Phi)/a$. The adjustment of the pressure gauge is generally such that p_a is taken as the reference pressure (i.e., zero). For a 20-bar material, assuming that $\Phi = 0$, and $\gamma = 72$ dynes/cm, $a \simeq 1.44 \times 10^{-4}$ cm. The pores/capillaries in an actual ceramic are interconnected, highly irregular in shape, and flow rates through the ceramic are affected by the surface tractions between the fluid and the ceramic material, so that actual pore openings for a 20-bar ceramic, for example, can be larger than 10^{-4} cm. When the damp ceramic tensiometer tip is inserted into the soil sample of interest, and a water film exists between the tip and the soil, some amount of water leaves the column due to capillary action of the soil, and the pressure measured by the gauge of Figure 7.1 is below zero–generally in units of centibars. The lower limit of matric potential of which a tensiometer is

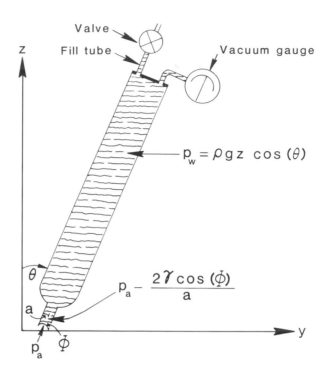

Figure 7.2: The pressure at any point in the column of water is $\rho g z \cos(\theta)$ where ρ is the density of the water, g is the acceleration due to gravity, θ is the angle between the tube and the vertical, and a is the radius of the capillary.

theoretically capable is the negative of the local atmospheric pressure, while the largest (least negative) value is the local atmospheric pressure minus the standard pressure at sea level. Thus, for example, at a location at which the atmospheric pressure is $12.5\,\text{lb/in.}^2$ instead of $14.7\,\text{lb/in.}^2$, the largest matric potential reading possible would be 100 centibars $\times 12.5/14.7 = -85$ centibars due to the fact that the pressure gauge is set to zero in air at each location before use. As a practical matter, it is important to avoid air bubbles in the tensiometer–either the response will be slow, or in more extreme cases, the instrument will cease to function altogether. As a further practical matter, if the instrument is stored with the column filled, the tip should be kept wrapped in damp sponge to prevent evaporation and the consequent formation of air bubbles.

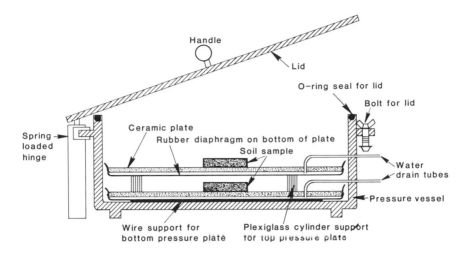

Figure 7.3: Cross section of a typical pressure plate apparatus. Each ceramic pressure plate is sealed at its edges by a rubber diaphragm and is penetrated by a tube to drain water forced through the plate under pressure from the diaphragm to the outside of the pressure vessel while the pressure, usually from a tank of nitrogen gas, is being maintained.

7.2.2   Pressure Plate Apparatus

We now describe a piece of equipment capable of measuring matric potentials on the drying branch of the moisture characteristic in the range of a few centibars to about -15 bars, namely the pressure plate apparatus shown in Figure 7.3. The procedure for using the apparatus is as follows. The ceramic plate is saturated with water, preferably distilled or deionized, by being submerged for 24 h. In the traditional procedure, splits of a disturbed and possibly reconstituted soil sample are contained on the surface of the pressure plate in areas formed by rubber bands about $\frac{3}{8}$ in. high and saturated. A map is drawn showing the location of each sample, or split as the case may be, on the top of the pressure plate. The plate is then placed in the pressure vessel and the metal drain tube passing through the ceramic plate is connected to an outlet–from the pressure vessel–tube via flexible, stiff-walled (to withstand the applied pressure inside the vessel without collapsing) tubing. The lid is bolted down and an increment of gas pressure, usually from a nitrogen tank, is applied. Since a fully charged nitrogen tank contains the gas at a pressure of approximately 2000 lb/in.^{2}, and 15 bars, the nominal upper limit of the pressure vessel, is 210 lb/in.^{2}, care must be taken, via a suitable system of pressure gauges and gas regulator valves, not to subject the system to excessive pressure.

When water outflow has ceased, the pressure is released, one of the splits is removed from the pressure plate and weighed, and the process repeated at higher pressure. The applied pressure in each case balances the capillary forces holding the water in the soil sample splits after water outflow has ceased so that the matric potential of the water in the sample is the negative of the applied pressure. If the splits are eventually oven-dried and weighed, and could be weighed when fully saturated, and their volumes measured, it would be possible, in principle, using this technique, to obtain, in addition to the drying branch of the moisture characteristic hysteresis loop, the bulk densities and grain densities of the samples as well. Since, however, the integrity of each split is destroyed upon its removal from the porous plate, a variation of the traditional technique is required to accomplish the objective. A suitable variation is to place the initially saturated soil sample in a brass sleeve on its own ceramic disk, such as the sleeve and disk from a Tempe cell, instead of using a rubber band. Then a given sample can be pressurized, removed, weighed, along with the sleeve and disk, replaced in the pressure vessel, and the process repeated to yield, after finally oven drying the sample, the bulk density, grain density, porosity, and matric potential as a function of degree of saturation for the drying branch of the moisture characteristic. The equation set is

$$^{(0)}\epsilon V_B = V_p \tag{7.1}$$

$$\frac{M_B}{V_B} = {}^{(0)}\rho_b \tag{7.2}$$

$$M_w = V_p \, {}^{(1)}\theta \, {}^{(1)}\rho \tag{7.3}$$

$$_g\rho = \frac{M_B}{(1 - {}^{(0)}\epsilon)V_B} \tag{7.4}$$

where $^{(0)}\epsilon$ is the porosity of the sample in the brass sleeve, assumed to be the same to within experimental error whether the sample is wet or dry (a swelling clay would be an exception to this), V_B is the bulk volume of the sample, assumed to be the same as the sleeve, which could contain a field core, as well as a recompacted sample, M_B is the mass of the dry sample,

$^{(0)}\rho_b$ is the bulk density, M_w is the mass of water in the sample, at any given time, V_p is the pore volume of the sample, $^{(1)}\theta$ is the degree of saturation of the sample at any time, $_g\rho$ is the grain density of the sample, which is the mass of the dry sample divided by the volume of the soil in the sample, and $^{(1)}\rho$ is the density of water. Experiment yields M_B, V_B, M_w, and, for $^{(1)}\theta = 1$, $^{(0)}\epsilon$, directly, assuming that $^{(1)}\rho$ is known. The equilibrium degree of saturation at any applied pressure (negative of $^{(1)}\Psi$) is calculated from M_w to yield the desorption branch of the hysteretic moisture characteristic. It should be noted that the texture of a reconstituted sample, and hence its hydraulic conductivity, porosity, permeability, and bulk density, are never the same as those of an undisturbed core of the same material. In general, the bulk density of the reconstituted sample is smaller than that of the core, while the porosity and hydraulic conductivity of the reconstituted sample are larger than in undisturbed cores. The inherent limitations of the pressure plate apparatus on the range of matric potentials that can be measured, and the limitation of collecting desorption data only can be removed, in principle, by the vacuum oven scheme discussed next.

7.2.3 Vacuum Oven Techniques

A vacuum oven, shown in cross section in Figure 7.4, differs from an ordinary oven for baking soil samples in that (1) a rubber vacuum seal, lightly greased, is attached to the door, (2) provision is made for attaching a mechanical vacuum pump, called a "fore pump" in vacuum parlance, to the oven to allow the interior to be evacuated, (3) provision of a port for measurements of the gas pressure inside the oven, in addition to a built-in vacuum gauge, and (4) provision of a valve to admit air to the evacuated oven. In this scheme, the soil sample is saturated, weighed, the oven temperature is set, and the vapor pressure adjusted to the first of an ever-decreasing set of values by a judicious combination of pumping and partial closure of the metered leak valve between the trap and the oven. When an equilibrium vapor pressure is reached as judged by constant vacuum gauge readings (a vacuum capacitance manometer with digital readout capable of pressure measurement in the range $10^{-3} \le p \le 760$ torr, where 1 torr is the pressure required to support a column of mercury 1 millimeter high, proves quite satisfactory in practice), the sample is weighed, the vapor pressure readjusted, and the process repeated until the lowest value of degree of saturation of the soil sample that can be attained at that temperature is reached. The process is then reversed and the soil sample is weighed at increasing values of vapor pressure until it is saturated once again. The temperature of the oven is then increased and the desorption–wetting cycle repeated.

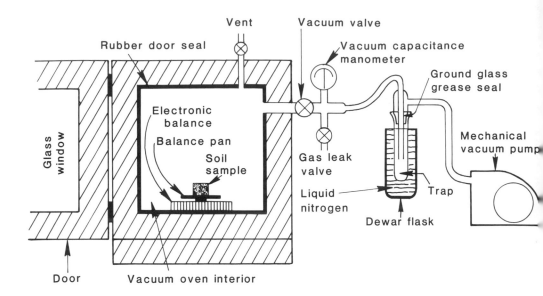

Figure 7.4: Vacuum oven shown schematically in cross section containing an electronic balance on which is the soil sample of interest in a teflon mesh container and attached to a vacuum gauge, a liquid nitrogen dewar and trap and a mechanical vacuum pump.

The matric potential of the water in the soil containing impurities, Ψ, is computed from the equilibrium vapor pressure in the oven, $\bar{p}_v$, via

$$\Psi = \Psi_0 \ln\left[\frac{\bar{p}_v(T)}{p_0(T)}\right] \qquad (7.5\text{a})$$

where $p_0(T)$ is the equilibrium pressure of water over a flat water surface that is at the same absolute temperature, T, and contains the same chemical impurities as the water in the soil sample, and

$$\Psi_0 = \frac{\rho(T)kT}{m} \qquad (7.5\text{b})$$

where, as in Chapter 1, $\rho(T)$ is the density of soil solution at absolute temperature T, k is Boltzmann's constant, and m is the weighted average of a water molecule and the masses of the dissolved impurities. The quantity

$\bar{p}_v(T)$ is seen to have the same physical interpretation as given in Chapter I. To obtain moisture characteristic curves for temperatures below room temperature, the entire set up can be placed in a freezer storage room and the oven temperature controls used to set the temperature from freezing upwards. As the sample temperature is increased, the residual degree of saturation of the soil sample, $^{(1)}\theta_0(T)$, decreases, while the maximum degree of saturation is still $^{(1)}\theta = 1$. Lower values of $^{(1)}\theta_0(T)$ mean lower (more negative) values of matric potential as well so that desorption matric potential curves extend further toward the low-saturation end for higher temperatures and as temperature increases, given values of the matric potential correspond to ever lower values of $^{(1)}\theta$. Scanning curves can also be developed at each oven temperature using this method.

7.2.4 Thermocouple Psychrometer

The thermocouple psychrometer makes use of a thermocouple, generally of chromel–constantan, to measure the average ambient vapor pressure, $\bar{p}_v$, in a soil sample. This vapor pressure is related to the matric potential via (7.5). Thermoelectricity is an example of the coupled effects of the flows of heat and electricity, in the sense of nonequilibrium thermodynamics, as previously discussed in Chapter 6. We consider the arrangement of thermocouple junctions shown in Figure 7.5. We assume that the three reference junctions are kept at the same constant temperature. If an electric current is forced through the chromel–constantan junction by connecting copper leads 2 and 3 in the figure to a seat of emf (electromotive force, usually measured in volts), then, depending on the direction of current flow and the amount of current, that junction cools, while heat is evolved from the copper–chromel reference junction and the copper–constantan reference junction. This absorption and liberation of heat is called the *Peltier effect* after Jean C. A. Peltier, who discovered it in 1834. Conversely, if a temperature difference exists between the copper–constantan junction and the copper–copper and copper–constantan reference junctions, then an emf is set up that drives a current through the circuit formed by connecting leads 1 and 2 in the figure to a voltmeter, reading in microvolts, such that Peltier heating and cooling act to tend to equalize the temperature at junctions externally maintained at dissimilar temperatures. This inverse of the Peltier effect is called the *Seebeck effect* after its discoverer in 1821, Thomas J. Seebeck. A third effect occurs along with the first two, namely the *Thompson effect*, which can be described as follows. In a wire made of a single material, such as the segments between each of the junctions shown in Figure 7.5, a temperature difference between the ends causes an imbalance in the distribution of conduction electrons in the wire such that relative depletion

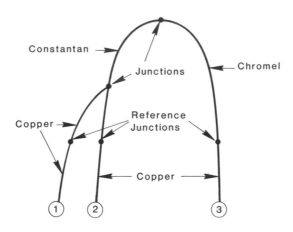

Figure 7.5: Two thermocouple junctions–the chromel–constantan junction is the one on which water vapor is condensed by Peltier cooling, while the copper–constantan junction is used to measure ambient temperature.

occurs at the warmer end and a corresponding increase occurs at the cooler end. This in turn causes an emf to be set up in such a direction as to tend to equalize the conduction-electron distribution in the metal. This emf is called the *Thompson emf* after its discoverer, Sir William Thompson (Lord Kelvin). We consider now the thermocouple junctions shown in Figure 7.6.

The Seebeck emf, E_{12}, due to unequal temperatures at the junctions of materials 1 and 2 and taken to appear across the microvoltmeter as measured by the microvoltmeter, is the sum of the Peltier emfs at the two junctions. At each junction of two metals, 1 and 2, the heat energy absorbed or released per unit charge passing through the junction is customarily written π_{12}. The Thompson emfs occur in the copper and constantan wires, where the amount of heat absorbed or liberated per coulomb passing through an element of wire of length dl, at unit temperature of wire of material j, is customarily written as σ_j, which is sometimes called the *specific heat of electricity*.

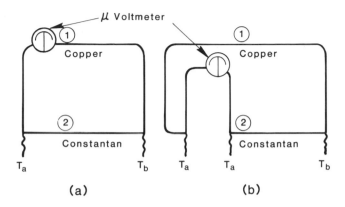

Figure 7.6: Two copper-constantan thermocouple junctions showing (a) a non-isothermal meter connection in the copper loop, and (b) an isothermal meter connection.

We express these considerations mathematically by adding up the voltage drops around the thermocouple circuit, which sum, since it is around a single loop, equals zero (this is Kirchhoff's first law of circuit analysis; Kirchhoff's second law of circuit analysis is that the sum of currents into and out of a point in a circuit equals zero). Thus we have, beginning with the junction at temperature T_a and proceeding around the circuit clockwise

$$\pi_{12}(T_a) + \int_{T_a}^{T_b} \sigma_1(T)\, dT - \pi_{12}(T_b) + \int_{T_b}^{T_a} \sigma_2(T)\, dT + E_{12} = 0 \tag{7.6}$$

or

$$E_{12} = \pi_{12}(T_b) - \pi_{12}(T_a) + \int_{T_a}^{T_b} [\sigma_2(T) - \sigma_1(T)]\, dT \tag{7.7}$$

An additional relation for a thermocouple junction is

$$\pi_{12} = T\frac{dE_{12}}{dT} \tag{7.8}$$

This relation can be evaluated at the temperature of either junction. If it is evaluated at the nonreference junction, the quantity $E_{12}(T_b)$ is called the *sensitivity* of the thermocouple, or the thermoelectric power, ϵ_{12}, so that

$$\epsilon_{12} = \frac{\pi_{12}(T_b)}{T_b} \tag{7.9}$$

From the definition of the Peltier emf, the rate at which Peltier heat W is gained or lost at the junction is

$$\frac{dW}{dt} = \pi_{12}(T_b)i \tag{7.10}$$

where i is the instantaneous current flowing through the junction and t is time. (Thompson heat due to flow of current in the wires leading to the junction is evolved, but will not be considered further.) If r_{12} is the resistance of the junction, then the power dissipated due to the Joule heating that occurs when a current i flows through it is

$$P_{12} = i^2 r_{12} \tag{7.11}$$

so that the total power, P_t, evolved/absorbed at the junction is

$$P_t = \pi_{12}(T_b)i + r_{12}i^2 \tag{7.12}$$

Thus, for small currents (8 mA for some thermocouples) the flow of current in the proper direction can cause cooling at the junction to occur, while for larger currents the Joule heating overwhelms the Peltier cooling. This sets a practical limit on the lowest temperature that can be reached. The thermocouple psychrometer can be operated in either the psychrometric mode, so called, or the dew-point mode. In the psychrometric mode the thermocouple junction is cooled below the dew point of the water vapor in the porous medium, thus inducing water vapor to condense onto it. The cooling current is then switched off and the water that has condensed onto the junction is allowed to evaporate, thereby cooling the junction to a temperature below ambient but above the dew-point temperature. If the difference in temperature between ambient, usually measured before the initial Peltier cooling, and the temperature to which the junction is cooled by evaporation is expressed as a difference between the Seebeck emf for each temperature, ΔE_{12}, then empirically it is found that for a chromel–constantan thermocouple the matric potential of the water in the soil is given by

$$\Psi = -2.13 \times 10^6 \, \Delta E_{12} \text{bars/V} \tag{7.13}$$

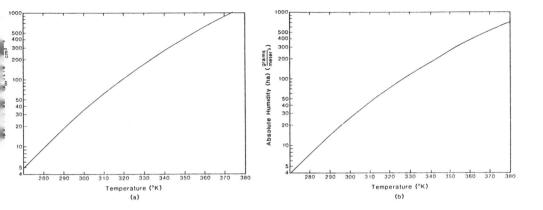

Figure 7.7: (a) The vapor pressure of water over a flat water surface where the water is assumed to contain no impurities. (b) Calculated from the data of (a), the absolute humidity of water vapor in air at saturation.

at an ambient temperature of 25 °C with ΔE_{12} measured in volts. At this point a slight elaboration of the concepts of dew point and relative humidity of water vapor in a porous medium is in order.

The dew point as actually defined is the temperature, generally below ambient, at which water vapor in air becomes saturated. Figure 7.7(a) shows the vapor pressure curve for water vapor over a flat water surface. The presence of dissolved impurities in the water would, of course, as discussed in Chapter 4, lower the vapor pressure at each temperature. When the ideal gas law is applied to the data of Figure 7.7(a) in the form

$$\frac{mN}{V} = \frac{p_v m}{kT} \tag{7.14}$$

where m is the mass of a water vapor molecule and the rest of the symbols are as previously defined, to calculate the corresponding mass of water vapor per unit volume of air, which is the absolute humidity at saturation, h_a (for this application $h_a = p_v m/kT$), we obtain the curve of Figure 7.7(b). (The term *saturation* as used in this context means that if the temperature is lowered, water vapor in the air will condense until the lower saturation vapor pressure at the lower temperature is reached.) If now the amount

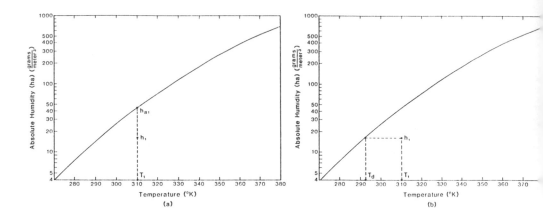

Figure 7.8: (a) A point of absolute humidity h_1 at temperature T_1 below the saturated vapor pressure at that temperature, h_{a1}. The ratio h_1/h_{a1}, expressed as a percent, is called the *relative humidity*. (b) The effect of lowering the temperature of the air and water vapor at a point. T_d is called the *dew point*.

of water vapor in the air at a given temperature is below the curve of Figure 7.7(b), as depicted in Figure 7.8(a), lowering the ambient temperature to the point at which the given amount of water vapor lies on the vapor saturation curve, as depicted in Figure 7.8(b), means that the air has been cooled to the dew point as mentioned previously. In a porous medium, the curve $h_a(T)$ is lowered from that shown in Figure 7.7(b) due to two effects. The first is that water in soil always contains dissolved impurities, and the second, also discussed in Chapter 2, is that the vapor pressure of water in a porous medium is lowered below that just above a flat surface due to the concave shape and the air–water menisci interfaces. Thus, the concepts of dew point and relative and absolute humidity can be applied in a porous medium, but it must be recognized that the terms do not have quite the same meaning as in their normal context. In fact the average water vapor pressure "seen" by a thermocouple psychrometer probe is the weighted–by relative frequencies of groups of menisci sizes and shapes–average of vapor pressure contributions from all menisci in a manner similar to Dalton's law of partial pressures. Thus, the concepts of dew point and relative humidity as applied to thermocouple psychrometers operated in either the psychrometric or dew point modes are to be understood as given previously.

We now describe the dew point mode of operation of a thermocouple psychrometer. In this mode of operation the chromel–constantan thermocouple junction is cooled by a current consisting of discrete pulses. If the initial temperature to which the junction is cooled is below the dew point of the water vapor in the porous medium, then, during the portions of the pulse cycle when no current is flowing, water condenses on the junction. Conversely, if the junction is cooled initially to a temperature above the dew point, then water evaporates from the junction, absorbing heat of evaporation to do so, thus lowering the temperature of the junction toward the dew point. In practice an electronic feedback circuit is required to sense changes in temperature of the junction and automatically adjust the pulsed current to maintain the junction at the dew point of the water vapor in the porous medium. Empirically it is found that if ΔE_{12} is the difference in Seebeck emf between E_{12} of the (Peltier) chromel–constantan junction at the ambient temperature and the value of E_{12} at the junction at the dew point, then the matric potential of the water in the soil as reflected in the pressure of the vapor evolving from the air–water interfaces (menisci) is

$$\Psi \simeq -1.33 \times 10^6 \, \Delta E_{12} \text{bars/V} \qquad (7.15)$$

where ΔE_{12} is in volts. Since the Peltier junction is maintained at the dew point electronically, it is generally held that the dew point method of operation is more accurate than the psychrometric mode. We now explore the relation between ΔE_{12} and Ψ in more detail by computing $\partial E_{12}/\partial \Psi$ as follows. If all independent variables other than temperature are held constant, we can write

$$\frac{\partial E_{12}}{\partial \Psi} = \frac{\partial E_{12}}{\partial T} \frac{\partial T}{\partial \Psi} = \frac{\partial E_{12}}{\partial T} \frac{\partial \Psi}{\partial T} \qquad (7.16a)$$

where $\partial E_{12}/\partial T$ is given by (7.8) and $\partial \Psi/\partial T$ is given by

$$\frac{\partial \Psi}{\partial T} = \frac{\partial}{\partial T}\left[\Psi_0 \ln\left(\frac{\bar{p}_v}{p_{0v}}\right)\right] = \frac{\Psi}{T} + \frac{1}{p_{0v}}\frac{\partial p_{0v}}{\partial T} + \frac{1}{p_{0v}}\frac{\partial \bar{p}_v}{\partial T} \qquad (7.16b)$$

where all the symbols are as previously defined in Chapter 1. At this point it is customary in some treatments to approximate $\partial \bar{p}_v/\partial T$ by Eq. (4.5). This is correct only for calibration situations in which the psychrometer probe is suspended in an atmosphere of vapor at a known pressure and

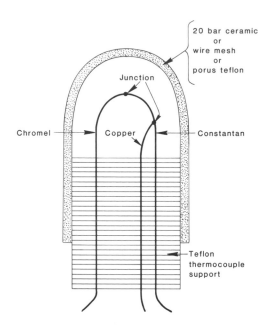

Figure 7.9: A possible design of thermocouple psychrometer probe showing the thermocouple junctions, teflon support, and the protective porous cap over the thermocouples, which may be made of ceramic, stainless steel wire mesh, or teflon with many small holes drilled through the walls.

temperature generated by an aqueous solution, having a flat surface, of known composition, often NaOH, and known concentration. In practice a series of such solutions at different concentrations is used, Ψ is computed from

$$\Psi = \Psi_0 \ln \left(\frac{\bar{p}_v}{p_o} \right) \tag{7.17}$$

where $\bar{p}_v$ is taken to be the vapor pressure of the solution, computed approximately from (3.86a) or (3.90), and p_0 is taken to be the vapor pressure of pure water at the some temperature. The values of Ψ so obtained are then correlated with the measured values of ΔE_{12} for a given thermocouple for specific solutions at known temperatures, though 25 °C is a generally accepted standard temperature, to calibrate a given thermocouple psychrometer. Tables of Ψ as a function of solution type and concentration given in the literature are computed using the scheme outlined. In a porous medium Eq. (4.5) cannot approximate $\partial \bar{p}_v / \partial T$. Instead, a porous medium model-dependent form must be used in an explicit theoretical relation for the change in E_{12} with respect to Ψ. This model-dependent form for a bundle of capillaries with a radius frequency distribution is given by the derivative with respect to temperature of Eq. (1.3).

Figure 7.9 shows a cross section of a possible thermocouple psychrometer probe. If stainless steel mesh or porous teflon is used for the protective cap, then the size of the pore opening in the cap is generally larger than the mean free path of water vapor molecules in the atmosphere, which is approximately 4.18×10^{-6} cm (Kennard, 1938, p. 149), and the water vapor molecules can diffuse freely through. If, on the other hand, the cap is made from 20-bar ceramic, then a nominal pore radius opening in it is about 7.2×10^{-6} cm, or about twice the mean free path of water vapor in the atmosphere. Thus, the equilibration of the vapor pressure inside, since the cooled Peltier junction is acting as a vapor sink, can take from several minutes to several hours. Conversely, when the thermocouple junction is heated (Joule heating by passing a current through it) to dry it prior to another measurement, several minutes must be allowed to elapse for the high-pressure water vapor generated within the protective cap to diffuse out. A heating cycle of a few seconds followed immediately by a measurement cycle is thus not satisfactory, particularly in a arid unsaturated zone environment, since if insufficient time is allowed for the evaporated water vapor to diffuse out of the protective cap, the subsequent measurement of matric potential will yield an incorrectly high value. As a final practical matter, the ultimate useful life of a thermocouple psychrometer probe is limited by the deposition, over time, of salts dissolved in the water onto the thermocouple junction when it is heated between measurement cycles. If the psychrometer probe is sufficiently accessible so that the protective cap can be removed and the thermocouple junction cleaned by gently spraying it with distilled water, then the probe can last almost indefinitely. If, however, probes are inaccessible, then their useful lives range from days to weeks in a humid greenhouse environment to 7–10 years in the field in an arid environment.

7.3 MOISTURE CONTENT OF SOILS

7.3.1 General Discussion

The moisture content of soils may be determined using either the vacuum oven or the pressure plate apparatus as described. Also, once the moisture characteristic has been measured, values of matric potential can be used to infer, depending on the amount of hysteresis shown by the soil, corresponding values of degree of saturation as well. Instruments also exist that

measure the dielectric constant(s) of earth materials, and with suitable calibration can be used to infer soil moisture contents, although here again,
the degree to which this can be done depends on the degree of hysteresis displayed by the moisture characteristic. The emerging–in the study of
porous media–technique of nuclear magnetic resonance can be used to characterize the moisture content of small laboratory samples and, in common
with the vacuum oven, can do so in a manner that is independent of the
hysteretic nature of the moisture characteristic for the medium. The principal potential utility of nuclear magnetic resonance in studies of porous
media, however, is in characterizing the actual pore volume structure of
a laboratory sample, and in elucidating the chemical interactions between
water containing impurities adsorbed onto pore/capillary walls and thus an
introductory section is devoted to this technique though, in practice, iron
in the soil samples can interfere with the efficiency of this measurement
technique. In field work, one technique for nondestructive soil moisture
determination has become fairly widespread. This technique is based on
the scattering of neutrons by the protons in the water in soil, and to this
method we now turn.

7.3.2   Neutron Moisture Gauge

The principle underlying the neutron moisture gauge is the following. "Fast"
neutrons, that is, neutrons having kinetic energies of up to 10 MeV, from a
radioactive source are scattered essentially elastically from the protons in
the water in the soil. The scattering process causes the scattered neutrons
to lose some of their initial kinetic energy. In close proximity to the source
is a neutron detector made in such a way as to be sensitive to thermal (kinetic energies of the order of $3kT/2$ where k is Boltzmann's constant and
T is absolute temperature) neutrons backscattered toward it. The number
of thermal neutrons detected per unit time is a measure of the degree of
saturation of the soil near the source and detector. The neutron moisture
gauge must be calibrated for each soil locality where it is to be used. A
possible calibration setup is shown in Figure 7.10. The soil surrounding
the access tube is sampled and the samples baked to determine directly the
ambient soil saturation(s). The radius of influence, r_i, of the probe can be
estimated from the relation

$$r_i = \frac{d_f}{{}^{(0)}\epsilon\,{}^{(1)}\theta} \left[1 + f(a)g({}^{(1)}\theta)\right] \tag{7.18}$$

where d_f is the fast diffusion length for neutrons in water, 5.75 cm, ${}^{(0)}\epsilon$ is
the porosity of the soil, ${}^{(1)}\theta$ is the degree of saturation of the soil by

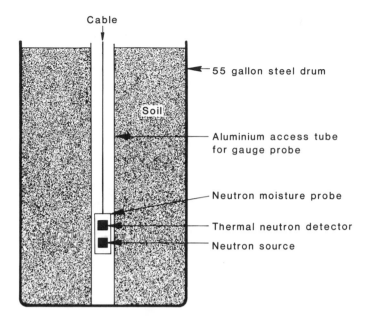

Figure 7.10: A possible calibration setup for a neutron moisture gauge consisting of an aluminum access tube in the soil to be studied under a range of artificially induced and directly measured saturations.

water, $f(a)$ is a slowly increasing dimensionless function of the activity of the neutron source in the gauge probe, and $g(^{(1)}\theta)$, also dimensionless, depends on the soil type and degree of saturation and may become negative if the soils of interest contain appreciable amounts of lithium, cadmium, and/or boron, which are significant absorbers of neutrons. Both $f(a)$ and $g(^{(1)}\theta)$ are measured for each specific field situation of interest by calibrating probes of different source strengths as functions of soil saturation and comparing the results. A typical value of the inside diameter of the aluminum access tube is 1.9 inches for most moisture gauge probes that contain both the source and the detector. The gauge can be calibrated for iron access tubing as well, but PVC (polyvinyl chloride) or fiberglass are not acceptable as tubing materials due to their high hydrogen content. The usual method of using a disturbed soil sample in a 55 gallon drum (for which r_i is generally larger than the radius of the drum) with an aluminum access tube coaxial with the drum as a calibration setup falls short of what is actually needed in practice, which is calibration in the field, and so its use is not elaborated here.

We now consider nuclear fundamentals in more detail. We consider only neutrons, protons, and electrons as atomic and molecular constituents.

A free proton, one not in the nucleus of an atom, has a rest mass of 1.672×10^{-27} kg and a charge of $+1.602 \times 10^{-19}$ C while a free neutron has a rest mass of 1.675×10^{-27} kg and is electrically neutral. A free proton is stable, but a free neutron has an average lifetime of 11.7 min and decays into a proton, electron (beta particle), and a neutrino (denoted by ν) according to

$$_0n^1 \longrightarrow {}_1p^1 + {}_{-1}\beta^0 + \nu \tag{7.19}$$

where the superscripts are atomic mass numbers, and the subscripts are the charges of the particles in units of the charge on an electron. The sum of the superscripts and subscripts on each side of the equation must be equal for a correctly written nuclear reaction. Nuclear energies are millions of times larger than chemical binding energies. A convenient and customary unit for measuring atomic and nuclear energies is the electron volt, (eV) which is the magnitude of the charge on an electron, -1.602×10^{-19} C multiplied by 1 V or, 1 eV $= 1.602 \times 10^{-19}$ C $\times$ 1 V $= 1.602 \times 10^{-19}$ J $= 1.602 \times 10^{-12}$ erg, where 1 J $= 1$ kg meter/sec^2 $= 1$ C V, one thousand electron volts is 1 keV, and an energy of one million electron volts is written as 1 MeV.

The radius of a nucleus is given approximately by

$$R_m = r_0 A^{\frac{1}{3}} \tag{7.20}$$

where A is the mass number of the atomic nucleus of approximate radius R_m, the sum of the number of neutrons, N, and protons, Z, in the nucleus, that is

$$A = N + Z \tag{7.21a}$$

and

$$r_0 = 1.4 \times 10^{-15} \text{ m} \tag{7.21b}$$

Nuclei having the same numbers of protons (the number of protons determines the atomic species) but different numbers of neutrons are called *isotopes*. Nuclei having different numbers of protons but the same numbers of neutrons are called *isotones*, while nuclei having identical mass numbers are called *isobars*. The nuclear binding energies can be computed from the differences in mass between the masses of the free protons and neutrons required to make up a given nucleus and the mass of the nucleus itself. This possibility is contained in the relation

$$E = mc^2 \tag{7.22}$$

from Einstein's special theory (as distinct from the general theory) of relativity. In this equation, E is the total relativistic energy of a body, which would be a particle of relativistic mass m, and $c \simeq 3 \times 10^8$ m/sec, is the speed of light in vacuum. The relativistic mass of a body as perceived by an observer is larger than its rest mass if a state of relative motion exists between the observer and the body. Specifically, if a body is moving past an observer with velocity of magnitude v along a straight line path, then the relation from special relativity, between the rest mass of the body, m_0, which is the mass of the body perceived by the observer when the observer and the body are at rest relative to each other, and m is

$$m = \frac{m_0}{\left(1 - \frac{v^2}{c^2}\right)^{\frac{1}{2}}} \tag{7.23}$$

The rest energy, E_0 of the body is thus written as

$$E_0 = m_0 c^2 \tag{7.24}$$

and thus, on combining (7.19) and (7.20), we have

$$E = \frac{E_0}{\sqrt{1 - \frac{v^2}{c^2}}} \tag{7.25}$$

Thus, mass can be expressed in energy units, and vice versa. All the masses discussed in Chapters I–VI have been taken to be rest masses and it is only in this section that the distinction between rest mass and relativistic mass will be made. The kinetic energy, or energy of motion of a body of rest mass m_0, is given by

$$E_k = E - E_0 \tag{7.26}$$

which, by expanding the square root in (7.22) in a binomial series and retaining only the first two terms, valid if $v \ll c$, reduces in that case to the usual nonrelativistic expression for kinetic energy

$$E_k \simeq \frac{1}{2} m_0 v^2 \tag{7.27}$$

The binding energy per nucleon of an atom can be computed by knowing to high precision the masses of the nuclei of interest as well as the masses of its constituents, assumed in this approximation to be neutrons and protons. The results of such computations are available for the stable isotopes and are plotted in Figure 7.11 as the binding energy per nucleon as a function of mass number.

We now discuss aspects of unstable isotopes, that is, those that spontaneously decay. If N_0 is the number of radioactive atoms of a given species present at time $t = 0$, then the change in the number of nuclei, $N(t)$, as a function of time is given by

$$\frac{dN}{dt} = -\lambda_d N \tag{7.28a}$$

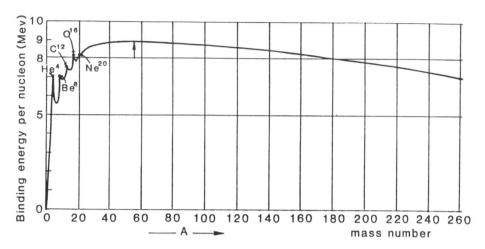

Figure 7.11: The binding energy per nucleon for the stable isotopes in the periodic table. The average binding energy is about 8 MeV per nucleon.

where λ_d is called the *decay constant*. The quantity $-dN(t)/dt$ is called the *activity* of the radioactive species and is measured in Curies, where 1 curie = 1 Ci = 3.7×10^{10} disintegrations/ second. Equation (7.25) may be integrated to yield

$$N(t) = N_0 e^{-\lambda_d t} \tag{7.28b}$$

If $T_{\frac{1}{2}}$ is the time required for half the atoms of a given species to decay, the half-life of the species, then

$$\frac{1}{2} = \frac{N(T_{\frac{1}{2}})}{N_0} = e^{-\lambda_d T_{\frac{1}{2}}} \tag{7.29a}$$

or, on taking the natural logarithm of both sides and rearranging

$$\lambda_d = \frac{\ln(2)}{T_{\frac{1}{2}}} = \frac{0.693}{T_{\frac{1}{2}}} \tag{7.29b}$$

The average lifetime, $\bar{t}$, of each of the atoms of a decaying species, by contrast, is

$$\bar{t} = \int_0^\infty t\, N(t)\, dt = \frac{1}{\lambda_d} \tag{7.30}$$

We now consider the concept of specific activity (SpA), or activity per gram or per mol of radioactive species. From (7.28a) we see that the activity of a sample of $N(t)$ atoms is $\lambda_d N(t)$. If n is the number of atoms per gram of a given radioactive species, then $\lambda_d n$ is the specific activity, ultimately in units of curies per gram, of the radioactive species. To compute n we write $n = L/A$ where $L = 6.02 \times 10^{23}$ particles per mol, Avogadro's number, and A, the atomic mass number, equals, by definition, the number of grams per mol. Thus, the specific activity, in units of disintegrations per second per gram is given by $\lambda_d L/A$. If the half-life of the species is expressed in seconds, we convert from disintegrations per second to curies via

$$\text{SpA} = \frac{\lambda_d L}{A} \frac{1 \text{ Ci}}{3.7 \times 10^{10} \text{ disintegrations/sec}} = \frac{1.128 \times 10^{13}}{A T'_{\frac{1}{2}}} \frac{\text{Ci}}{\text{g}}$$

where $T_{\frac{1}{2}}$ is given in seconds. In other units we have

$$SpA = \frac{1.879 \times 10^{11}}{A T_{\frac{1}{2}}} \frac{\text{Ci}}{\text{g}} \tag{7.31a}$$

for $T_{\frac{1}{2}}$ in minutes

$$SpA = \frac{1.305 \times 10^{8}}{A T'_{\frac{1}{2}}} \frac{\text{Ci}}{\text{g}} \tag{7.31b}$$

for $T_{\frac{1}{2}}$ in days

$$SpA = \frac{3.573 \times 10^{5}}{A T_{\frac{1}{2}}} \frac{\text{Ci}}{\text{g}} \tag{7.31c}$$

and

$$SpA = \frac{9.789 \times 10^{2}}{A T_{\frac{1}{2}}} \frac{\text{Ci}}{\text{g}} \tag{7.31d}$$

for $T_{\frac{1}{2}}$ in years. Multiplying SpA by the density of the species at a given temperature yields the activity per unit volume of the material at that temperature. Since the production of neutrons in the neutron moisture gauge generally begins with the alpha decay of $_{95}\text{Am}^{241}$, we now consider the energetics of alpha decay. If M_{0p} is the rest mass of the parent nucleus, P, M_{0d}, the rest mass of the daughter nucleus, D, and $M_{0\alpha}$, the rest mass of the alpha particle, α, then we write for the decay of the parent nucleus, where Q is the disintegration energy of the reaction, or the Q value of the reaction (decay in this example),

$$P \rightarrow D + \alpha + Q \tag{7.32}$$

If the parent nucleus is initially at rest, so that its kinetic energy, E_{kp}, is taken to be zero, conservation of mass–energy in the sense of Eq. (7.19) yields

$$M_{0p}c^2 = (M_{0d} + M_{0\alpha})c^2 + E_{kd} + E_{k\alpha} \tag{7.33}$$

where E_{kd} is the kinetic energy of the daughter, and $E_{k\alpha}$ is the kinetic energy of the α particle. We identify

$$Q = E_{kd} + E_{k\alpha} \tag{7.34}$$

Since the parent nucleus was assumed to be initially at rest, thus having zero momentum, from conservation of momentum for the decay, which means that the momentum of the parent is equal to the sum of the momenta of daughter and the α particle, we have

$$0 = M_{0d}v_d + M_{0\alpha}v_\alpha \tag{7.35a}$$

or,

$$M_{0d}v_d = -M_{0\alpha}v_\alpha \tag{7.35b}$$

where v_d and v_α are the velocities of the daughter and the α particle, respectively. Squaring each side of (7.35) and multiplying by $\frac{1}{2}$ yields

$$M_{0d}\left(\frac{1}{2}M_{0d}v_d^2\right) = M_{0\alpha}\left(\frac{1}{2}M_{0\alpha}v_\alpha^2\right) \tag{7.36}$$

If the velocities of the α particle and the daughter are low enough so that the nonrelativistic form of the kinetic energy can be used to good approximation, we can write

$$E_{k\alpha} = \frac{1}{2}M_{0\alpha}v_\alpha^2 \tag{7.37}$$

and

$$E_{kd} = \frac{1}{2}M_{0d}v_d^2 \tag{7.38}$$

so that (7.36) becomes

$$M_{0d}E_{kd} = M_{0\alpha}E_{k\alpha} \tag{7.39}$$

If M_{0p} has atomic number A_p and $M_{0\alpha}$ has atomic number 4, then measuring M_{0d} and $M_{0\alpha}$ in amu yields from (7.39) with $A_d = A_p - 4$,

$$E_{kd} = \left(\frac{4}{A_p - 4}\right)E_{k\alpha} \tag{7.40}$$

From (7.33) and (7.40) we have

$$Q = E_{kd} + E_{k\alpha} = \left(\frac{4}{A_p - 4}\right)E_{k\alpha} + E_{k\alpha} \tag{7.41a}$$

or

$$E_{k\alpha} = \left(\frac{A_p - 4}{A_p}\right)Q \tag{7.41b}$$

and, from (7.40) and (7.41b)

$$E_{kd} = \frac{4Q}{A_p} \tag{7.42}$$

The next reaction to consider is the capture of an alpha by $_4\text{Be}^9$, namely

$$_2\text{He}^4 + {}_4\text{Be}^9 \rightarrow {}_6\text{C}^{13*} \rightarrow {}_6\text{C}^{12*} + {}_0\text{n}^1 + Q \tag{7.43}$$

The asterisk on the $_6\text{C}^{12*}$ indicates that in this reaction the isotope of carbon is produced in an excited (higher energy than the ground state) state, which decays to the ground state by gamma emission. The $_6\text{C}^{13}$ is an intermediate excited compound nucleus. The total energy produced by the reaction is the disintegration energy, Q, plus the kinetic energy of the incoming alpha relative to the $_4\text{Be}^9$ nucleus, which is assumed to be at rest. The other reaction by which neutrons are produced in the Am-Be source is the photodisintegration reaction

$$\gamma + {}_4\text{Be}^9 \rightarrow {}_4\text{Be}^8 + {}_0\text{n}^1 + Q \tag{7.44}$$

which can occur because the gamma rays emitted from the $_6\text{C}^{12*}$ isomeric transition to the ground state have energies in the range of 1-9 MeV while the threshold energy for photodisintegration to occur is 1.666 MeV. The efficiency of production of neutrons by an Am-Be source is not 100%, however. In fact for a 10 mCi (1 mCi $= 1 \times 10^{-3}$ Ci) source, where the source strength designation of 10 mCi refers to the Am disintegrations occurring in the source, the production of neutrons is about 2.5×10^4 neutrons/sec as a net result of the processes mentioned.

We now consider the scattering and consequent moderation of neutrons that can occur in the body of a neutron moisture gauge, as shown in Figure 7.12, as a classical rather than a quantum-mechanical scattering problem as follows. The scattering and consequent energy loss of a neutron having mass m_{01} per collision with a nucleus of mass m_{02} are most conveniently considered in a coordinate system in which the center of mass of the two particles is the origin. The center of mass, $\vec{R}$, of a collection of particles having masses $m_1, m_2, \cdots, m_n$ is the weighted, by particle mass, average of the positions of the particles. Thus,

$$\vec{R} = \frac{\sum_{i=1}^n \vec{r}_i m_i}{\sum_{i=1}^n m_i} \tag{7.45}$$

where $\vec{r}_i$ is the (vector) position of the ith mass. For two masses, we have $n = 2$ and

$$\vec{R} = \frac{\vec{r}_1 m_1 + \vec{r}_2 m_2}{m_1 + m_2} \tag{7.46}$$

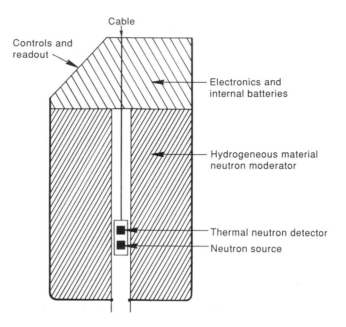

Figure 7.12: Schematic cross section of a neutron moisture gauge showing the polyethylene shield, which serves as a moderating medium to thermalize the fast neutrons from the source.

The coordinate, $\vec{r}$, of m_1 relative to m_2 is

$$\vec{r} = \vec{r}_1 - \vec{r}_2 \tag{7.47}$$

We now write for the positions of particles 1 and 2 referred to their common center of mass as the origin

$$\vec{r}_{1c} = \vec{r}_1 - \vec{R} \tag{7.48}$$

and

$$\vec{r}_{2c} = \vec{r}_2 - \vec{R} \tag{7.49}$$

and call these the center-of-mass (CM) coordinates of particles 1 and 2, respectively. We can express the center-of-mass coordinates in terms of the coordinate $\vec{r}$ as

$$\vec{r}_{1c} = \frac{m_2}{m_1 + m_2}\vec{r} = \frac{m_R}{m_2}\vec{r} \tag{7.50}$$

$$\vec{r}_{2c} = -\frac{m_R}{m_2}\vec{r} \tag{7.51}$$

where m_R is called the *reduced mass* and is given by

$$m_R = \frac{m_1 m_2}{m_1 + m_2} \tag{7.52}$$

The velocities of the two particles in CM coordinates are

$$\dot{\vec{r}}_{1c} = \frac{m_R}{m_1} \dot{\vec{r}} \tag{7.53}$$

$$\dot{\vec{r}}_{2c} = -\frac{m_R}{m_2} \dot{\vec{r}} \tag{7.54}$$

where the dot denotes differentiation with respect to time. From (7.53) and (7.54) we see that the momenta of the two particles, $m_i \dot{\vec{r}}_{ic}$, $(i = 1, 2)$, are equal in magnitude and oppositely directed so that their total momentum in CM coordinates is zero, both before scattering takes place and afterwards, since momentum is always conserved in collisions whether kinetic energy is or not.

The collision process between an incoming particle of mass m_1, which in this case will be taken to be a neutron, and a target nucleus of mass m_2, assumed to be at rest in the laboratory, is sketched in the laboratory system of coordinates in Figure 7.13 and in CM coordinates in Figure 7.14. After the collision, particle 1 moves off at an angle θ_1 with respect to the initial direction of motion, with a velocity $\vec{v}_1$ and kinetic energy E_1, while particle 2 moves off at angle θ_2, with a velocity $\vec{v}_2$, and kinetic energy E_2.

The vector $\vec{r}$ still goes from particle 1 to particle 2 after the collision as shown. Since m_2 is assumed to be at rest before the collision, $\dot{\vec{r}}_{20} = 0$, and the velocity of the center of mass in the laboratory becomes

$$\vec{V}_{cm} = \frac{m_1}{m_1 + m_2} \dot{\vec{r}} = \vec{v}_{10} \frac{m_1}{m_1 + m_2} \tag{7.55}$$

In the laboratory system, the incoming particle scatters at an angle θ_1 with its original direction while in the CM system the incoming particle scatters at an angle Φ_1 with its original direction. The relation between the velocity of the incoming particle before and after scattering can be obtained from

$$\vec{V}_{cm} = \dot{\vec{R}} = \dot{\vec{r}}_1 - \frac{m_2}{m_1 + m_2} \dot{\vec{r}} \tag{7.56a}$$

with

$$\vec{v}_1 = \dot{\vec{r}}_{1c} + \vec{V}_{cm} \tag{7.56b}$$

which states that the final velocity of m_1 in the laboratory frame is the vector sum of its velocity in the center-of-mass system and the velocity of the center of mass of the two particles in the laboratory system. Note also that since m_2 is assumed to be initially at rest, the time derivative of the

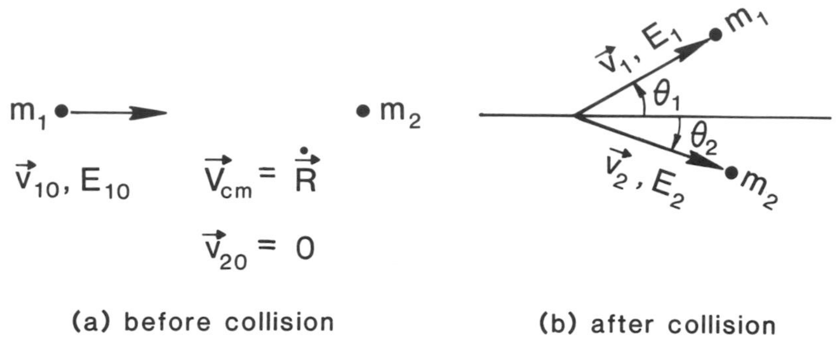

(a) before collision   (b) after collision

Figure 7.13: An incoming particle of mass m_1 having an initial velocity $\vec{v}_{10}$ and kinetic energy E_{10}. The center of mass of the two-particle system is moving toward the right with velocity $\vec{V}_{\text{cm}} = \dot{\vec{R}}$.

relative coordinate, $\dot{\vec{r}}$, equals the velocity of m_1 before scattering (its initial velocity), $\vec{v}_{10}$. Thus we have, with the help of (7.52a),

$$\vec{V}_{\text{cm}} = \vec{v}_1 - \frac{m_2}{m_1 + m_2}\dot{\vec{r}} = \vec{v}_1 - \frac{m_R}{m_1}\vec{v}_{10} \qquad (7.57)$$

so that

$$\vec{v}_1 = \frac{m_R}{m_1}\vec{v}_{10} + \vec{V}_{\text{cm}} \qquad (7.58)$$

Equation (7.58) is depicted graphically in Figure 7.15. Applying the law of cosines to the velocity triangle shown in Figure 7.15 yields

$$v_1^2 = \left(\frac{m_R}{m_1}\right)^2 v_{10}^2 + V_{\text{cm}}^2 + 2\frac{m_R}{m_1}v_{10}V_{\text{cm}}\cos(\Phi_1) \qquad (7.59)$$

where

$$v_1^2 = \vec{v}_1 \cdot \vec{v}_1 \qquad (7.60)$$

etc. The ratio of the kinetic energy of m_1 after scattering to its kinetic energy before scattering is given by, with the help of (7.55) and (7.56),

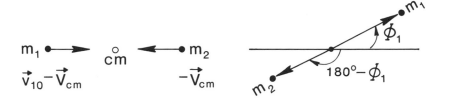

(a) before collision   (b) after collision

Figure 7.14: The target nucleus, assumed to be at rest in the laboratory system, moves to the left with velocity $-\vec{V}_{cm} = \dot{\vec{R}}$. Since the total momentum in the CM system after the collision is zero as it was before the collision, the particles move away from each other in opposite directions.

(note that $\vec{V}_{cm}$ is the same both before and after the collision, since no force external to the two-particle system is applied to it) in the nonrelativistic approximation as

$$\frac{E_1}{E_{10}} = \frac{\left(\frac{1}{2}\right) m_{o1} v_1^2}{\left(\frac{1}{2}\right) m_{o1} v_{10}^2}$$

$$= \left(\frac{m_2}{m_1 + m_2}\right)^2 + \left(\frac{m_1}{m_1 + m_2}\right)^2 + 2\frac{m_1 m_2}{(m_1 + m_2)^2} \cos(\Phi_1) \quad (7.61)$$

If we measure m_1 and m_2 in terms of mass numbers, using $m_1 = 1$ and $m_2 = A$, (7.61) becomes

$$\frac{E_1}{E_{10}} = \frac{A^2 + 1 + 2A \cos(\Phi_1)}{(A+1)^2} \quad (7.62)$$

For a collision in which the scattered particle is hit "head on" and moves forward in the laboratory frame, $\Phi_1 = 180°$ and (7.62) becomes

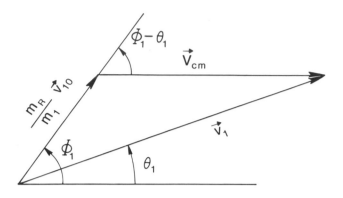

Figure 7.15: Graphical representation of the vector equation $\vec{v}_1 = m_R \vec{v}_{10}/m_1 + \vec{V}_{cm}$.

$$\frac{E_1}{E_{10}} = \frac{A^2 - 2A + 1}{(A+1)^2} = \frac{(A-1)^2}{(A+1)^2} \tag{7.63}$$

Thus, for a "head on" collision we have for the change in energy of the incident neutron

$$\Delta E = E_{10} - E_1 = E_{10}\left[1 - \frac{(A-1)^2}{(A+1)^2}\right] = E_{10}\frac{4A}{(A+1)^2} \tag{7.64}$$

which is a maximum for $A = 1$, corresponding to $_1\mathrm{H}^1$ nuclei.

Since (7.63) represents the maximum change in incident neutron energy per collision, we may write

$$E_{1m} = E_{10}\, x \tag{7.65}$$

where

$$x = \left(\frac{A-1}{A+1}\right)^2 \tag{7.66}$$

Thus, the difference between the initial neutron energy, E_{10}, and the range of possible energies that the neutron can have after one collision is $E_{10}(1-x)$ and varies from E_{10}, corresponding to a head on collision with a hydrogen nucleus, for which $x = 0$, to very close to zero for elastic collisions of neutrons with nuclei having large values of A. Thus, the probability that after one head on collision the neutron energy will be in the range E to $E + dE$ is

$$p(E)\, dE = \frac{dE}{E_{10}(1-x)} \tag{7.67}$$

The logarithmic decrement of the neutron energy, ξ, is by definition for one collision

$$\xi = \ln\left(\frac{\overline{E}_{10}}{E_1}\right) \tag{7.68}$$

where the bar represents the average over $p(E)\, dE$, or

$$\xi = \int_{xE_{10}}^{E_{10}} \ln\left(\frac{E_{10}}{E}\right) p(E)\, dE \tag{7.69}$$

where the range of integration is over the difference between the initial neutron energy and the possible energy left after one collision. Setting $y = E/E_{10}$ yields $dy = dE/E_{10}$; when $E = E_{10}, y = 1$, and when $E = xE_{10}$, $y = x$, so that

$$\xi = \frac{1}{1-x} \int_x^1 \ln\left(\frac{1}{y}\right) dy = \frac{1}{1-x} \int_1^x \ln(y) dy$$

$$= \frac{1}{1-x}[1 - x + x\ln(x)] = 1 + \frac{x\ln(x)}{1-x} \tag{7.70}$$

Substituting (7.66) into (7.70) and rearranging yields

$$\xi = 1 - \frac{(A-1)^2}{2A} \ln\left(\frac{A+1}{A-1}\right) \tag{7.71a}$$

or for $A \geq 10$

$$\xi = \frac{6}{2 + 3A} \tag{7.71b}$$

By taking the limit as $A \to 1$, we find $\xi = 1$, which means that

$$1 = \ln\left(\frac{\overline{E}_{10}}{E_1}\right) \tag{7.72}$$

or

$$\overline{E}_1 = \frac{E_{10}}{e} \tag{7.73}$$

where e is the base of natural logarithms. Since ξ is the average logarithmic decrement in neutron energy per collision, the number of collisions, N_c, needed to reduce a neutron of initial energy E_{10} to a final energy of E_{1f} is given by

$$N_c = \frac{\ln\left(\frac{E_{10}}{E_{1f}}\right)}{\xi} = \frac{\ln\left(\frac{E_{10}}{E_{1f}}\right)}{1 - \frac{(A-1)^2}{2A}\ln\left(\frac{A+1}{A-1}\right)} \tag{7.74}$$

Thus, for example, the average number of collisions with silicon, with A taken to be 27, needed to thermalize a 10 MeV neutron is

$$N = \frac{\ln(10 \times 10^6 \,\mathrm{eV}/0.025\,\mathrm{eV})}{1 - \frac{(27)^2}{56}\ln\left(\frac{29}{27}\right)} \simeq 284$$

We have considered the scattering due to one nucleus. We now consider the effect of the number density, ρ_n, of scattering nuclei of mass number A. If ρ_n is the number of scatters per cm^3, then

$$\rho_n = \frac{L\rho}{A} \tag{7.75}$$

where L is Avogadro's number and ρ is the mass density of the scattering material of atomic mass number A as previously and we write

$$\sigma = \rho_n \sigma_s = \frac{L\rho}{A}\sigma_s \tag{7.76}$$

where σ_s is the microscopic scattering cross section and σ is called the macroscopic scattering cross section and represents the probability per cm thickness of scattering medium that a neutron traversing it will be scattered. For a composite medium, such as polyethylene, σ is the sum of σ_H, with H denoting hydrogen, and σ_C, C denoting carbon, the macroscopic scattering cross sections computed for each constituent separately. In earth materials σ is the sum of the σ_i for each atomic constituent weighted by relative abundance. The neutrons thermalized by elastic scattering have a Maxwellian distribution of velocities, as discussed for gas molecules in Chapter 4 and a concomitant energy distribution given by

$$n(E)\,dE = \frac{2\pi n}{(\pi kT)^{3/2}}e^{-E/kT}E^{1/2}\,dE \tag{7.77}$$

where n is the number of thermalized neutrons per unit volume. The neutrons thermalized by the polyethylene when the probe is in the body of the instrument, can be read by the detector to yield a standard, baseline thermal neutron count, and the instrument can be calibrated in terms of the ratio of the neutrons counted following thermalization by a given soil type with varying degrees of saturation to the standard count. In both cases the thermalized neutrons diffuse toward and away from the detector. Thus, fewer than half of the neutrons actually thermalized may come into contact with the detector so that detector sensitivity is important. Two types of neutron detectors are in common use in neutron moisture gauges. We now discuss each of these in turn.

7.3.3 Neutron Detectors

The detection reaction for what is called a $_2He^3$ detector is

$$_2He^3 + _0n^1 \rightarrow _1H^1 + _1H^3 + Q \tag{7.78}$$

where the positively charged $_1H^1$ and $_1H^3$ ions are accelerated toward the wall of the tube and detected as pulses. The detector used in this case is a Geiger–Muller (GM) tube filled with He^3 and operated at such a voltage for its dimensions as to be used as a proportional counter. A cross section of a GM tube is shown in Figure 7.16. The voltage applied to a GM tube, relative to its dimensions, determines the mode in which the tube will operate, as shown in Figure 7.17. An alternative neutron detection reaction is

$$_5B^{10} + _0n^1 \rightarrow _2He^4 + _3Li^7 + Q \tag{7.79a}$$

and the instrument in this case is called a BF_3 counter. Both the alpha particle, the $_2He^4$, and the lithium are detected. The GM tube, using BF_3 gas in this case, can be operated in the proportional counter region for this means of neutron detection. Elastic scattering is not the only fate befalling neutrons leaving the radioactive source. Some of the neutrons can undergo nuclear reactions with the aluminum access tubing and with some of the elements present in the earth materials as well. Some of the possible reactions of neutrons with the access tubing are

$$_0n^1 + _{13}Al^{27} \rightarrow _{11}Na^{24} + _2He^9$$

$$_0n^1 + _{13}Al^{27} \rightarrow _{13}Al^{28} + \gamma$$

where

$$_{13}Al^{28} \rightarrow _{14}Si^{28} + _{-1}e^0$$

where $_{-1}e^0$ is an electron. Also,

$$_0n^1 + _1H^1 \rightarrow [_1H^3]^* \rightarrow _1H^3 + \gamma \tag{7.79b}$$

which is the formation of an isomer of $_1H^3$, denoted by $[_1H^3]^*$, followed by a transition to the ground state via gamma emission, is possible as well.

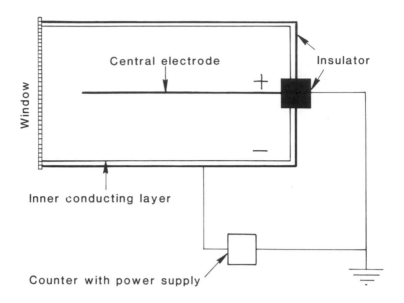

Figure 7.16: Cross-section of a $G - M$ tube used as a proportional counter for detecting slow neutrons. The $_2He^3$ is a gas inside the tube at a pressure between one and ten atmospheres.

well. The probability that this nuclear reaction will occur increases with the degree of soil saturation and competes with the thermalization process discussed. Thus, the relation between neutron gauge counts of thermal neutrons and degree of soil saturation is nonlinear, and underscores the need for calibration of each gauge in the soil and range of saturation in which it is to be used.

7.4   BULK DENSITY OF SOILS

7.4.1   General Discussion

The bulk density of soils can be determined in the laboratory using the vacuum oven technique with soil cores as discussed. For reconstituted soil, as is occasionally encountered in the use of the pressure plate apparatus, a measurement of bulk density can be made as well, after the reconstituted sample is dried. In the field, an available technique is the gamma–gamma $(\gamma - \gamma)$ density probe, which we now discuss.

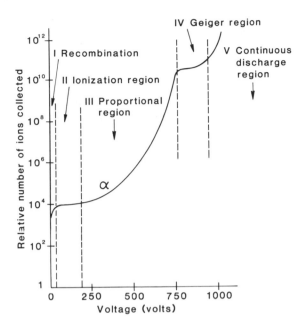

Figure 7.17: Response characteristics of a Geiger–Muller tube as a function of applied voltage relative to the dimensions of the tube.

7.4.2 Gamma–Gamma Density Probe

A γ – γ density probe is shown schematically in Figure 7.18. The nuclear source, which can be $_{55}Cs^{137}$, is the parent of a gamma source according to the reactions

$$_{55}Cs^{127} \rightarrow [_{56}Ba^{137}] + _{-1}e^0 \tag{7.80}$$

and

$$[_{56}Ba^{137}] \rightarrow {}_{56}Ba^{137} + \gamma \tag{7.81}$$

where the γ has an energy of about 0.66 MeV. The probe is calibrated to measure the combined density of the soil and the water the soil pores contain by detecting a fraction of the gammas emitted by the source that have been scattered by the electrons of the atoms making up the soil and the waters in it. Since the density of the electrons in the volume of soil being sampled affects the scattering, higher density means more scattering, and since in general more dense materials have higher values of Z and hence more electrons per atom, the degree to which gamma rays are scattered is a measure of the density of the material in an approximately spherical volume of radius 5 cm around the location of the nuclear source. There are

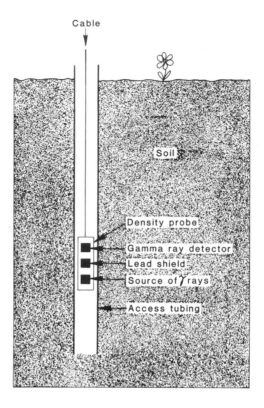

Figure 7.18: A $\gamma - \gamma$ density probe showing the source, detector, aluminum access tube, and surrounding soil.

three physical mechanisms by which gamma rays are scattered by electrons. These are (1) the photoelectric effect, (2) Compton scattering, and (3) pair production. The photoelectric effect consists of the ejection of an orbital electron from an atom, which may be part of a molecule (constituent of a mineral) with the consequent loss by the gamma-ray photon of the kinetic energy imparted to the ejected electron plus its binding energy. The photoelectric effect is the dominant energy loss mechanism at lower gamma-ray energies–up to about 0.05 MeV for aluminum and up to about 0.5 MeV for lead. At higher gamma-ray energies, in the range of 0.05 to 15 MeV for aluminum and 0.5 to 5 MeV for lead, for example, Compton scattering is the dominant gamma-ray energy loss mechanism. In Compton scattering a gamma-ray photon interacts with an electron, loses energy, and thus increases in wavelength according to

$$\lambda - \lambda_0 = \frac{h}{m_0 c}[1 - \cos(\beta)] \qquad (7.82)$$

where λ_0 is the wavelength of the original gamma-ray photon and is given by

$$\lambda_0 = \frac{hc}{E_{\lambda_0}} \tag{7.83}$$

where E_{λ_0} is the energy of the unscattered photon, possibly in MeV, h is Planck's constant, equal to 6.62×10^{-34} J sec $= 1.602 \times 10^{-13}$ MeV sec, c is the speed of light in vacuum, m_0 is the rest mass of the electron, equal to 9.11×10^{-31} kg, and β is the angle through which the incident photon is scattered measured relative to its initial direction of motion. Equation (7.82) can be rearranged to yield for the energy of a scattered photon, E_λ,

$$E_\lambda = \frac{E_{\lambda 0}}{1 + \frac{E_{\lambda 0}}{E_0}[1 - \cos(\beta)]} \tag{7.84}$$

where $E_0 = m_o c^2$ is the rest energy of the electron. The quantity $h/m_o c$ is called the *Compton wavelength*, λ_c, and for an electron is about 2.426×10^{-8} m. The kinetic energy imparted to the scattered electron by this process, E_k, is

$$E_k = E_{\lambda_0} - E_\lambda \tag{7.85}$$

A measure of the probability that a given gamma-ray photon will undergo a single Compton scattering is the cross section per electron σ_e, given by (Kaplan, 1955, p. 332)

$$\sigma_e = 2\pi \left(\frac{e^2}{E_0}\right)^2 \left[\frac{1 + e_r}{e_r^2}\left(\frac{2(1 + e_r)}{1 + 2e_r} - \frac{1}{e_r}\ln(1 + 2e_r)\right)\right.$$

$$\left. + \frac{1}{2e_r}\ln(1 + 2e_r) - \frac{1 + 3e_r}{(1 + 2e_r)^2}\right] \tag{7.86}$$

where $e_r = E_{\lambda 0}/E_0$, and e is the charge on an electron. The absorption coefficient for Compton scattering, μ_{ci}, is given for a homogeneous isotropic medium of atoms of species i having atomic number Z_i, mass number A_i, and mass density ρ_i by

$$\mu_{ci} = \frac{\rho_i L Z_i}{A_i}\sigma_e \tag{7.87}$$

where L is Avogadro's number. For an absorber of a geometry such that rescattering of gamma rays back into the beam can be assumed not to occur, the total absorption coefficient for Compton scattering, μ_{cT}, may be given by

$$\mu_{cT} = \sum_{i=1}^{N} f_i \mu_{ci} \tag{7.88}$$

where f_i is the fraction of the atoms of species i in the path of the beam of gamma rays. The decrease in intensity of the beam is then given by

$$I = I_0 e^{-\mu_{cr} x} \tag{7.89}$$

where I_0 is the initial intensity of the beam and I is its intensity after it has traversed a thickness x of absorber. If a body of absorbing material is sufficiently thick, however, gamma rays initially removed from the beam may be returned to the beam by further scattering. This effect is accounted for in calculation by introducing a factor, b_f, the so-called build-up factor as a multiplier of I_0 in (7.89).

7.4.3   Combined Density–Saturation Gauge

It is possible, by the proper choice of nuclear source and combination of detectors, to construct an instrument capable of measuring, again given suitable prior calibration, both bulk density and soil moisture content. This is accomplished by incorporating both gamma-ray and thermal neutron detectors into the instrument and either utilizing a nuclear source that produces sufficient gamma rays as well as neutrons, or two separate sources in the same probe. An example of the former is the older radium–beryllium neutron source, which produces gamma rays and neutrons, according to (7.45), while the alpha particles are produced by the decay

$$_{88}\text{Ra}^{226} \rightarrow \,_{86}\text{Rn}^{222} + \,_2\text{He}^4 + \gamma \tag{7.90}$$

The intense gamma radiation from a Ra-Be source requires care in shielding and radiation protection, as do the neutrons and gamma rays emitted as products of the reactions given, and it is this subject to which we now turn.

7.5   ASPECTS OF RADIATION SAFETY

The safe handling of radioactive materials, even those contained in permanently sealed capsules, so-called sealed sources, as used in the instruments discussed, requires some knowledge of permissible human exposure to radiation, the so-called body burden, radiation levels normally associated with a given source, shielded and unshielded, periodic tests for leaking sources, and routine radiation monitoring. We begin with customary terminology. A rad, or "radiation absorbed dose" is an absorbed energy of 100 ergs per gram of tissue. This definition is independent of the type of radiation incident on the tissue, but the biological effect is not. The relative biological effect (RBE) of radiation in damaging tissue depends the type of radiation and its energy–kinetic energy in the case of particles, and photon energy

Table 7.1: QF for various radiations

Radiation	Quality Factor (QF)
Gamma rays	1.0
X rays	1.0
Beta particles having $E_k > 0.03$ MeV	1.0
Beta particles having $E_k < 0.03$ MeV	1.7
Thermal neutrons	3.0
Fast neutrons	10.0
Protons	10.0
Alpha particles	20.0
Heavy ions	20.0

in the case of gamma rays. The maximum value of the relative biological effectiveness for a given type of radiation is called its quality factor (QF) and, like RBE, is a dimensionless number. The QF values for the various radiations of interest are given in Table 7.1.

The dose of x rays in air required to ionize it to the level of 2.58×10^{-4} coulombs of ionic charge per kilogram of dry air is called the Roentgen and is abbreviated R. For historical reasons, the dose of ionizing radiation to organisms is measured in units of "Roentgen-equivalent-man" or rem, even though radiations other than x rays are included in the definition. Hence, the dose is called the *dose equivalent*, abbreviated DE, or dose of other radiations equivalent to the given dose of x rays, and is given by

$$DE = D \times QF \qquad (7.91)$$

where DE is the dose equivalent in rems, D, dose, is the energy deposited in the tissue in rads, and QF is the quality factor. For one internal (to the body) radiation dose, perhaps from ingesting water containing radioactive contamination, or breathing contaminated dust, QF is multiplied by DF, the "distribution function," to account for the non uniform distribution of a radionuclide in the tissues and organs of the body. In this connection we note that alpha and beta emitters are more serious internal (to the body) radiation hazards than external (to the body) hazards because their relatively short range results in highly localized deposition of energy in tissue. Neutrons and gamma rays, on the other hand, are more serious external rather than internal hazards due to their high penetrating power - i.e the neutrons or gamma rays from an ingested neutron or gamma source tend, relative to alpha and beta particles, to leave the body rather than cause ionization. The maximum permissible exposure limits for radiation

Table 7.2: Maximum exposures for radiation workers

Body part	13 week exposure limit (millirems)
Whole body, eyes, gonads, skull	1250
Kidneys, spleen, lungs, liver	5000
Skin of whole body	7500
Hands, arms, feet, ankles	18,750

workers are given in Table 7.2. A complete listing of maximum permissible body burdens by organ, and type and energy of radiation, is given in "Maximum Permissible Body Burdens and Maximum Permissible Concentrations of Radionuclides in Air and in Water for Occupational Exposure," U.S. Department of Commerce, National Bureau of Standards, Handbook 69, August, 1963.

The estimation of a possible dose of radiation depends on the circumstances of the particular situation being considered, and some simplifying assumptions are probably unavoidable. In order to illustrate some of the concepts involved we now consider the potential exposure from an unshielded sealed 10 mCi (10 millicurie) AmBe neutron source held in an operator's hand–something never to be done in an actual field or laboratory situation except in the most extreme circumstances (the designation of 10 mCi refers to the activity of the alpha emitter that is part of the neutron source and not to the number of neutrons emitted per second). We assume spatially isotropic neutron emission from the source of 2.5×10^4 neutrons/sec with an average kinetic energy of 4.5 MeV per neutron. We further assume the source to be cylindrical of length 5 cm and diameter 5 cm. The assumption that an average neutron energy can be used in calculations is a significant oversimplification, since the energies of the emitted neutrons actually depend in part on the energies of the alpha particles emitted from the americium, or other alpha particle source, and range from thermal (0.025 eV) to about 10 MeV. We assume further that the operator's hand fully encloses the cylinder and that the flesh surrounding the cylinder is about 4.5 cm thick and made of pure water–again considerably oversimplifying the situation. The fast diffusion length is the average straight line distance traveled by an initially fast neutron before it is thermalized by elastic collisions with surrounding nuclei. For water, the fast diffusion length is 5.75 cm which would say that no neutrons are thermalized by the operator's hand–an oversimplification resulting from neglecting the energy spread–and since the absorption cross sections of hydrogen and oxygen for fast neutrons are quite low, the calculation for the hand ends at this point. We now assume that the source is held at a distance of 1.75 ft

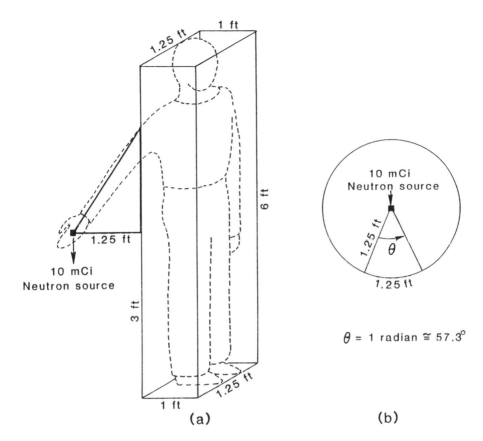

Figure 7.19: (a) An operator approximated as being 6 ft tall, 1 ft thick, and 1.5 ft wide, holding (b) a neutron source 1.75 ft away from his body at a height of 3 ft above the ground.

from the center of the body of the operator, as shown in Figure 7.19(a) with the idealized dimensions given there and continue the calculation. In top view the source is held vertically at the center of a circle of radius 1.75 ft with the operator subtending an angle of approximately 41°. We now assume that the distance of the operator from the source and the area of the operator are such that the fraction of the neutrons emitted from the source that strike the operator is about 0.11, which is the ratio of 41° to 360°, as shown in Figure 7.19. Since the thickness of the operator's body is assumed to be greater than the fast diffusion length of neutrons in water, we assume that all the neutrons incident on the operator's body are thermalized. This is a flux of 2.7×10^3 neutrons/sec representing an average energy deposition per unit time, upon thermalization, in the operator's body of about 1.95×10^{-2} erg/sec (1.602×10^{-6} erg $=1$ MeV). If we assume that the body of the operator is composed entirely of water and that the thermal neutrons are uniformly distributed throughout it, then the mass of the operator would be 2.8×10^4 g so that the dose rate in rads per second throughout the operator's body would be

$$\frac{1 \text{ rad} \times 1.95 \times 10^{-2} \text{ erg/sec}}{\frac{100 \text{ ergs}}{\text{g}} \times 2.8 \times 10^4 \text{ g}} = 6.95 \times 10^{-9} \text{ rad/sec}$$

The equivalent dose rate, DR, is then

$$DR = \frac{6.95 \times 10^{-9} \text{ rad}}{\text{sec}} \times \frac{10 \text{ rem}}{\text{rad}} = 0.25 \frac{\text{mr}}{\text{h}}$$

where the quality factor of 10 for fast neutrons has been used and mr stands for millirem. By way of comparison we note that an average rate of occupational exposure of 1250 mr/13 weeks is the equivalent of an exposure of 0.573 mr/hr, assuming a constant rate of exposure for 24 hours a day, 7 days per week for 13 weeks. We see that the estimated rate of exposure from the energy loss due to thermalization of the neutrons is about half the allowed rate of occupational exposure using the oversimplifying assumptions given. In addition, the neutron capture mechanism of (7.79b) which produces deuterium and 2.26 MeV gamma rays, making the operator's body a source of gamma rays, operates in this case because the thermal diffusion length for neutrons, which is the thickness of material required to attenuate a beam of thermal neutrons by a factor of $1/e$, where e is the base of natural logarithms, is 2.88 cm for water. We thus assume that the entire thermal neutron flux is available for capture reactions and that the microscopic cross section, σ_n, for neutron capture is 0.33 barns, where 1 barn $= 10^{-24}$ cm^2. The associated macroscopic cross section, σ_m, is

$$\sigma_m = \rho_n \sigma_n \tag{7.92}$$

where ρ_n is the number of atoms of the target material per unit volume given by (7.75). The fraction of thermal neutrons producing gamma rays via the capture reaction discussed, is

$$\frac{I}{I_0} = e^{-\sigma_m d} \tag{7.93}$$

where d is the thickness of the absorber, taken to be one half foot (15.24 cm) in this case (compare Figure 7.19), while the appropriate value of $\rho_n \sigma_n$ is estimated by computing ρ_n for water using (7.75) and then multiplying by σ_n

and is $\sigma_m = 5.5 \times 10^{-3}\,\mathrm{cm}^{-1}$ so that $I/I_0 \simeq 0.92$. This means that the flux of generated gamma rays can be estimated to be $I \simeq 2.5 \times 10^3$ gammas/sec. Multiplying I by the average energy of each gamma ray, further multiplying by 1 rad divided by 100 ergs/g multiplied by the mass of the operator yields the maximum potential exposure in rad/sec (this quantity, while in the same units as absorbed dose rate, is in reality a maximum exposure rather than an actual dose rate, since some of the gamma rays may escape with little interaction) which is 3.2×10^{-9} rad/sec which corresponds to a dose rate of 1.2×10^{-2} mr/hr for a total assumed dose/exposure of 0.26 mr/hr. Moving the source closer to the body changes this picture since

$$I_1 R_1^2 = I_2 R_2^2 \tag{7.94}$$

in which, if I_1 is the intensity of radiation a distance R_1 from a source of radiation, then I_2 is the intensity of the radiation at a point is a distance R_2 from the source of the radiation. This is the so-called *inverse square law* and is derived by considering the area on the surface of a sphere subtending a constant solid angle with its vertex and the source at the center as its radius expands from R_1 to R_2. If the source is moved from a distance of 1.75 ft from the operator's body to a distance of 0.58 ft, the intensities discussed are increased by a factor of 9, which, when multiplied by 0.26 mr/h yields 2.34 mr/h or about half the legal whole body exposure rate for radiation workers. The entire scenario given has been oversimplified and is given to illustrate concepts and calculations only. It is seen, however, that if a given neutron/gamma gauge is used in such a way that the source is shielded either in the body of the instrument or in an access tube in the soil except for periodic–generally every 6 months–wipe testing to check the sealed source for leaks, which procedure will be discussed shortly, such use is relatively safe as far as maximum permissible whole-body exposure is concerned.

The current version of the accepted formula for the maximum permissible lifetime exposure for a nonradiation worker, DE_m, is

$$DE_m = 0.5(y - 18)\ \mathrm{rem/yr} \tag{7.95}$$

where y is an individual's age in years greater than 18 and the exposure limit for nonradiation workers of 0.5 rem per year has been taken into account. This formula and the previous discussion assume that low levels of exposure are safe. Individuals vary greatly in sensitivity to radiation. Pregnant women are assumed to have much lower tolerance limits to exposure than those given, for example, and in fact if the genetic code is altered in the nuclei of reproducing cells by single-particle collisions, it is seen that there is no necessarily minimum safe (threshold) dose of radioactivity. Further, "safe" doses, even though conservative in nature in the

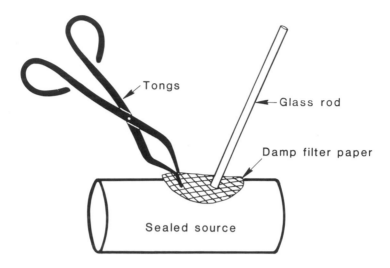

Figure 7.20: A sealed source is wiped with damp filter paper in such a way that the operator does not touch either the filter paper or the source directly.

context in which they arise, are based in part on immediately observable somatic effects when, in fact, the various effects may be delayed in their appearance, sometimes for years following the exposure. This is so because even if exposure to radiation has ceased, the cellular effects of the dose, if any, remain behind and lead to later, more easily observed damage.

Wipe tests on sealed sources are conducted as follows. In Figure 7.20 is shown a sealed source, depicted as a cylinder, with a filter paper that has been dampened with distilled water, held with tongs, and carefully pushed over the surface of the source with a disposable glass rod. The common use of a wooden dowel rod to move damp filter paper over the surface of a sealed source is susceptible to absorption of contaminated moisture from the filter paper into the wood of the dowel, and hence this practice is not recommended. Further, surgical gloves and a film badge should be worn as well. If, after analysis, the filter paper is found to contain between 0.001 and 0.005 μCi (1 μ Ci = 10^{-6} curies) of activity, a further wipe test should be conducted, and if an activity in that range is observed a second time, the gauge should be removed from service and sent to its manufacturer for repair or replacement of the source. If an activity of 0.005 μCi or above is observed, the gauge should be removed from service immediately and sent for repair or replacement of the source.

The need for shielding intense, high-energy beta sources does not generally arise in applications of radioactivity to earth science measurements. However, it should be noted that lead shields for beta particles can lead to the production of x rays, called bremsstrahlung, by a process of energy loss in which high-energy beta particles interact with the electric field of the nucleus of an atom and undergo a change in direction and a consequent loss of energy. This energy appears as bremsstrahlung [braking (slowing) rays]. The fraction f_β of the initial energy of a beam or other source of beta particles converted to bremsstrahlung may be estimated via

$$f_\beta = 3.5 \times 10^{-4} Z E_m \tag{7.96}$$

where Z is the atomic number of the absorbing material and E_m is the maximum energy of a given beta particle in the beam. Since the efficiency of x ray production increases linearly with Z, lower-Z shielding material for beta particles is preferable. In practice, aluminum, for which $Z = 13$, is the highest-Z material generally useful for beta shielding purposes. The use of mild steel 55-gallon drums for shielding high-energy beta particle emitters can be questioned on these grounds, since Z for iron is 26, thus doubling the bremsstrahlung production over that due to aluminum, according to (7.96), for a given source of beta particles having sufficiently high energies. In a health physics calculation using (7.96), a conservative dose estimate would be made by computing f_β and then assuming that each gamma-ray photon had the same energy as did each beta particle prior to scattering.

The shielding of gamma-ray sources can be accomplished, on the other hand, by using high-Z materials, such as lead, iron, or even concrete. Figure 7.21 shows the factors by which gamma rays are attenuated by lead, iron, and concrete. The buildup factor, b_f, due to initially scattered gamma rays, which are rescattered in such a way as to contribute energy to the exiting flux of radiation, is not taken into account. The curves of Figure 7.21 are calculated from $I/I_0 = e^{-\mu_l x}$ where μ_l is the linear attenuation coefficient for a given material and gamma ray photon energy and x is the thickness of the shielding material.

Shielding against fast neutrons requires first a layer of hydrogenous material thicker than the fast diffusion length characteristic of it to thermalize the neutrons, then a strong absorber of thermal neutrons such as cadmium (see Table 7.3 for the absorption cross sections for thermal neutrons for

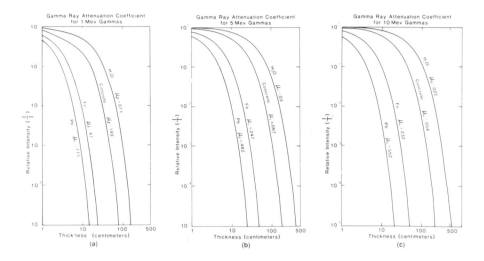

Figure 7.21: Attenuation factor $\frac{I}{I_o}$-fractional transmission of incident intensity-of gamma rays of selected energies for lead, iron, and concrete.

selected materials), and finally a lead or other high-Z shield for the gamma rays generated by the neutron capture reactions in the hydrogenous material and the strong neutron absorber.

The computation of the radiation exposure of the hypothetical gauge operator could have been refined by using the data of Table 7.3 and the total macroscopic cross section, σ_T, computed for the elements making up the standard man via

$$\sigma_T = L \sum_{i=1}^{n} N_i \sigma_i \qquad (7.97)$$

where L is Avogadro's number, N_i is the number of atoms of species i per unit volume, and σ_i is the microscopic absorption cross section for thermal neutrons for atoms of species i. The gamma production due to the neutron flux from a moisture gauge can be very roughly estimated for a given spherical volume of earth surrounding the source when it is in its access tube using the relative abundance data of Table 7.3 along with (7.96)

Table 7.3: Selected elements in the standard man and the earth's crust.

Element (symbol)	Standard man	Earth's crust	σ_i
Hydrogen (H)	0.6282	0.6143	0.33
Oxygen (O)	0.2572		0.0002
Carbon (C)	0.0949		0.0034
Nitrogen (N)	0.01356		1.90
Calcium (Ca)	0.00237	0.0277	0.43
Phosphorus	0.00204		0.19
Sulfur (S)	0.000494		0.51
Sodium (Na)	0.000413	0.0217	0.53
Potassium (K)	0.000324		2.10
Chlorine (Cl)	0.000268		33.0
Magnesium (Mg)	0.000130	0.0251	0.063
Iron (Fe)	6.48×10^{-6}	0.0226	2.53
Copper(Cu)	1.42×10^{-7}		
Manganese (Mn)	3.29×10^{-8}		13.30
Iodine (I)	2.14×10^{-8}		13.30
Silicon (Si)		0.2106	7.16

and

$$I = I_0 e^{-\sigma_T r} \qquad (7.98)$$

where r is the radius of the given spherical volume surrounding the source. In Table 7.3 is given the relative abundance in the earth's crust of selected elements, the fraction of the "Standard Man" that each element listed comprises, and the microscopic absorption cross-section σ_i in barns for thermal neutrons for certain of the elements.

7.6 CHARACTERIZATION OF PORE STRUCTURE

7.6.1 General Discussion

The experimental study of the pore structure of porous media has been approached in various ways over the years. The three methods that have been used in most recent times are (1) mercury injection porosimetry, which requires an assumed geometrical model for its interpretation as will be discussed shortly; (2) the use of thin sections with a petrographic microscope; and (3) scanning electron microscopy, which in a sense, is a variation of (2). While scanning electron microscopy is of value in quantifying pore opening shapes to some degree, neither it nor the other two techniques just mentioned have so far proven capable of characterizing the interior pore structure of a porous medium sample though repeated shaving of porous medium samples combined with scanning electron microscope observations has been suggested as a means of directly characterizing the interior structure of competent, in the geologic sense, initially intact porous medium samples.

A well-established technique with other, prior applications, is available however, and as an emerging method for the laboratory characterization of the actual pore structure in a porous medium sample, the characterization of the location(s) of fluid in a sample of porous medium as a function of degree of saturation of the medium, the direct observation of vapor diffusion within the sample as it occurs, and the characterization of the chemical environment of species adsorbed of the interior pore walls shows considerable promise. This technique is called *nuclear magnetic resonance* and will now be briefly discussed.

7.6.2 Nuclear Magnetic Resonance

Neutrons and protons possess an intrinsic magnetic moment. This is a quantum-mechanical attribute and has no analog in classical physics (Derivations of the various quantum-mechanical results we shall state are beyond the scope of this work and so will not be given here). We recall as a point of reference from classical physics, however, that the magnitude of the magnetic moment, μ_m, generated by a charge q moving in a circular orbit of radius r_c and area πr_c^2 is $\mu_m = q\pi r_c^2/T_q$ where T_q is the period of the circular orbit of the charged particle. Neutrons and protons also possess intrinsic angular momenta, and we recall as another point of reference from classical physics that the angular momentum, $\vec{L}$, of a particle of rest mass m_0 moving in an orbit characterized by radius vector $\vec{r}$ with velocity $\vec{v}$ is given in vector notation by $\vec{L} = m_0(\vec{r} \times \vec{v})$. The angular momentum of protons and neutrons is an intrinsic property of the particles themselves, independent of their motion through space. Electrons also have an intrinsic angular momentum and intrinsic magnetic moment as well. In the case of water the spin angular momenta (spins) of the electrons cancel each other so that the spins and magnetic moments remaining are those due to the protons comprising the nuclei of the hydrogen atoms in the water molecules. The projection of the angular momentum of a proton along a particular direction, the z direction, for example, can take on only certain discrete values and is thus said to be quantized. This is also a result from quantum mechanics and has no classical analog. We characterize the z component of angular momentum by the angular momentum quantum number, m_z. The z component of angular momentum, S_z, is called spin angular momentum,

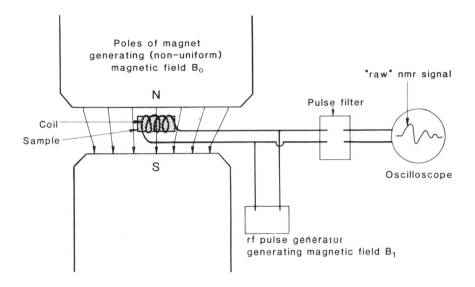

Figure 7.22: A sample of partly saturated porous medium in an aluminum sample holder, which is contained in a coil of wire connected to a radio-frequency pulse generator and a detector, depicted here as an oscilloscope, for detecting the nmr signal.

or spin, for short, and is given by

$$S_z = m_z \hbar \qquad (7.99a)$$

where $\hbar = h/2\pi$ and h is Planck's constant. The magnetic moment of a proton is

$$\mu_p = \frac{g_p e \hbar m_z}{2 m_p} \qquad (7.99b)$$

where m_p is the mass of a proton ($m_p \simeq 1.673 \times 10^{-27}$ kg), g_p is called the gyromagnetic ratio which for a proton is 2×2.79, and e is the charge on a proton ($e \simeq 1.6 \times 10^{-19}$ C). It can be shown that the two possible values of m_z for a proton are $\pm\frac{1}{2}$. We now consider water in the pores of a sample of porous medium, as shown in the simplified experimental setup depicted in Figure 7.22. The nmr signal from the water in the sample is generated as follows. The constant magnetic field B_0 in the direction depicted by the downward arrows in Figure 7.22 exerts a torque, $\vec{\tau}_p$, on the magnetic moments of the protons and is given for an individual proton by

$$\vec{\tau}_p = \vec{\mu}_p \times \vec{B} \tag{7.100}$$

The tendency is for the magnetic moments of the protons to become aligned parallel to the direction of $\vec{B}$. This tendency is opposed, however, by the angular momentum of each proton (A point of reference is the tendency for the angular momentum of a spinning gyroscope wheel to oppose a change in the angle between its axis and the local direction of the gravitational field, and instead precess around an axis in the direction of the external field). In the case of a proton, the spin angular momentum vector of each proton precesses around the direction of $\vec{B}$. The precession frequency, f_p, is given by

$$f_p = \frac{g_p}{2\pi}\left(\frac{e}{2m_p}\right)B_0 \tag{7.101}$$

where f_p is in Hertz, abbreviated Hz, ($1\ \text{Hz} = 1\ \text{sec}^{-1}$), B_0 is the magnitude of $\vec{B}_0$ in Teslas ($1\ \text{T} = \frac{1\,\text{N}}{\text{A}\,\text{m}} = \frac{1\,\text{W}}{\text{m}^2} = 10^4\ \text{G}$), and the rest of the symbols are as previously defined. For $B_0 = 1\ \text{T}$, $f_p \simeq 4.25 \times 10^7\ \text{Hz} = 42.5\ \text{kHz}$. The application of a second magnetic field, $\vec{B}_1$, at a right angle to $\vec{B}_0$ and generated by a radio-frequency pulse of frequency f_p applied to the coil surrounding the sample holder will generate an additional torque acting on the magnetic moments of the protons and cause the angles the magnetic moments of the protons make with the direction of $\vec{B}_0$ to change even further and in fact to precess around the axis formed by the originally precessing magnetic moment vector, so that the net trajectory of the magnetic moment vector is the sum of two precessions.

A field, $\vec{B}_1$, of sufficient strength to cause the directions of the magnetic moments to make an angle of 90° with the direction of $\vec{B}_0$ is called a 90° field, while a field, $\vec{B}_1$, of sufficient strength to flip the spins completely so that they make an angle of 180° with $\vec{B}_0$ is called a 180° field. When the field $\vec{B}_1$ is removed, the directions of the magnetic moments of the protons tend to return to those prevalent when only the field $\vec{B}_0$ was present. This rotation of magnetic moments past the turns of wire in the coil generates a current in it and this time-varying current is the nmr signal.

Since the chemical and nuclear (nuclear spin–spin interactions) (electronic shielding from the effects of $\vec{B}_0$) environments of the water molecules in the porous medium affect the rate of "relaxation" of the directions of the moments, and concomitantly the spins of the protons toward their pre-$\vec{B}_1$ directions, as well as the exact frequency f_p, at which resonance occurs, the "raw" nmr signal consists of a damped sine wave of varying frequency. In practice, the radio-frequency (rf) pulse generating $\vec{B}_1$ is varied to "sweep through" the various values of f_p of possible interest. Pulses of varying duration ("length") and amplitude (pulse "height") are made up of components of different amplitudes and frequencies and thus pulses of varying lengths can affect protons precessing at a range of frequencies. Also, the infinite Fourier transform of the raw data is generally taken so that the various resonance frequencies are displayed directly as signal amplitude versus signal frequency plots (frequency domain) rather than as a plot of signal amplitude versus time (time domain). In the emerging applications of this technique to the characterization of the bulk pore structure of porous media, the spatial locations of agglomerations of water molecules in the pores are "mapped" by using a field $\vec{B}_0$ that is nonuniform with vertical distance through the sample, thus requiring the use of applied rf signals, $\vec{B}_1$, of concomitantly varying values of f_p vertical position in the sample and yielding a microscopic view of the highly irregular three-dimensional pore structure of the sample. If redistribution of water in partially saturated samples is allowed, the successive equilibrium locations of water in the sample will be those at varying average degrees of saturation and various equilibrium temperatures. It is also possible, as mentioned, to follow the diffusion of water vapor, and other vapors as well, depending on their nmr response, directly through the interior of a porous medium sample. This gives the possibility of studying molecular sieving (the relative rates of transport of different molecular species through a porous medium) from a fundamental point of view and hence explaining the results of laboratory experiments characterizing macroscopic diffusion of gases through porous medium samples. The resolution of an nmr apparatus as here described is about 10^{-2} cm. Iron in the soil matrix can interfere with the nmr response to water in the soil pores, and its possible presence must be considered in the application of this technique. We now discuss a final point, namely the energy state of the protons in the sense of thermodynamics. The oscillating shape of the "raw" data wave form represents spin energy being added to the field of the rf coil when a group of proton spins flips "up" and energy taken from the field when a group of proton spins flips "down" to the lower of the two possible quantized energy states. The ratio of the number of protons per unit sample volume in the upper energy state, N_+, in which the tendency of the proton spins is to be aligned antiparallel to $\vec{B}_0$, to the number of protons per unit sample volume in the lower energy state, N_-, in which the tendency of the proton spins is to be aligned parallel to $\vec{B}_0$, is given by

$$\frac{N_+}{N_-} = e^{-2\mu_p B_o / kT} \tag{7.102}$$

where k is Boltzmann's constant, T is the absolute temperature, and the other symbols are as previously defined. This relation means that as the protons try to come to thermal equilibrium by exchanging energy with the electron spins (in a crystal this would be termed an exchange of energy with the crystal lattice), the tendency of the thermal vibrations is that $N_- > N_+$. We further see that at higher water temperature, a larger magnetic field is required to make nmr observations, and from (7.101) we see that a higher-frequency rf pulse is concomitantly required.

7.7   UNSATURATED HYDRAULIC CONDUCTIVITY

A number of methods for measuring unsaturated hydraulic conductivity have been suggested in the literature. The method suggested here does not depend on a solution of the flow equation for a given set of conditions, which would be the normal procedure for developing an equation set for use with field or laboratory data as is done to arrive at the various pumping test methods used for saturated flow in confined aquifers. It does, however, rely on some particular assumptions, as follows. We consider the one-dimensional horizontal flow situation depicted in Figure 7.23. The left end of the sample is assumed to be kept saturated ($^{(1)}\theta = 1$), while the right end is assumed to be at the residual degree of saturation, $^{(1)}\theta = {}^{(1)}\theta_0$.

We assume a constant horizontal flux that continues until an increase in the degree of soil saturation at the right-hand end is noted, at which time the experiment is considered completed. We further assume that $\partial\, ^{(1)}\Psi / \partial\, ^{(1)}\theta$ for the main wetting branch of the moisture characteristic is known so that the thermocouple psychrometer measurements of $^{(1)}\Psi$ along the length of the tube yield values of $^{(1)}\theta$ as a function of distance x along the tube. Beginning with the expression for the x component of Darcy velocity for unsaturated flow

$$^{(1)}J_x = -\,^{(1)}K(\,^{(1)}\theta)\frac{\partial}{\partial x}\left(\frac{^{(1)}\Psi}{^{(1)}\rho g} + z\right) \tag{7.103}$$

where all the symbols are as previously defined, and assuming a particular functional form for $^{(1)}K(\,^{(1)}\theta)$, namely

$$^{(1)}K(\,^{(1)}\theta) = \,^{(1)}K_s e^{-(1-\,^{(1)}\theta)\beta} \tag{7.104}$$

where $0.2 < \,^{(1)}\theta \le 1$, and $^{(1)}K_s$ is the saturated hydraulic conductivity, assumed to have been measured in a separate experiment, we find for the quantities $\partial\, ^{(1)}\Psi / \partial x$ and $\partial\left(\partial\, ^{(1)}\Psi / \partial x\right) / \partial x$, after some rearrangement,

$$\frac{\partial\, ^{(1)}\Psi}{\partial x} = -\frac{^{(1)}J_x\,^{(1)}\rho g}{^{(1)}K(\,^{(1)}\theta)} \tag{7.105a}$$

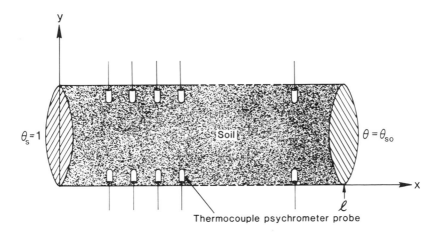

Figure 7.23: A horizontal tube of circular cross section and radius small compared with its length having thermocouple psychrometer probes embedded along its length in the soil sample it contains.

$$\frac{\partial\left(\frac{\partial\,^{(1)}\Psi}{\partial x}\right)}{\partial x} = \frac{^{(1)}J_x\,^{(1)}\rho g}{^{(1)}K(\,^{(1)}\theta)}\beta\frac{\partial\,^{(1)}\theta}{\partial x} \tag{7.105b}$$

The quantities $\partial^{(1)}\Psi/\partial x$, $\partial^2\,^{(1)}\Psi/\partial x^2$, and $\partial^{(1)}\theta_v/\partial r$, which last is negative since $^{(1)}\theta$ decreases as x increases in this case, are assumed to be known from analysis of the psychrometer data. Thus, taking the ratio of (7.105a) and (7.105b) evaluated at a particular place along the tube, x_0, yields,

$$\frac{\left.\frac{\partial^2\,^{(1)}\Psi}{\partial x^2}\right|_{x=x_0}}{\left.\frac{\partial\,^{(1)}\Psi}{\partial x}\right|_{x=x_0}} = -\beta\left.\frac{\partial\,^{(1)}\theta}{\partial x}\right|_{x=x_0} \tag{7.106}$$

Equation (7.106) is now solved for β, which is then substituted into (7.104) to complete the measurement. It is assumed in this development that β is a constant for the range of values of $^{(1)}\theta$ in the experiment. This assumption may not always be correct for the full range of $^{(1)}\theta$ and thus a number of measurements of β at different ranges of $^{(1)}\theta$, effectively yielding $\beta(\,^{(1)}\theta)$, may be required. The permeability of a porous medium, K_p, is related to its saturated conductivity when filled with medium i, $^{(i)}K_s$, by

$$K_p = \frac{^{(i)}\eta}{^{(i)}\rho g} \, ^{(i)}K_s \qquad (7.107)$$

where $^{(i)}\eta$ is the viscosity of the fluid filling the pores, $^{(i)}\rho$ is its density, and g is the acceleration due to gravity. In principle, the permeability, while it is actually measured by measuring K_s, is a property of the porous medium independent of the fluid it contains.

7.8 MERCURY INJECTION POROSIMETRY

We now consider a geometrical property of a porous medium, namely the distribution of the volumes of its pores parameterized on the average, effective radii of the cross-sectional area(s) of their connected openings, which are *assumed* to be circular. We begin with what has been the current picture of mercury injection as a means of measuring experimentally, subject to certain assumptions, the distribution of pore volumes parameterized on average, effective, connected pore/capillary entry radii. Since the angle of capillarity for mercury in contact with soil is greater than 90°, pressure must be exerted on the mercury to force it into the soil pores/capillaries. If the assumptions are made that the pore/capillary openings are circular in cross section and small enough in radius so that the liquid mercury–mercury vapor interface has the shape of a section of a hemisphere, (2.82) can be used as the starting point to compute the pore radius into which a given applied pressure will force mercury as follows. Identifying z_0 as the distance the top of the mercury hemisphere would be depressed below the free mercury surface into which a capillary of radius ξ has been pushed, multiplying this result by $(\rho_l - \rho_v)g$, and setting the result equal to $p_v - p_l$, where p_v is the pressure of the mercury vapor on one side of the vapor–liquid interface and p_l is the pressure of the liquid just under the surface on the other, yields

$$p_v - p_l = (\rho_l - \rho_v)g z_0 = \frac{C}{\xi} - A_1\xi \qquad (7.108a)$$

where

$$A_1 = \left(\frac{2}{3\cos^3(\varphi)} [1 + \sin^3(\varphi)] - \frac{1}{\cos(\varphi)} \right) (\rho_l - \rho_v)g \qquad (7.108b)$$

$$C = 2\gamma \cos(\varphi) \tag{7.108c}$$

and $\xi = -r_a \cos(\varphi)$ for $\varphi > \pi/2$. In mercury injection as it is applied to pore volume determinations, $p_l >> p_v$ so that Eq. (7.108a) becomes, upon multiplying by ξ and dividing by p_l

$$\xi \simeq \frac{-C}{p_l} + \frac{A_1 \xi^2}{p_l} \tag{7.109}$$

Note in passing that $A_1 \xi$ is generally neglected compared to C/ξ, which is valid if ξ is small enough. To estimate a suitably small value of ξ, we assume that $\varphi = 140°$, $\gamma = 480$ dyn/cm (the values for clean mercury in contact with clean glass), $\rho_l = 13.6$ g/cm^3, and $g = 980$ cm/sec^2. This gives $A_1 \simeq -3.1915 \times 10^4$ dyn/cm^2 and $C \simeq -7.354 \times 10^2$ dyn/cm. If we use the criterion that to be neglected, the second term should be 1% or less of the first, we have $\xi \simeq 1.52 \times 10^{-2}$ cm $= 0.152$ mm which, just using the first term alone, and the same values as above, corresponds to an injection pressure of 4.85×10^4 dyn/cm^2 or 4.85×10^{-2} bars. Equation (7.109) can be solved for ξ as a function of applied mercury pressure, p_l, via Lagrange's expansion (Whittaker and Watson, 1969, p. 133) which development we now carry out. We recall that if we are given a function for ζ in the form

$$\zeta = \alpha + \beta \Phi(\zeta) \tag{7.110a}$$

where α and β are parameters independent of ζ, and $\Phi(\zeta)$ is a given function of ζ, then a function of ζ, $f(\zeta)$ can be expressed in terms of $\Phi(\zeta)$ by

$$f(\zeta) = f(\alpha) + \sum_{n=1}^{\infty} \frac{\beta^n}{n!} \frac{d^{n-1}}{d\alpha^{n-1}} \{f'(\alpha)[\Phi(\alpha)]^n\} \tag{7.110b}$$

In the special case in which $f(\zeta) = \zeta$, (7.110b) becomes

$$\zeta = \alpha + \sum_{n=1}^{\infty} \frac{\beta^n}{n!} \frac{d^{n-1}}{d\alpha^{n-1}} \{[\Phi(\alpha)]^n\} \tag{7.111}$$

which amounts to solving (7.110a) for ζ. To solve (7.109) for ξ we thus identify ζ with ξ, α with $-C/p_l$, β with A_1/p_l and $\Phi(\xi) = \xi^2$. Substituting these quantities into (7.111) yields

$$\xi = \frac{-C}{p_l} + \sum_{n=1}^{\infty} \frac{A_1^n}{n! p_l^n} \frac{d^{n-1}(\alpha^{2n})}{d\alpha^{n-1}} \Bigg|_{\alpha = \frac{-C}{p_l}} \tag{7.112}$$

Recalling that

$$\frac{d^p(x^q)}{dx^p} = x^{q-p} p! \binom{q}{p} \tag{7.113a}$$

where

$$\binom{q}{p} = \frac{q!}{p!(q-p)!} \tag{7.113b}$$

and setting $p = n - 1$, $q = 2n$, and $x = \alpha$, we have for the derivative term of (7.112)

$$\frac{d^{n-1}(\alpha^{2n})}{d\alpha^{n-1}} = \alpha^{n+1}(n-1)! \binom{2n}{n-1} \tag{7.114}$$

Substituting (7.114) into (7.112) and evaluating the result at $\alpha = -C/p_l$ yields

$$\xi = -\frac{C}{p_l} - \frac{C}{p_l} \sum_{n=1}^{\infty} \frac{(-A_1 C)^n}{n p_l^{2n}} \binom{2n}{n-1} \tag{7.115}$$

for the radius of a circular opening penetrated by mercury at an injection pressure p_l. We now define the pore/capillary volume distribution $D_c(\xi)$ as being the pore volume contained in the range of mercury entry radii ξ to $\xi + d\xi$ so that the total pore volume of the sample, V_0, is given by

$$V_0 = \int_{r_m}^{R} D_c(\xi') \, d\xi' \tag{7.116}$$

where r_m is the smallest measurable pore radius in the sample and R is the largest. Dividing $D(\xi)$ by V_0 yields a normalized (to one) pore volume frequency distribution, $d_c(\xi)$, so that

$$\int_{r_m}^{R} \frac{D_c(\xi')}{V_0} \, d\xi' = \int_{r_m}^{R} d_c(\xi') \, d\xi' = 1 \tag{7.117}$$

Setting V_1 equal to the volume of the sample filled from radius R at pressure P to radius ξ at pressure p_l we have

$$V_1 = \int_{R}^{\xi} D_c(\xi') \, d\xi' \tag{7.118}$$

When the slope of the data curve is divided by $\partial\xi/\partial p_l$, which is computed by differentiating Eq. (7.115), and the result plotted against ξ, computed for each value of p_l, using Eq. (7.115), the result is a graph of $D_c(\xi)$ versus ξ (Figure 7.24), as can be seen from Eq. (7.119)

$$\frac{\partial V_1}{\partial p_l} = \frac{\partial V_1}{\partial \xi} \frac{\partial \xi}{\partial p_l} = D_c(\xi) \frac{\partial \xi}{\partial p_l} \tag{7.119}$$

The sequence of data analysis steps is shown schematically in Figure 7.24. We now express the average, effective, pore/capillary radius frequency distribution, $f_c(\xi)$, for a "bundle of capillaries" model in terms of $D_c(\xi)$

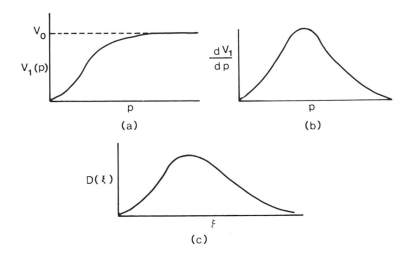

Figure 7.24: Sequence of data analysis steps involved in a mercury injection experiment. (a) Experimental data; (b) form of the derivative of the curve shown in (a); (c) $D_c(\xi)$ as a function of ξ, with ξ computed from p_l using Eq. (7.115).

as follows. Let N_c be the total number of pores/capillaries in the sample of porous medium being considered and $n_c(\xi)\,d\xi$ be the number of pores/capillaries in the sample having average, effective radii in the range ξ to $\xi + d\xi$. Then

$$N_o = \int_{r_m}^{R} n_c(\xi)\,d\xi \qquad (7.120)$$

If now we write $v_c(\xi)$ for the volume of a pore/capillary of average, effective radius ξ, which can take on different functional forms, as well as different values in different ranges of ξ, we have for the pore volume of the sample in the range ξ to $\xi + d\xi$,

$$D_c(\xi)\,d\xi = n_c(\xi)v_c(\xi)\,d\xi \qquad (7.121)$$

so that the expression for the total pore volume of the sample [Eq. (7.116)] becomes

$$V_0 = \int_{r_m}^{R} n_c(\xi)v_c(\xi)\,d\xi \qquad (7.122)$$

and V_2, the pore volume in the sample having average, effective radii in the range $r_m < \xi < r$, takes the form

$$V_2 = \int_{r_m}^{R} n_c(\xi) v_c(\xi) \, d\xi \tag{7.123}$$

We now write for $f_c(\xi)$, the fraction of pores/capillaries having average, effective radii in the range ξ to $\xi + d\xi$

$$f_c(\xi) \, d\xi = \frac{n_c(\xi)}{N_c} \, d\xi \tag{7.124}$$

so that, from (7.120) and (7.124)

$$1 = \int_{r_m}^{R} \frac{n_c(\xi)}{N_c} \, d\xi = \int_{r_m}^{R} f_c(\xi) \, d\xi \tag{7.125}$$

from which we say that $f_c(\xi)$ is normalized to 1 on the interval $r_m < \xi < R$. From (7.120) and (7.121) we have

$$N_c = \int_{r_m}^{R} \frac{D_c(\xi)}{v_c(\xi)} \, d\xi \tag{7.126}$$

so that, from (7.124), (7.126), and (7.124)

$$f_c(\xi) \, d\xi = \frac{\frac{D(\xi) \, d\xi}{v_c(\xi)}}{\int_{r_m}^{R} \frac{D(\xi)}{v_c(\xi)} \, d\xi} \tag{7.127}$$

The utility of $f_c(\xi)$ is that it is the frequency distribution over which averages of quantities that depend on pore/capillary radius are computed. We now write the expression for θ_s, temporarily neglecting spontaneous redistribution, as

$$\theta_s = \frac{\int_{r_m}^{r} v_c(\xi) f_c(\xi) \, d\xi}{\int_{r_m}^{R} v_c(\xi) f_c(\xi) \, d\xi} \tag{7.128}$$

where it is assumed that all the pores/capillaries having average effective radii in the range $r_m \le \xi \le r$ are filled with fluid.

Poiseuille's law affords a direct means of measuring values of $n_c(\xi)$ at discrete values of ξ, computed from discrete values of applied injection pressure using Eq. (7.115), in the "bundle of capillaries" model when the capillaries are circular in cross section. For a tube of circular cross section with radius ξ, Poiseuille's law is given by

$$q = \frac{\pi \xi^4 (p_2 - p_1)}{8 \eta \ell} \tag{7.129}$$

where q is the volume of liquid per unit time flowing through a tube of length ℓ under the influence of a pressure difference across ℓ of $p_2 - p_1$, $\pi \simeq 3.1416$, and η is the dynamic viscosity of the liquid in the tube. The first term on the right hand-side of Eq. (7.115) gives, to lowest approximation, the pressure required to force mercury through the tube against capillary action, if $\varphi > \pi/2$, which is

$$p_2 \simeq \frac{-2\gamma \cos(\varphi)}{\xi} \tag{7.130}$$

where p_1 in this case is the vapor pressure of mercury at room temperature, $\simeq 10^{-3}$ Torr, and may be neglected. Now,

$$q(\xi, t) = \pi \xi^2 \frac{dh(\xi, t)}{dt} \tag{7.131}$$

where $h(\xi, t)$ is the liquid-filled length of tube at time t, and $dh(\xi, t)/dt$ is the average velocity of liquid in the tube, neglecting the initial acceleration of the liquid needed to start it moving. Identifying $h(\xi, t)$ with ℓ of Eq. (7.129), substituting (7.131) and (7.130) into (7.129), separating variables, h and t, and integrating yields

$$h(t) = A_0(\xi, \varphi)t^{1/2} \tag{7.132a}$$

where

$$A_0(\xi, \varphi) = \left(-\frac{\gamma \xi \cos(\phi)}{2\eta} \right)^{1/2} \tag{7.132b}$$

Thus

$$q = \frac{\pi \xi^2}{2} A_0(\xi, \varphi)t^{-1/2} \tag{7.133}$$

At a given value of injection pressure, or equivalently the radius of the ith size range of capillaries, ξ_i), if $Q(\xi_i, t)$ is the total volume per unit time of mercury being injected into the sample as determined from the slope of a curve of volume injected at a given pressure as a function of time, as shown for a ceramic sample in Figure 7.25, then

$$n_c(\xi_i) = \frac{Q(\xi_i, t)}{q(\xi_i, t)} \tag{7.134}$$

Thus, normalizing a plot of $n_c(\xi_i)$ versus ξ_i via

$$f_c(\xi_i) = \frac{n_c(\xi_i)}{\sum_{i=1}^{m} n_c(\xi_i)} \tag{7.135}$$

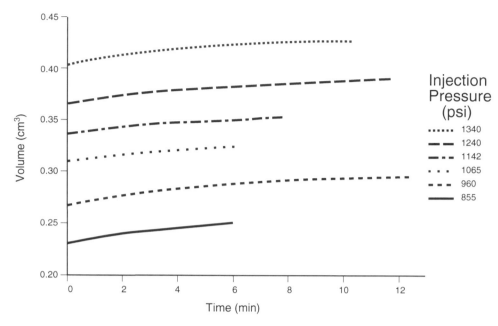

Figure 7.25: Volumes of mercury injected into a ceramic sample as functions of time at constant injection pressures.

where m is the number of discrete pressures at which injected volume–time measurements have been made, and fitting a curve to the points so computed yields a plot of $f_c(\xi)$ versus ξ. We note, by way of contrast between the data of Figure 7.25 and the results of the usual mercury injection experiment, that in the usual experiment, the data consist only of the injection pressures and the volumes injected when no more mercury will enter the sample at a given pressure–with no time-dependent data being collected. The analysis methodology for the data from a "standard" mercury injection experiment was discussed in connection with Figure 7.24. A significant limitation of this analysis is the assumption of the existence of pore openings of circular cross section, which have probably never been observed in earth materials. Rather, pore openings of irregular shapes are observed to occur in earth materials, as well as in ceramics and concrete. Scanning electron microscopy has been recently applied to ceramics to determine actual pore opening shapes and sizes and is needed because pressure–volume data alone do not determine the shapes of pore openings. Further, different pore opening shapes of the same area can require different mercury injection pressures. For example, if the pore openings are line fractures, idealized as smooth-walled structures, of length l and width d, then from Eq. (2.5) we have for the radius of curvature parallel to the fracture, r_2, say, $r_2 \to \infty$ and for the radius of curvature of the entering mercury–mercury vapor interface, if the opening is sufficiently narrow so that the cross section of the entering mercury is approximately hemispherical, $r_1 = d/2$. Thus, the injection pressure measured is approximately

$$p_l = \frac{-2\gamma \cos(\varphi)}{d} \tag{7.136}$$

where $p_1 = p_v$, $p_2 = p_l$, and $p_l >> p_v$ in this case. If $d/2$ is equal to ξ of (7.115), then the pore opening areas are ld for the idealized fracture and $\pi d^2/4$ for the corresponding cylindrical capillary, respectively. Further, if the length of the capillary of radius $d/2$ equals d_ℓ, and the depth of the microfracture equals d_ℓ as well, then the volume of mercury injected is $d_\ell^2 d$ for the fracture and $\pi d^2 d_\ell/4$ for the capillary, a ratio of capillary to fracture volumes of $\pi d/4d_\ell$. Since $d << d_\ell$, much more mercury is seen to be injected into the fracture under the given conditions than would be injected at the same pressure into the capillary having comparable dimensions. For a wedge-shaped fracture of length d_ℓ, depth d_ℓ, opening width d, and half-angle α, the comparable capillary to fracture volume is $\pi d/2d_\ell$. Thus discrete, possibly large steps in the plot of injected volume versus injection pressure may indicate: (1) small capillaries transmitting mercury to larger-sized internal microcracks or pores, or (2) if the rate of injection does not have a corresponding spike, a large internal volume of capillaries having approximately the same average, effective radius.

7.9 CENTRIFUGATION

We now consider the centrifugation of a soil core to remove the water contained therein. In what follows, we shall neglect the coupled nonequilibrium thermodynamic effects that would be expected to occur. The distances from the axis of rotation used below are shown in Figure 7.26. We begin with

$$d^{(i)}\mu = \frac{\partial^{(i)}\mu}{\partial r}dr + \frac{\partial^{(i)}\mu}{\partial P}dP + \frac{\partial^{(i)}\mu}{\partial T}dT$$
$$+ \frac{\partial^{(i)}\mu}{\partial \Psi}\frac{\partial \Psi}{\partial r}d\Psi + \frac{\partial^{(i)}\mu}{\partial^{(i)}n}d^{(i)}n = 0 \tag{7.137}$$

If the soil core is at or below field capacity, then $\partial^{(i)}\mu/\partial P$, a hydrostatic pressure term, drops out, as previously, and at constant temperature (refrigerated centrifuge) $dT = 0$. Further [compare Eq. (3.168)]

$$\frac{\partial^{(i)}\mu}{\partial r} = {}^{(i)}w(\omega^2 r) \tag{7.138}$$

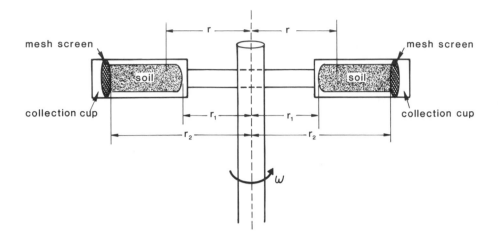

Figure 7.26: Centrifugation of a soil core showing soil sample holder and soil solution collection cup.

where $\omega^2 r$ is the angular acceleration of an element of mass a distance r from the axis of rotation of the centrifuge and ω is the angular frequency of rotation given by

$$\omega = 2\pi f \qquad (7.139)$$

where f is the frequency of rotation in revolutions per second or $\sec^{-1}$. [The force per unit mass in "gs" is given by $\omega^2 r/g$ and increases with (perpendicular) distance from the axis of rotation.] Also, we note that

$$\frac{\partial\, ^{(i)}\mu}{\partial\Psi} = {}^{(i)}v \qquad (7.140)$$

The pressure P per unit length of soil sample on the fluid in the soil pores (force per unit volume) is

$$\frac{dP}{dr} = \rho\omega^2 r \qquad (7.141)$$

where ρ is the mass density of the soil solution and can be expressed as

$$\rho = \frac{w}{v} \qquad (7.142)$$

where w is the mass per mol of soil solution and v is the volume per mol of soil solution as previously. For a fairly dilute solution such as normally occurs in soils, ρ is approximately constant even at varying chemical concentrations. We now recall that

$$\Psi = \bar{p}_l - p_{0l} = \frac{\rho_l kT}{m} \ln\left(\frac{\bar{p}_v}{p_{0v}}\right) \tag{7.143}$$

with all quantities as previously defined. Since $\bar{P}_l$, the average pressure of the water in the soil, is balanced by the pressure applied by centrifugation and since p_{0l} is constant for a soil solution with given temperature and chemistry, we have

$$\frac{dp}{dr} = -\frac{d\Psi}{dr} = \rho\omega^2 r \tag{7.144}$$

where it is assumed that the core does not contain "dry spots" that are of higher matric potential than $-P$. Using the above we have

$$\frac{d\,^{(i)}n}{^{(i)}n} = -[\,^{(i)}w - \,^{(i)}v\rho]\frac{\omega^2 r}{R_g T}\,dr \tag{7.145}$$

Integrating from $^{(i)}n_1$, the concentration of a solution species i in the soil column at distance r_1 from the axis of rotation to r, to a larger distance from the axis, where the concentration of species i in the soil solution is $^{(i)}n$, yields

$$\ln\left(\frac{^{(i)}n}{^{(i)}n_1}\right) = -\left(^{(i)}w - \,^{(i)}v\rho\right)\frac{\omega^2}{2R_g T}(r^2 - r_1^2) \tag{7.146}$$

or

$$^{(i)}n = \,^{(i)}n_1 e^{-(\,^{(i)}w - \,^{(i)}v\rho)\omega^2(r^2 - r_1^2)/2R_g T} \tag{7.147}$$

Equation (7.147) gives the relative mol fractions of soil–water solution species i as a function of distance from the axis of rotation and angular frequency of rotation. In the sample holder shown in Figure 7.26 soil water can flow out of the core into the collection cup so that matric potentials in the soil core will, if water can flow at all, be such as to have a gradient equal to the applied force per unit volume. At very high values of force per unit mass, 20,000 gs for example, easily attainable with current ultracentrifuges, the texture of the soil sample itself may well be altered via compression of the soil sample against the screen of the sample holder. The physical consequence of this is that centrifugation of unsaturated soil samples increases both the relative concentration of solutes near the screen, but also the absolute concentration of those constituents having solubilities that increase with applied pressure–calcium carbonate for example. Conversely, for some solution components, increased pressure will cause a decrease in concentration of that component in solution, with consequent precipitation of it onto the material of the pore walls. The same effects occur in reverse, except to a much smaller degree, when the soil solution is in a porous medium due to the reduction in pressure from that just under a flat surface arising from the capillary effects of the soil solution "wetting" the soil and thus forming concave menisci at any air–water interfaces that occur as given in Eq. (3.121). Thus, for example, calcium carbonate concentration in the soil solution would be lower in a porous medium in which the capillary effects were more pronounced–smaller pores/capillaries–than in one with larger pore openings.

SELECTED REFERENCES

Cember, H., *Introduction to Health Physics*, Pergamon Press, New York, 1969.

Gadian, D. G., *Nuclear magnetic resonance and its applications to living systems*, Oxford University Press, New York, 1982.

Kaplan, I., *Nuclear Physics*, Addison -Wesley Publishing Co., Inc., Cambridge, MA, 1955.

König, L. A., et al., "1961 Nuclidic Mass Table", *Nuclear Physics*, Vol. 31, pp. 18–42, 1962.

Whittaker, E. T., and G. N. Watson, *Modern Analysis*, 4th ed., Cambridge University Press, New York, NY, 1969.

Radiological Health Handbook, U.S. Dept. of Health, Education and Welfare, Public Health Service, Bureau of Radiological Health, Washington D.C., 1970.

PROBLEMS

7.1 Using Eqs. (7.46)–(7.49) and (7.52) as required, derive Eqs. (7.50) and (7.51) of the text.

7.2 Show that the velocity of the center of mass of a two-particle system, V_{cm}, when one particle is initially at rest, is

$$V_{cm} = \frac{m_1 \dot{\vec{r}}}{m_1 + m_2}$$

where m_2 is the mass of the particle initially at rest ($\dot{\vec{r}}_{20} = 0$), m_1 is the mass of the incoming particle, and $\vec{r}$ is the position of particle 1 measured relative to the position of particle 2.

7.3 Using the relations $\vec{V}_{cm} = \dot{\vec{R}}$, $\vec{v}_1 = \dot{\vec{r}}_1 = \dot{\vec{r}}_2 + \dot{\vec{r}}$, and other relations as required, show that (a)

$$\vec{v}_1 = \vec{V}_{cm} + \frac{m_R}{m_1}\vec{v}$$

and that (b)

$$\vec{v}_2 = \vec{V}_{cm} + \frac{m_R}{m_2}\vec{v}$$

where m_R is the reduced mass of particles 1 and 2 given in (7.52) of the text, and $\vec{v}$ is the relative velocity of the two particles.

7.4 Show that the total kinetic energy of the two particles after scattering, E_T, namely

$$E_T = \frac{1}{2}m_1 v_1^2 + \frac{1}{2}m_2 v_2^2$$

is equal to

$$E_T = \frac{1}{2}(m_1 + m_2)V_{cm}^2 + \frac{1}{2}m_R v^2$$

where $V_{cm}^2 = \vec{V}_{cm} \cdot \vec{V}_{cm}$ and $v^2 = \vec{v} \cdot \vec{v}$.

7.5 Show that the total angular momentum, $\vec{L}$, of particles m_1 and m_2 after scattering has occurred, namely

$$\vec{L} = m_1 \vec{r}_1 \times \vec{v}_1 + m_2 \vec{r}_2 \times \vec{v}_2$$

can be expressed as

$$\vec{L} = (m_1 + m_2)\vec{R} \times \vec{V}_{cm} + m_R \vec{r} \times \vec{v}$$

where $\times$ is the usual cross product, and the rest of the quantities are as previously defined.

7.6 Derive Eq. (7.59) of the text by applying the law of cosines to the velocity triangle of Figure 7.15.

7.7 Derive Eq. (7.61) of the text.

7.8 Show, by evaluating $d(\Delta E)/dA = 0$, where ΔE is given by Eq. (7.64) of the text, that ΔE is a maximum for $A = 1$.

7.9 By considering Figure 7.15, deduce the relations [note Eq. (7.55)]

$$\tan(\theta_1) = \overline{de} \,/\, \overline{ad} = \overline{bc} \,/\, \overline{ad}$$

$$\overline{bc} = \frac{m_R}{m_1} v_{10} \, \sin(\Phi_1)$$

$$\overline{ad} = \overline{ab} + V_{\text{cm}} = \frac{m_R}{m_1} v_{10} \cos(\Phi_1) + \frac{m_R}{m_2} v_{10}$$

to show that

$$\tan(\theta_1) = \frac{\sin(\Phi_1)}{\frac{1}{A} + \cos(\Phi_1)}$$

with $m_1 = 1$ and $m_2 = A$ measured in amu, which is the relation between the scattering angle in laboratory coordinates, θ_1, and the scattering angle in center-of-mass coordinates, Φ_1.

8

STATISTICAL
TREATMENT OF DATA

8.1 INTRODUCTION

The application of statistics in the experimental aspects of flow in unsaturated porous media is to separate signals–values of the parameter(s) of interest–from some lack of information, or what may be termed *noise*, in the broadest sense. Noise can be a spread of experimental measurements of a time-independent parameter, an attribute of members of a population, or an extraneous time-varying signal superposed on the parameter(s) of interest. The effects of instrumentation, convolved with the actual physical phenomena to produce the measured experimental results, are also sources of lack of information, or noise. The rate at which time-dependent phenomena are sampled can induce noise if the rate is not appropriate to the phenomena being studied, and reconstruction of the original phenomena from the discrete data points comprising the sample must be carried out in a statistically optimum way to reduce the loss of information, manifested as noise, inherent in the process. Further, statistical methods may be used for testing hypotheses made concerning measured data. Such hypotheses might concern the drift of measuring equipment with time, the effects of a particular experimental procedure, relative to an alternate procedure, quality control in an equipment construction process, etc. The theory of extreme values–realizations of infrequent events–depends on the relative probabilities of the occurrence of each event of interest. Environmental monitoring and assessment often require applications of unsaturated flow methodology, and assessment of environmental contamination levels has become a common application of statistical methodology and hence valuable to the unsaturated flow practitioner involved in an environmental assessment. The branch of statistical analysis featuring the explicit use of frequency distributions and analysis of data in terms of the parameters on which they depend is called *parametric statistics*. If the frequency distribution functions of attributes of the systems of interest are not specified so that the parameters characterizing the distributions are not known, we term the body of methods available in that case *nonparametric statistics*. Though both of these branches of statistical analysis have their uses, we will be concerned primarily with parametric statistics in which the distribution functions of attributes of members of a population are either assumed or inferred from data.

8.2 POPULATIONS AND SAMPLES

In statistical analysis, the term *population* refers to the collection of entities (members) having attributes amenable to measurement. It is the task of statistical analysis to determine, to some level of confidence, this concept to be made precise shortly, the distribution of values of attributes over the members of the population by determining the distributions of the attributes in sets of sample members of the population selected, presumably in a random way, for measurement of the attributes. Populations may have members with attributes varying with time and/or attributes that are presumed to be constant in time by the observer. They may have a finite or infinite number of members. Samples of these populations may be taken with or without replacement of the sample in the population with distinct statistical implications for each sampling strategy. If the attributes of interest are time dependent, then an important aspect of sampling strategy is the sampling frequency, or the rate at which samples of members of the population having the attribute(s) of interest are taken. If the population consists of values of a parameter, such as matric potential, which may vary with time in a given location, and vary with location at a given field site, then the process of sampling the population is time and space dependent, with possible uncertainties contributing to the total uncertainty–noise–from uncertainties in determining locations and times, as well as instrumental uncertainties superposed on the distribution of values of the attribute in the population itself.

8.3 NONPARAMETRIC STATISTICS

8.3.1 Mann–Whitney Test of Ordinal Data

We begin a test designed to determine, at some level of confidence, whether or not two sets of data are drawn from the same population. This particular test is an example of a group of tests that do not assume any particular functional form for the underlying distribution(s) of attribute(s) over the population. This branch of statistical analysis is called *nonparametric statistics* and the particular test we shall discuss is called the *Mann-Whitney test* of ordinal (ranked) data (Mann and Whitney, 1947). This test is a variation and extension of the test given first by Wilcoxen (Wilcoxen, 1945), the differences being in the test statistic and the extension from equal numbers of data points in two sets of data being compared to determine if, statistically speaking, they can be said to come from the same population, to unequal numbers of data points. Let n be the number of data points, consisting of values $x_1, \cdots, x_n$, in the larger of the two data sets being compared, and m be the number of data points in the smaller of the two data sets that consists of the values $y_1, \cdots, y_n$. As just noted, the case $m = n$ was the one considered by Wilcoxen (1945). The two data sets are then combined and the resultant data set is ranked in the order of numerical values of the data points. The Mann–Whitney test statistic, u_n, is obtained by counting, in the combined, ranked set, the number of x_i following each successive y_i and adding the results. The relation between this statistic and the statistic u_m, computed by counting for each successive x_i the number of times y_i of higher rank occur in the combined data set and adding the results, is

$$u_n = nm - u_m \qquad (8.1)$$

Further, if the sum of the ranks of the x_i in the combined data set is r_n and the sum of the ranks of the y_i in the combined data set is r_m, we have the relations

$$u_n = nm + \frac{n(n+1)}{2} - r_n \qquad (8.2a)$$

and

$$u_n = nm + \frac{m(m+1)}{2} - r_m \qquad (8.2b)$$

Eliminating u_m between (8.1) and (8.2b) and then eliminating u_n between the result and (8.2a) yields, upon rearrangement,

$$r_m + r_n = \frac{(n+m)(n+m+1)}{2} \qquad (8.3)$$

We note that the statistic w, used by Wilcoxen, is

$$w = u_n + \frac{m(m+1)}{2} \qquad (8.4)$$

which is equal to r_m of (8.2b) as can be seen by combining (8.1) and (8.2b). The statistic u_n can be used to test the null hypothesis, namely that two sets of data being considered have the same frequency distribution so that, at a certain level of significance, $1 - \alpha$, which concept will be made precise shortly, the probability, p, of computing from measured data a value of u_n that is less than or equal to that found by considering all possible occurrences of x_i having ranks larger than those of y_i leading to a value of $u = u_d$, is

$$p(u_i \leq u_d) = \alpha \qquad (8.5)$$

where u_i stands for either u_n or u_m, whichever is less in a given experimental situation. If the value of u_i calculated from the ranked data sets is larger than u_d at the level of significance $1 - \alpha$, then the null hypothesis is rejected and it is said that at the $(1 - \alpha)$th level of significance a difference exists between the distribution(s) of the attribute(s) in the population(s) from which the two samples were drawn. Table D.1 contains values of

$$p(u_d, n, m) = p(u_i \leq u_d) = \alpha \qquad (8.6)$$

for various ranges of u_d, n, and m, subject to the condition that $m \leq n$. Since the method of calculation of these tables appears not to be widely discussed in the statistical literature, and since a variation of one of the literature methods was employed to compute the tables given in this work, a discussion of the method of calculation will now be given. The method used here follows essentially that of Fix and Hodges (1955) though the tables arrived at by them are of different form and content and more general than those generally used. Table D.1 is of the customary form.

The quantity $p(u_d, n, m)$ is related to a certain standard partition function arising in the theory of numbers as follows. If $p(u_d, m)$ is the number of ways of constructing sets of numbers, the sum of which is u_d, subject to the condition that each set, or partition, as it is called, is made up of no more than m parts, (see Gupta, 1958, for a discussion of this and other standard partitions) then, in essentially the notation of Fix and Hodges (1955),

$$p_0(u_d, m) = A_0(u_d, m) - A_0(u_d - 1, m) \qquad (8.7)$$

where the $A_0(u_d, m)$ are computed via the recursion relation

$$A_0(u_d, m) = A_0(u_d, m - 1) + A_0(u_d - m, m) \qquad (8.8a)$$

with the boundary and initial conditions, respectively,

$$A_0(0, m) = 1 \qquad (8.8b)$$

$$A_0(u_d, 1) = u_d + 1 \qquad (8.8c)$$

with the convention that

$$A_0(u_d, m) = 0 \qquad (8.8d)$$

for $u_d < 0$. We note in passing that this recursion relation is the same as the one used to compute successive values of $p_0(u_d, m)$ (Gupta, loc. cit.), but that the initial and boundary conditions, respectively, in this case are $p_0(0, m) = 1$, $p_0(n, 0) = 0$, and $p_0(n, 1) = 1$ for $n \geq 1$. We now compute the quantity

$$A(u_d, n, m) = \sum_{k \geq 0} (-1)^k A_k \left(u_d - kn - \frac{k(k+1)}{2}, m - k \right) \qquad (8.9)$$

where, for $k \geq 1$ (Fix and Hodges, loc. cit.)

$$A_k(x, m - k) = \sum_{\ell \geq 0} p_0(\ell, k) A_0(x - \ell, m - k) \qquad (8.10)$$

In (8.9) and (8.10) the sums are carried out; that is, k and ℓ are increased from zero until just before the first arguments of A_k and A_0, respectively, become negative. The relation between $A(u_d, n, m)$ and $p(u_d, n, m)$ is (Fix and Hodges, loc. cit.)

$$p(u_d, n, m) = A(u_d, n, m) / \binom{n + m}{m} \qquad (8.11)$$

where

$$\binom{n + m}{m} = \frac{(n + m)!}{m!(n + m - m)!} = \frac{(n + m)!}{n!m!} \qquad (8.12)$$

is the usual binomial coefficient and $n! = n(n - 1)(n - 2) \cdots$ as usual. The calculational steps leading to the production of Table D.1 may be summarized as follows: (1) values of $A_0(u_d, m)$ covering the ranges of interest of u_d, m, and n, with $m \leq n$, are computed using (8.8a)–(8.8d); (2) the corresponding values of $p(u_d, m)$ are then computed using (8.7); (3) the corresponding values of A_k are computed using the results generated in (8.10); (4) (8.9) is then used with the results of (8.10) to compute values of $A(u_d, n, m)$ that are each divided by the appropriately valued binomial coefficients as shown in (8.11) to yield the values of $p(u_d, n, m)$ for the ranges of u_d, n, and m of interest. Since only the condition $u \leq u_d$ is being tested, this particular form of the Mann–Whitney test is called a *one-tailed* (tail of the unspecified frequency distribution) *test*. The quantity u_d is called the *critical value* of u and is the lower boundary, so far as values of u go, of the rejection region of the unspecified frequency distribution as shown schematically in Figure 8.1 for a one-tailed test.

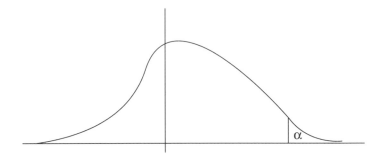

Figure 8.1: An unspecified frequency distribution, assumed to be normalized so that the area under the entire curve equals one, with the shaded area, α, taken to be the region in which the hypothesis to be tested is rejected for a one-tailed test.

We now consider an example of the Mann-Whitney test. Two psychrometer probes, designated A and B, have been calibrated in the laboratory in the vapor over a 0.1 molar NaCl solution at 20°C to read -4.5 bars. Later, the probes and associated electronics are checked for drift by carrying out a series of measurements under the calibration conditions. The measurements are shown in Table 8.1. It is the goal of the drift test to discover whether or not the two probes agree in their readings in the given experimental situation to within acceptable experimental error. A related question, which is an extension of this experiment, is whether or not the probes agree over a range of matric potential values. We begin the analysis by constructing a table of the combined data ranked in order of size. The results of this step are shown in Table 8.2.

We note that there were three values of matric potential that were the same in each of the lists. These are called *ties* in the combined rank scores and are dealt with by computing the average of the ranks the tied values would occupy and assigning the average value to each of the ranked items. The ranks that were averaged do not appear in the final ranked list of values. For example, the value -4.55 bars appears in both the A list and the B list. These values, if not averaged, would occupy ranks 4 and 5 in the list with consequent ambiguity as to which rank was assigned to the A or B value if averaging were not carried out. The original ranks 4 and 5 do not appear in the final list. Similarly, ranks 6 and 7 and ranks 9 and

Table 8.1: Psychrometer data to check for drift

Probe A (bars)	Probe B (bars)
-4.51	-4.54
-4.56	-4.53
-4.55	-4.47
-4.60	-4.49
-4.58	-4.52
-4.54	-4.55
-4.52	-4.46
-4.50	
-4.48	

10 have been averaged and are missing from the final ranked list as well. We now compute r_n, the sum of the ranks of the x_i in the combined list, the A values in this case (recall that our convention is that if the data lists are of unequal lengths, the x_i are the members of the longer list), to yield $r_n = 63.5$ and similarly for r_m, the sum of the B ranks in the combined list, to yield $r_m = 72.5$. These values can be checked by substitution into Eq. (8.3). We now compute the u_n and u_m values from Eqs. (8.2a) and (8.2b), respectively, to yield $u_n = 44.5$ and $u_m = 18.5$. These values can be checked by substitution into Eq. (8.1). Table D.1 is arranged such that values of $p(u_i \leq u_d) = \alpha$ are given for values of $m \leq n$ for given values of u_d and n. Thus, for convenience, we consider the smaller of the u_n and u_m, which is u_m in this case, and upon rounding 18.5 up to 19, since the tables exist only for integer values of u_i, we find from Table D.1h, that for $n = 9$, $m = 7$, and $u_i = 19$, the probability of $u_i \leq u_d$ for $\alpha = 0.05$ is zero since u_d in this case lies between 15 and 16 for a one-tailed test at the 95% confidence level, and so the value of $u_i = 19$ is in the rejection region. We thus reject the hypothesis that at the $(1 - \alpha) \times 100\%$ or 95% confidence level, the two psychrometer probes agree in their readings. We note finally that the average or expectation value of u, denoted by $E(u)$, is $E(u) = n_1 n_2/2$, and the variance of u, denoted by σ_u^2, is given by $\sigma_u^2 = n_1 n_2 (n_1 + n_2 + 1)/12$. We shall shortly discuss [Eq. (8.102)] the Gaussian distribution having a mean of zero and a standard deviation equal to one. The variate associated with that distribution, denoted by z, is computed from the mean(s) and standard deviation(s) of Gaussian distributions for which the mean(s) are not necessarily equal to zero and the standard deviations are not necessarily equal to one from the relation $z = [u - E(u)]/\sigma_u$, which is called the z score of u in this case. Using the z score allows the distribution of u to be approximated by a Gaussian distribution if it is desired to do so to compute

Table 8.2: Combined data sets ranked in numerical order

Matric potential (bars)	Rank	Probe
-4.60	1	A
-4.58	2	A
-4.56	3	A
-4.55	4.5	A
-4.55	4.5	B
-4.54	6.5	A
-4.54	6.5	B
-4.53	8	B
-4.52	9.5	A
-4.52	9.5	B
-4.51	11	A
-4.50	12	A
-4.49	13	B
-4.48	14	A
-4.47	15	B
-4.46	16	B

confidence intervals for $E(u)$, for example. We now consider a nonparametric test for periodicities and/or trends in the data, namely the runs test.

8.3.2   Runs Test for Ordinal Data

We now consider a test for the randomness of the values in a sequence of ordinal data expressed as what might be described as a series of either–or values of a particular variable of interest. An example of such data is a set of deviations of time series or other data from their mean. We consider data containing n_1 values of outcome of type one and n_2 values of outcome of type two. If n_1 is greater or less than n_2 by one unit, then the maximum possible number of runs, R_m, where a run is an uninterrupted sequence of outcomes of a particular kind, is

$$n_1 + n_2 = R_m \tag{8.13}$$

The number of ways of choosing ℓ_1 distinguishable items from a group of n_1 items is given by $C_{\ell_1}^{n_1}$, where $C_{\ell_1}^{n_1}$ is the usual binomial coefficient given by

$$C_{\ell_1}^{n_1} = \frac{n_1!}{\ell_1!(n_1 - \ell_1)!} \tag{8.14}$$

Thus, ℓ_1 runs, where $\ell_1 \leq n_1$, can be observed in $C_{\ell_1-1}^{n_1-1}$ ways, and similarly, the number of ways in which ℓ_2 runs out of a possible total of n_2 runs can be observed is $C_{\ell_2-1}^{n_2-1}$. Thus, the probability of having ℓ_1 runs and ℓ_2 runs simultaneously, $p(\ell_1, \ell_2)$, is

$$p(\ell_1, \ell_2) = \frac{C_{\ell_1-1}^{n_1-1} C_{\ell_2-1}^{n_2-1}}{C_{n_1}^{n_1+n_2}} \tag{8.15}$$

where $C_{n_1}^{n_1+n_2} = C_{n_2}^{n_1+n_2}$ is the maximum number of ways in which runs may occur. The probability of having a total number of runs R, out of a possible maximum of R_m, is equal to the sum of $p(\ell_1, \ell_2)$ over ℓ_1 and ℓ_2 subject to the condition

$$\ell_1 + \ell_2 = R \tag{8.16}$$

where R is given in advance. Since the number of runs of attribute 1 are interspersed with those of attribute 2, we have by inspection the further relations for the case in which R is odd, which we denote by R_o,

$$\ell_1 = \ell_2 - 1 \tag{8.17a}$$

or

$$\ell_1 = \ell_2 + 1 \tag{8.17b}$$

These relations are mutually exclusive, and each contributes a term to the sum over $p(\ell_1, \ell_2)$ to yield the probability $p(R_o, n_1, n_2)$, of the number of possible runs being equal to R_o. Combining (8.17a) and (8.17b) with (8.16) yields, for R odd,

$$\ell_1 = \frac{R_o - 1}{2} \tag{8.18a}$$

$$\ell_2 = \frac{R_o + 1}{2} \tag{8.18b}$$

and

$$\ell_1 = \frac{R_o + 1}{2} \tag{8.19a}$$

$$\ell_2 = \frac{R_o - 1}{2} \tag{8.19b}$$

Thus, for R odd we have for the probability, $p_o(n_1, n_2, R_o)$, of having R_o runs out of a total of n_1 outcomes of type 1 and n_2 outcomes of type 2

$$p_o(n_1, n_2, R_o) = \frac{C_{\frac{R_o-3}{2}}^{n_1-1} C_{\frac{R_o-1}{2}}^{n_2-1} + C_{\frac{R_o-1}{2}}^{n_1-1} C_{\frac{R_o-3}{2}}^{n_2-1}}{C_{n_1}^{n_1+n_2}} \tag{8.20}$$

For the case in which R is even, we denote R by R_e and see by inspection

$$\ell_1 = \ell_2 = R_e/2 \tag{8.21}$$

There are two mutually exclusive cases here also, since the sequence of data can begin either with results of type 1 or of type 2. Thus, for R_e, the quantity $p(\ell_1, \ell_2)$ can be cast in the form $p_o(n_1, n_2, R_e)$ and is the probability of having R_e runs in a set of data having n_1 results of type 1 and n_2 results of type 2 and is given by

$$p_e(n_1, n_2, R_e) = 2 \frac{C^{n_1-1}_{\frac{R_e}{2}-1} C^{n_2-1}_{\frac{R_e}{2}-1}}{C^{n_1+n_2}_{n_1}} \tag{8.22}$$

We now consider the cumulative probability distribution, $P(n_1, n_2, R)$, of the probabilities given in (8.20) and (8.22) given by

$$P(n_1, n_2, R) = \sum_{R'=2}^{R_e} p_e + \sum_{R'=3}^{R_o} p_o \tag{8.23}$$

where the values of R' in the two sums indicated in (8.23) are over even and odd values, respectively. This cumulative probability distribution is the same as the one that would be calculated by summing $p(\ell_1, \ell_2)$ over ℓ_1 and ℓ_2 subject to the condition (8.16). Values of R that are small and in some cases values of R that are large indicate nonrandomness in the data and hence would lead to rejection of the hypothesis that the data, which could consist of deviations from a mean, for example, are random. A one-tailed test considers an acceptance region having values of $P(n_1, n_2, R)$ less than or equal to a given value, say, b. Similarly, a two-tailed test has an acceptance region for the hypothesis of randomness of $P(n_1, n_2, R) \geq a$ and $P(n_1, n_2, R) \leq b$, for $a < b$. From a table giving $P(n_1, n_2, R)$ as a function of n_1, n_2, and R, it is possible to choose the values of $P(n_1, n_2, R)$ separating the regions of acceptance and rejection of the hypothesis of randomness, the critical values, $P(n_1, n_2, R_c)$, of R_c, n_1, and n_2. By way of example, we consider the deviations from the mean of the values of matric potential given for psychrometer probe A in Table 8.1. These deviations from the mean are given in Table 8.3. Deviations above the mean are taken to be events of type 1, while deviations below the mean are taken to be events of type 2. From Table 8.3 we see that $n_1 = 5$, $n_2 = 4$, and $R = 3$. We find from Table D.2a, computed using the relations given in Eqs. (8.20), (8.22), and (8.14), that $P(5, 4, 3) = 0.071$, and $P(5, 4, 2) = 0.016$. This indicates that the hypothesis that the deviations from the mean are random errors of measurement would be considered to be correct at the 91.3% confidence level, computed as $100\% - [100\% \times (0.016 + 0.071)]$.

Table 8.3: Deviations from the mean as type 1 or type 2 events

Deviation	Event Type
-0.0278	2
0.0222	1
0.0122	1
0.0622	1
0.0422	1
0.0022	1
-0.0178	2
-0.0378	2
-0.0577	2

8.4 PERMUTATIONS AND COMBINATIONS

We now consider the notion of probability in more detail. We begin in an intuitive way by studying permutations and combinations in order to compute the probabilities of the occurrence of events in certain simple, well-defined situations and remind ourselves that statistics, with which we will be primarily concerned, is the attempt to infer the probabilities applicable in a given situation from sample data.

We first consider the number of ways of selecting m objects at a time from a total of n objects, or, put another way, the number of combinations of n objects taken m at a time. We then take as a first definition of the probability of an event occurring the number of ways it can occur in a given experiment divided by the total number of equally likely possible outcomes. We then compute the probability of given outcomes for particular examples.

8.4.1 Permutations

The number of ways of arranging n distinguishable objects taken m at a time in which the order in which the objects are arranged is taken into account is called the *number of possible permutations* of n objects taken m at a time, P_m^n, and is given by

$$P_m^n = \frac{n!}{(n-m)!} \tag{8.24}$$

where $n!$ has been previously defined. Note that the total number of permutations of n objects taken n at a time is

$$P_n^n = n! \tag{8.25}$$

Further, the total number of possible permutations, $P^n_{n_1,n_2,...,n_\ell}$, of n objects made up of n_1, n_2, $\ldots$, n_ℓ groups of identical objects is

$$P^n_{n_1,n_2,...,n_\ell} = \frac{n!}{n_1!n_2!\cdots n_\ell} \tag{8.26}$$

where

$$n_1 + n_2 \cdots + n_\ell = n \tag{8.27}$$

For example, the number of ways of arranging the letters in the word "array" can be computed using $n_1 = 2$ (two a's), $n_2 = 2$ (two r's), $n_3 = 1$ (one y) to yield

$$P^5_{2,2,1} = \frac{5!}{2!2!1!}$$

8.4.2   Combinations

A combination of n objects taken m at a time is a selection of m objects out of n without regard to the order in which the objects are arranged. The total number of possible combinations of n objects taken m at a time, C^n_m, is, as noted previously,

$$C^n_m = \frac{P^n_m}{m!} = \frac{n!}{m!(n-m)!} \tag{8.28}$$

For example, the possible combinations of the numbers 1, 2, and 3, are 12, 13, and 23. Here, 12 and 21 are the same combination but not the same permutation. The total number of combinations, $_TC^n_\ell$, of n objects taken successively $1, 2, \ldots, \ell$ at a time is

$$_TC^n_\ell = \sum_{i=1}^{\ell} C^n_i \tag{8.29}$$

In the special case in which $\ell = n$ it can be shown (Problem 8.5) that $_TC^n_n = 2^n - 1$. By way of example, we now compute the total number of ways of arranging five objects taken three at a time, $_TC^5_3$, to be

$$_TC^5_3 = \sum_{i=1}^{3} C^5_i = C^5_1 + C^5_2 + C^5_3 = 5 + 10 + 10 = 25 \tag{8.30}$$

8.5   PROBABILITY

We now take as our operational definition of the probability of the occurrence of an event the ratio of the number of times, m_1, the event is expected to occur in N trials, to the total number of possible outcomes in the N trials, namely, N. This operational definition and scenario presuppose a particular, known, set of circumstances, and concomitantly, knowledge of the characteristics of both the population and the sample as well. (Note in passing that the applications and utility of statistics, with which we will be primarily concerned, are to infer characteristics of attributes of the population from characteristics of the sample.) We can apply the combinatorial results just given to calculate probabilities of occurrence of specific events in specific cases using our operational definition as illustrated in the following example.

Six cards are drawn from a deck of 52 well-shuffled cards. To compute the probability that all 6 are red, we note that the number of ways in which 6 cards can be drawn from 52 (the number of ways of taking from 52 objects 6 at a time without regard to order) is C_6^{52}. The number of ways in which 6 red cards can be drawn from the 26 red cards in the deck is C_6^{26}. Thus the probability, p, of drawing six red cards from the deck is

$$p = \frac{C_6^{26}}{C_6^{52}} \simeq 0.0113$$

Since the probability of an event of some kind occurring in an experiment is equal to one, the probability that a given event will not occur, q, is given in terms of p by

$$q = 1 - p \tag{8.31}$$

The probabilities we have been considering are those of the occurrence of simple results, that is, results that cannot be further decomposed. We designate these results as R_i. Compound events are composed of simple events, and the occurrence of any of the simple events making up a compound event is sufficient for the compound event to be said to have occurred. Each simple event can be thought of as a point in a sample space, while collections of points can be thought of as compound events and indicated in a sketch of points in sample space by closed simple curves enclosing the simple results that comprise the compound event. These ideas are illustrated in Figure 8.2, in which the compound event E_1 consists only of the simple result R_1, the compound event E_2 consists of the simple events R_2 and R_8, the compound event E_3 consists of the simple events R_3, R_6, R_7, and R_9, and the compound event E_4 consists of the simple events R_5 and R_{10}. Thus, the event E_1 is said to have occurred if result R_1 is observed; the compound event E_2 is said to have occurred if any of the results R_2, R_4, or R_8 is observed; similarly the compound event E_3 is said to have occurred if any of the results R_3, R_6, R_7, or R_9 are observed, while E_4 is said to have occurred if either result R_5 or R_{10} is observed.

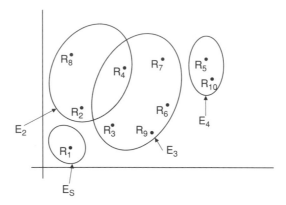

Figure 8.2: Illustration of a sample space in which a number of simple results, the R_i, are plotted as points while the compound events, the E_i, are depicted as closed curves each enclosing the R_i of which they are composed.

If each compound event is thought of as a set of points in sample space, then we may speak of the union and intersection of two sets as follows. The union of two sets, denoted by the symbol $\cup$, is the set consisting of all the points in both the original sets. Thus, in our example, $E_1 \cup E_2 = E_1 + E_2 = R_1 + R_4 + R_8$. The intersection of two sets, denoted by the symbol $\cap$, is the set of points common to both sets. In the example given we have $E_2 \cap E_3 = R_4$ since only the single result R_4 is common to both sets. Some authors use the notation $E_i E_j$ for $E_i \cap E_j$. However, this notation is the same as that for multiplication in algebra, and hence its use will be avoided here. If two sets have no points in common, then they are said to be disjoint, and $E_i \cap E_j = \emptyset$, where $\emptyset$ denotes the empty set, that is, one having no members. In the example just given, $E_1 \cap E_2 \cap E_4 = \emptyset$.

8.5.1   Addition Rule of Probability

We now consider the addition rule of probability as follows. The probability of occurrence of either of two compound events, $P\{E_i \cup E_j\}$, is given by

$$P\{E_i \cup E_j\} = P\{E_i\} + P\{E_j\} - P\{E_i \cap E_j\} \qquad (8.32)$$

where the probability of occurrence of events common to both E_i and E_j, namely $E_i \cap E_j$, is subtracted from the sum of the first two terms in (8.32) to avoid double counting. In the example shown in Figure 8.2, the probability of occurrence of either event E_2 or E_3, $P\{E_2 \cup E_3\}$, is

$$P\{E_2 \cup E_3\} = P\{E_2\} + P\{E_3\} - P\{E_2 \cap E_3\}$$
$$= P\{R_2\} + P\{R_4\} + P\{R_8\} + P\{R_3\} + P\{R_4\}$$
$$+P\{R_6\} + P\{R_7\} + P\{R_9\} - P\{R_4\} \qquad (8.33)$$

in which $P\{R_4\}$ is eventually only counted once and the occurrence of any one of the simple events on the right-hand side of (8.33) is taken to mean that one or the other of the compound events E_2 or E_3 has occurred. Another example of the addition rule is as follows. A box of 40 well-mixed psychrometer probes contains 10 old probes and 30 new ones. Each probe contains the usual two thermocouples, thermocouple A, used for vapor pressure measurements, and thermocouple B, used for measurements of ambient temperature. It is known that thermocouple A is defective in 2 of the 10 old probes and also in 5 of the new ones, while thermocouple B is defective in 1 of the old probes and in 2 of the new ones. If it is assumed that no probes have both thermocouples A and B defective, what is the probability of selecting a psychrometer probe and getting either a new one or one having a defective A thermocouple? The information given is summarized in Table 8.4. Denoting the result of drawing a new probe by E_2, we see that $P\{E_2\} = \frac{30}{40} = \frac{3}{4}$. Denoting the result of drawing a probe having a defective A thermocouple by E_1, we have $P\{E_1\} = \frac{7}{40}$. Denoting the result of drawing a probe that is both new and has a defective A thermocouple by $E_1 \cap E_2$, we have $P\{E_1 \cap E_2\} = \frac{5}{40}$. Thus, the probability of drawing a thermocouple that is either new or has a defective A thermocouple, $P\{E_1 \cup E_2\}$, is $P\{E_1 \cup E_2\} = P\{E_1\} + P\{E_2\} - P\{E_1 \cap E_2\} = \frac{3}{4} + \frac{7}{40} - \frac{5}{40} = 0.8$. As a second question we ask: What is the probability of choosing a probe having either a defective A thermocouple or a defective B thermocouple? Denoting the result of drawing a probe having a defective A thermocouple by E_1, as previously, and the result of drawing a defective B thermocouple by E_3, we have $P\{E_3\} = \frac{3}{40}$. Since, by assumption we have no probes having both thermocouples A and B defective, $E_1 \cap E_3 = \emptyset$, $P\{E_1 \cap E_3\} = 0$, and so, $P\{E_1 \cup E_3\} = P\{E_1\} + P\{E_1\} - P\{E_1 \cap E_3\} = \frac{7}{40} + \frac{3}{40} - 0 = \frac{1}{4}$.

Finally, we ask for the probability of choosing a nondefective thermocouple probe, either old or new. Denoting the probability of choosing a nondefective old probe by E_4, we have $P\{E_4\} = \frac{5}{40}$, while the probability of choosing a nondefective new probe is given by $E_5 = \frac{23}{40}$. Since the categories of new and old probes are mutually exclusive, $E_4 \cap E_5 = \emptyset$, $P\{E_4 \cap E_5\} = 0$, and we have for the probability of choosing a nondefective thermocouple, $P\{E_4 \cup E_5\}$, $P\{E_4 \cup E_5\} \simeq 0.675$.

The formulae giving the probabilities of observing at least one of three or more events can be written down as well. For example, by substituting

Table 8.4: Thermocouple probe data

Old Probes (10)	New Probes (30)
A defective (2)	A defective (5)
B defective (1)	B defective (2)
Neither A nor B defective (7)	Neither A nor B defective (23)

$E_j \cup E_k$ for E_j in (8.32) we find (see Problem 8.6) for the probability of observing at least one of the compound events E_i, E_j, or E_k

$$P\{E_i \cup E_j \cup E_k\} = P\{E_i\} + P\{E_j\} + P\{E_k\} - P\{E_j \cap E_k\}$$
$$- P\{E_i \cap E_j\} - P\{E_i\} + P\{E_i \cap E_j \cap E_k\} \quad (8.34)$$

In general, as can be shown by successive substitutions and induction

$$P\{E_1 \cup E_2 \cdots \cup E_n\} = \sum_{\{k\}}^{n} P\{E_k\} - \sum_{\{k\ell\}}^{n} P\{E_k \cap E_\ell\}$$

$$+ \sum_{\{k\ell m\}}^{n} P\{E_k \cap E_\ell \cap E_m\}$$

$$- \sum_{\{k\ell mn\}}^{n} P\{E_k \cap E_\ell \cap E_m \cap E_n + \cdots\} \quad (8.35)$$

In (8.35) the indicated summations are over the distinct combinations of integers in the sequence $1, 2, \ldots, n$, taken $1, 2, \ldots, n$ at a time where the numbers of possible combinations are given by Eq. (8.28). As an example of the use of (8.35), we compute $P\{E_1 \cup E_2 \cup E_3 \cup E_4\}$ as follows. From (8.35) we have

$$P\{E_1 \cup E_2 \cup E_3 \cup E_4\} = \sum_{\{k\}}^{4} P\{E_k\} - \sum_{\{k\ell\}}^{4} P\{E_k \cap E_\ell\} + \sum_{\{k\ell m\}}^{4} P\{E_k \cap E_\ell \cap E_m\}$$

$$- \sum_{\{k\ell mn\}}^{4} P\{E_k \cap E_\ell \cap E_m \cap E_n\} \quad (8.36)$$

We examine each of the indicated summations in (8.36) in turn. The first is over single integers. Thus,

$$\sum_{\{k\}}^{4} P\{E_k\} = P\{E_1\} + P\{E_2\} + P\{E_3\} + P\{E_4\} \quad (8.37)$$

The second sum is over the distinct combinations of four integers taken two at a time, which are, in this case, 12, 13, 14, 23, 24, and 34. Thus,

$$\sum_{\{k\ell m\}}^{4} P\{E_k \cap E_\ell E_m\} = P\{E_1 \cap E_2 \cap E_3\} + P\{E_1 \cap E_2 \cap E_4\}$$

$$+ P\{E_1 \cap E_3 \cap E_4\} + P\{E_2 \cap E_3 \cap E_4\} \qquad (8.38)$$

The combination of four integers taken four at a time is in this case 1234. Thus, we have one term only in the sum, and

$$\sum_{\{k\ell mn\}}^{4} P\{E_k \cap E_\ell \cap E_m \cap E_n\} = P\{E_1 \cap E_2 \cap E_3 \cap E_4\} \qquad (8.39)$$

Combining these results yields for the probability of occurrence of at least one of the events E_1, E_2, E_3, or E_4,

$$P\{E_1 \cup E_2 \cup E_3 \cup E_4\} = P\{E_1\} + P\{E_2\} + P\{E_3\} + P\{E_4\} - P\{E_1 \cap E_2\}$$

$$- P\{E_1 \cap E_3\} - P\{E_1 \cap E_4\} - P\{E_2 \cap E_3\}$$

$$- P\{E_2 \cap E_4\} - P\{E_3 \cap E_4\} + P\{E_1 \cap E_2 \cap E_3\}$$

$$+ P\{E_1 \cap E_2 \cap E_4\} + P\{E_1 \cap E_3 \cap E_4\}$$

$$+ P\{E_2 \cap E_3 \cap E_4\} - P\{E_1 \cap E_2 \cap E_3 \cap E_4\} \qquad (8.40)$$

If the compound events $E_1 \cdots E_n$ are mutually exclusive, meaning that if one of the events occurs the others cannot, and concomitantly the areas in sample space that they represent do not overlap (the sets of simple events comprising the compound events are disjoint), then

$$P\{E_1 \cup E_2 \cup \cdots \cup E_n\} = \sum_{i=1}^{n} P\{E_i\} \qquad (8.41)$$

8.5.2 Conditional Probability; Multiplication Rule

We now consider the notion of conditional probabilities. We write the probability, P, of occurrence of event E_2 given, that event E_1 has already occurred as $P\{E_2|E_1\}$ and say that $P\{E_2|E_1\}$ is the conditional probability of occurrence of E_2 given the occurrence of E_1. Further, we write for the probability of occurrence of both events E_1 and E_2, $P\{E_1 \cap E_2\}$,

$$P\{E_1 \cap E_2\} = P\{E_1\}P\{E_2|E_1\} \qquad (8.42)$$

This relation is called the *multiplication rule* in the theory of probability. If E_1 and E_2 are independent events, then

$$P\{E_2|E_1\} = P\{E_2\} \tag{8.43}$$

so, in this case,

$$P\{E_1 \cap E_2\} = P\{E_1\}P\{E_2\} \tag{8.44}$$

Further, we have for the probability of occurrence of all of three possible results, $P\{E_1 \cap E_2 \cap E_3\}$,

$$P\{E_1 \cap E_2 \cap E_3\} = P\{E_1\}P\{E_2|E_1\}P\{E_3|E_2 \cap E_1\} \tag{8.45}$$

where the event E_2 is conditional on the occurrence of event E_1, as expressed by the term $P\{E_2|E_1\}$, and the occurrence of E_3 is conditional on the prior occurrence of both E_1 and E_2, as expressed by the term $P\{E_3|E_2 \cap E_1\}$. If the probabilities of occurrence of E_1, E_2, and E_3 are independent, then

$$P\{E_1 \cap E_2 \cap E_3\} = P\{E_1\}P\{E_2\}P\{E_3\} \tag{8.46}$$

The probabilities of the occurrence of all of ℓ events are computed in a similar fashion, namely

$$P\{E_1 \cap E_2 \cdots \cap E_\ell\} = P\{E_1\}P\{E_2|E_3\}P\{E_3|E_2 \cap E_1\}$$
$$\cdots P\{E_\ell|E_1 \cap E_2 \cdots \cap E_{\ell-1}\} \tag{8.47}$$

We now consider an example of the multiplication rule as follows. An experimenter has a box of 10 well-mixed psychrometer probes, 3 of which are known to be defective. The probability of selecting 2 defective probes in a row, if the probes are not replaced in the box after being drawn, is computed as follows. Since the drawing of probes from the initial population of 10 probes is done without replacement and since the population is finite, the probability of drawing the the second defective psychrometer probe and hence the use of Eq. (8.42) instead of Eq. (8.43) is appropriate. (Note that since sampling a finite population with replacement leaves the probability of observing attributes of the population unchanged, this is the same as sampling an infinite population with or without replacement.) If E_1 is the probability of drawing the first defective psychrometer probe, then the probability of drawing two defective probes in succession is $P\{E_1 \cap E_2\} = P\{E_1\}P\{E_2|E_1\} = \frac{3}{10} \cdot \frac{2}{9} = \frac{2}{45}$. If the first probe is replaced before the second probe is drawn, then the probability of drawing the second probe is independent of the probability of drawing the first. Thus, in this case, $P\{E_2|E_1\} = P\{E_2\}P\{E_1\}$ and hence $P\{E_1 \cap E_2\} = \frac{3}{10} \cdot \frac{3}{10} = \frac{9}{100}$.

8.5.3   Bayes's Theorem

We now consider the relation between the probability of occurrence of E_1 given that E_2 has occurred in terms of the probability of occurrence of E_2 given that E_1 has occurred as follows. If in (8.42) we interchange subscripts we have

$$P\{E_2 \cap E_1\} = P\{E_2\}P\{E_1|E_2\} \tag{8.48}$$

Since $P\{E_1 \cap E_2\} = P\{E_2 \cap E_1\}$, we can equate (8.42) and (8.48) and rearrange to yield

$$P\{E_1|E_2\} = \frac{P\{E_1\}P\{E_2|E_1\}}{P\{E_2\}} \tag{8.49}$$

for the probability of occurrence of E_1, given that E_2 has occurred in terms of the probability of occurrence of E_2, given that E_1 has already occurred. This relation is Bayes's theorem.

This relation can be generalized as follows. If the E_i are all independent events, n in all, and $P\{\mathcal{E}\}$ is the probability of the occurrence of event $\mathcal{E}$, given that the event E_i has already occurred, then the total probability of observing the event $\mathcal{E}$, $P\{\mathcal{E}\}$, is given by

$$P\{\mathcal{E}\} = \sum_{i=1}^{n} P\{E_i\}P\{\mathcal{E}|E_i\} \tag{8.50}$$

If we now write $\mathcal{E}$ in place of E_2 and E_i in place of E_1 in Eq. (8.49) and substitute Eq. (8.50) into the result we have for the general form of Bayes's theorem

$$P\{E_i|\mathcal{E}\} = \frac{P\{E_i\}P\{\mathcal{E}|E_i\}}{\sum_{i=1}^{n} P\{E_i\}P\{\mathcal{E}|E_i\}} \tag{8.51}$$

We now consider an example of the use of Bayes's Theorem as follows. Meters for matric potential measurements are being manufactured. Three separate assembly lines are in operation. Following manufacture, the meters are mixed together for calibration. Assembly line 1 produces twice as many meters in a given length of time as assembly line 2. Assembly line 3 produces two and a half times as many meters in a given length of time as does assembly line 2. It is further found that 90% of the meters as received at the calibration station from assembly line number 1 are within an acceptable range of calibration while 95% of the meters from assembly line 2 and 80% of the meters from assembly line 3, respectively, are within acceptable calibration limits. We now ask for the probability that a meter found to be within acceptable calibration limits came from assembly line 3. If we denote by $\mathcal{E}$ the event of the meter being within acceptable limits, and by

E_i the event of the meter having come from assembly line i, we have for the probability of the meter having come from assembly line 3, $P\{E_3|\mathcal{E}\}$,

$$P\{E_3|\mathcal{E}\} = \frac{P\{E_3\}P\{\mathcal{E}|E_3\}}{P\{E_1\}P\{\mathcal{E}|E_1\} + P\{E_2\}P\{\mathcal{E}|E_2\} + P\{E_3\}P\{\mathcal{E}|E_3\}} \quad (8.52)$$

The probability of a meter coming from line 1, $P\{E_1\}$, expressed in terms of the probability of a meter coming from line 3, $P\{E_3\}$, is $P\{E_1\} = \frac{2.0}{2.5}P\{E_3\}$, while the probability of a meter coming from line 2, expressed in terms of the probability of a meter coming from line 3, is $P\{E_2\} = \frac{1}{2.5}P\{E_3\}$. Also, the probability of a meter from line 1 being within calibration limits as received, $P\{\mathcal{E}|E_1\}$, is, from the data given above, $P\{\mathcal{E}|E_1\} = 0.90$, while $P\{\mathcal{E}|E_2\} = 0.95$ and $P\{\mathcal{E}|E_3\} = 0.80$. Substituting these values into Eq. (8.52) yields

$$P\{E_3|\mathcal{E}\} = \frac{P\{E_3\} \times 0.80}{\frac{2.0}{2.5}P\{E_3\} \times 0.9 + \frac{1}{2.5}P\{E_3\} \times 0.95 + P\{E_3\} \times 0.80} \simeq 0.42$$

8.6 DISCRETE PROBABILITY DENSITY FUNCTIONS

We have previously adopted an operational definition of the probability of occurrence of an attribute of members of a population as being the frequency with which the attribute is observed. The frequencies with which ranges of values of the attribute are observed, called *class intervals*, can be displayed graphically and values of frequencies of observation of attributes of the population can be computed. The aggregate of these frequencies is called the *discrete probability density function*. An example of a discrete frequency distribution, or probability density function, determined experimentally from sample data is given by the data of Table 8.2 which are given in Table 8.5 grouped into classes–ranges of the dependent variable–which in this case is the matric potential. The data are grouped into classes as shown with a constant class interval of 0.02 bars. The absolute frequencies are the numbers of the members of each class, while the relative frequencies are computed by dividing each absolute frequency value by the sum of the absolute frequencies. The relative frequencies are the experimentally determined probabilities of observing a particular attribute of the population–in this case a particular value of matric potential from a population of values of matric potential. The data of Table 8.5 are shown graphically in Figure 8.3. If the values of the heights of the bars in the bar graph of Figure 8.3 are denoted by p_j, where the subscript j refers to the jth class interval, and x_j denotes the value of matric potential at the midpoint of the class interval, then averages of quantities over the probability density function

Table 8.5: Psychrometer data from Table 8.2 grouped into classes

Matric potential (bars)	Absolute frequency	Relative frequency
-4.46	3	0.1875
-4.47		
-4.48		
-4.49	3	0.1875
-4.50		
-4.51		
-4.52	5	0.3125
-4.52		
-4.53		
-4.54		
-4.54		
-4.55	3	0.1875
-4.56		
-4.56		
-4.58	2	0.1250
-4.60		

may be computed. If $f(x_j)$ is a function of the x_j, then the average of $f(x_j)$, $<f>$, over the probability density function may be computed as

$$<f> = \sum_{j=1}^{n} f(x_j)p_j \qquad (8.53)$$

where n is the number of class intervals. As an example of the use of Eq. (8.53) in this case, let us compute the average, $<x>$, of the observed matric potential values. In this case $f(x_j)$ equals x_j. These are the values of matric potential at the midpoints of the class intervals given in Table 8.5 and are computed as the arithmetic averages of the values of matric potential making up the classes. These values are given in Table 8.6. The p_j are the values of relative frequency from Table 8.5 and are reproduced in Table 8.6 for convenience. Thus,

$$<x> = \sum_{j=1}^{5} x_j p_j \simeq -4.526 \qquad (8.54)$$

We now consider a more elaborate example, namely the sample variance, denoted by s^2, which is

$$s^2 = <x^2> - <x>^2 \qquad (8.55)$$

We compute $<x^2>$ and $<x>^2$ as applications of Eq. (8.53) with

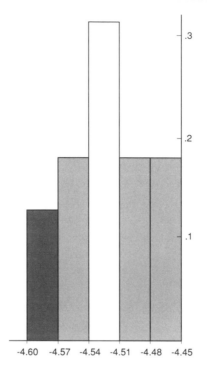

Figure 8.3: Discrete relative frequency function (experimentally determined probability density function) of matric potential measurements made in the process of calibrating a psychrometer probe.

$$< x^2 >= \sum_{j=1}^{5} x_j^2 p_j \simeq 20.48 \qquad (8.56a)$$

$$< x >^2= \left\{ \sum_{j=1}^{5} x_j p_j \right\}^2 \simeq 20.48 \qquad (8.56b)$$

to yield $s^2 \simeq 0$. The standard deviation of experimental values about the mean is

$$s = \left(< x^2 > - < x >^2 \right)^{\frac{1}{2}} \qquad (8.57)$$

In this case, $s = 0$, which is interpreted to mean that with this choice of grouping the data into classes, there is no significant deviation of matric potential values from the mean.

The mth moment of the distribution, $\alpha_m = < x^m >$, which is the average of x^m over the distribution, is given by

Table 8.6: Class intervals of Table 8.5

x_j (bars)	p_j
-4.470	0.1875
-4.500	0.1875
-4.530	0.3125
-4.557	0.1875
-4.590	0.1250

$$< x^m > = \sum_{j=1}^{n} (x_j)^m p_j \tag{8.58}$$

while the average of $(x- < x >)^m$ over the distribution, called the mth central moment, $\beta_m = < (x- < x >)^m >$, or moment about the mean, $< x >$, of the distribution, is given by

$$< (x- < x >)^m > = \sum_{j=1}^{n} (x_j- < x >)^m p_j \tag{8.59}$$

We have used the first moment, the mean, and the second moment, s^2, about the mean, namely the variance, in the example just given. Three quantities are commonly used as measures of central tendency of the probability density function. These are the mean, or first moment of the distribution, the median, x_a, which is the value of the abscissa that separates the graph of the probability density function into equal areas, and the mode, f_m, which is the maximum value of the probability density distribution. Denoting the class interval width by d_j, we have for the area, A, under the histogram

$$A = \sum_{j=1}^{n} d_j p_j \tag{8.60}$$

If $m < n$ is the value of j for which half the area of the histogram is computed, we have

$$\frac{A}{2} = \sum_{j=1}^{m} d_j p_j \tag{8.61}$$

and the median, x_a, is

$$x_a = \sum_{j=1}^{m} d_j \tag{8.62}$$

8.6.1 Binomial Distribution

We now consider a particular discrete theoretical distribution, the binomial distribution. The situation is one in which n trials of an experiment that has a probability p of the occurrence of a given outcome, called a *success*, and a probability $q = 1-p$, called a *failure*, given previously in Eq. (8.31), of the outcome not occurring. The probability, $p(m)$, of exactly m successes, where $0 \leq m \leq n$, out of n trials is

$$p(m) = \binom{n}{m} p^m q^{n-m} \tag{8.63}$$

The average number of successes, $< m >$, often denoted by μ for a theoretical distribution, is given by the average of m over $p(m)$, similarly to (8.54), for the experimental distribution, with $p(m)$ playing the role of the p_j and the x_j taking on the integer values of the variable m. Thus

$$< m >= \sum_{m=0}^{n} m \binom{n}{m} p^m (1 - p)^{n-m} \tag{8.64}$$

where $\binom{n}{m}$ is the usual binomial coefficient. We now evaluate $< m >$ explicitly as follows. Since the first term of the sum indicated in (8.64) is zero, we can begin the sum with the $m = 1$ term and expand the binomial coefficient to write

$$< m >= \sum_{m=1}^{n} m \frac{n(n - 1)!p}{(n - m)!m(m - 1)!} p^{m-1}(1 - p)^{n-m} \tag{8.65}$$

setting the dummy index of summation m equal to $k + 1$, which makes the sum run to $n-1$ instead of n, substituting in (8.65), and rearranging yields

$$< m > = np \sum_{k=0}^{n-1} \frac{(n - 1)!}{(n - 1 - k)!k!} p^k (1 - p)^{n-1-k}$$

$$= np \sum_{k=0}^{n-1} \binom{n - 1}{k} p^k (1 - p)^{n-1-k}$$

$$= np \left[p + (1 - p) \right]^{n-1} = np \tag{8.66}$$

where the binomial expansion

$$(a + b)^{\ell} = \sum_{k=0}^{\ell} \binom{\ell}{k} a^k b^{\ell-k} \tag{8.67}$$

has been used with the identifications $a = p$, $b = (1 - p)$, and $\ell = n - 1$. We now see that the binomial distribution is the summand of the binomial expansion of $(p + q)^n$, and that the the binomial distribution is so named because of its relation to the binomial expansion. The variance, σ^2, of a distribution, may written as

$$\sigma^2 =< m- < m >>^2=< m^2 - 2m < m > + < m >^2>$$

$$=< m^2 > - < 2 < m >^2> + < m >^2=< m^2 > - < m >^2 \quad (8.68)$$

Thus, evaluating $< m^2 >$ as $< m >$ was evaluated previously and substituting the result and (8.66) into (8.68), we have for the variance of the binomial distribution

$$\sigma^2 = np(1 - p) \quad (8.69)$$

and thus for the standard deviation, σ,

$$\sigma = \sqrt{np(1 - p)} \quad (8.70)$$

The probability, $^\ell p$, of at least ℓ successes out of n trials, is given by

$$^\ell p = \sum_{j=\ell}^{n} \binom{n}{j} p^j (1 - p)^{n-j} \quad (8.71)$$

where j is a dummy index of summation. Similarly, the probability, $_\ell p$, of at most (not more than) ℓ successes out of n trials is

$$_\ell p = \sum_{j=0}^{\ell} \binom{n}{j} p^j (1 - p)^{n-j} \quad (8.72)$$

We now consider an example of the use of the binomial distribution. The use of a psychrometer probe involves cooling an internal thermocouple to condense moisture onto its surface and later heating the thermocouple to dry it, by passing a current through it, in preparation for the next measurement. This process deposits salts onto the thermocouple surface over time, making the psychrometer probe increasingly less sensitive and less responsive until finally it fails altogether. This failure depends on the number of measurements made and the environment of the psychrometer probe. Very humid surroundings, such as the atmosphere of a greenhouse, are far more conducive to short psychrometer probe life than operations in an arid desert environment, for example. Assuming that a psychrometer probe has a probability $p = \frac{1}{500}$ of failure in a given environment, a "success" in our terminology as used so far, which corresponds to a failure in one out of 500 psychrometer readings, we can ask for the probability of no failures, $p(0)$, out of, say, $n = 250$ trials as follows. Substituting into Eq. (8.63), we have

$$p(0) = \binom{250}{0} \left(\frac{1}{500}\right)^0 \left(1 - \frac{1}{500}\right)^{250-0} \simeq 0.606$$

which is larger than the value 0.5 which might have been the intuitive "guess" in the absence of knowledge of one of the uses of the binomial distribution. We now compute the probability of at most one failure out of 250 trials, since one failure means the end of the experiment, as follows. Substituting into Eq. (8.72) yields

$$_1P = \sum_{j=0}^{1} \binom{250}{j} \left(\frac{1}{500}\right)^j \left(1 - \frac{1}{500}\right)^{250-j} = 0.91$$

8.6.2   Multinomial Distribution

The cases in which more than two outcomes are possible require the use of the multinomial distribution which we now briefly consider. We have seen that the summand in the expansion of $(p+q)^n$, given in Eq. (8.63), is the binomial distribution. Similarly, the summand in the expansion of $\left(\sum_{i=1}^{k} p_i\right)^n$, where the p_i are the probabilities of occurrence of the k possible individual events being considered and

$$\sum_{i=1}^{k} p_i = 1 \tag{8.73}$$

is

$$p(x_1, \ldots, x_k) = \frac{n!}{\Pi_{i=1}^{k} x_i!} \Pi_{i=1}^{k} p_i^{x_i} \tag{8.74}$$

and is the probability of observing values $x_1, \ldots, x_k$ of the variables they represent. In (8.74) the values of the x_i taken on by the parameters of interest are integers and

$$\sum_{i=1}^{k} x_i = n \tag{8.75}$$

We now consider an example of the use of the multinomial distribution as follows. Soil samples are being screened for transuranic radioactive isotopes as part of a site characterization and cleanup. The soil contamination level above which it is considered to be mixed transuranic waste, and hence subject to strict regulatory control, is 100×10^{-6} Curies/liter ($100\,\mu\,\mathrm{Ci}/\ell$). It is desired to to estimate the probabilities of exposure of workers to various levels of radiation based on the results of the screening process. It is found that on average 10% of the soil samples are found to have an activity level of exactly $100\,\mu\,\mathrm{Ci}/\ell$, 20% are found to have a level of activity less than $100\,\mu\,\mathrm{Ci}/\ell$, while 70% are found to have an activity greater than $100\,\mu\,\mathrm{Ci}/\ell$. If a worker takes 10 soil samples, the probability of particular numbers of the samples having levels of activity below x_1 at x_2 or above x_3, the regulatory differentiation value, is given by

$$p(x_1, x_2, x_3) = \frac{(x_1 + x_2 + x_3)!}{x_1! x_2! x_3!}(0.1)^{x_1}(0.2)^{x_2}(0.7)^{x_3} \qquad (8.76)$$

where

$$\sum_{i=1}^{3} x_i = 10 \qquad (8.77)$$

which is the number of samples, and each of the x_i can take on values ranging from 1 to 10 subject to the condition of Eq. (8.77). Equation (8.76) represents the situation considered and is used as follows. If values of $x_1 = 0$, $x_2 = 0$, and $x_3 = 10$ are chosen, corresponding to the the situation in which no samples at or below the regulatory cutoff are encountered by soil sampling workers, but instead all samples are mixed transuranic waste, we have for the probability of this occurrence,

$$p(0, 0, 10) = (0.7)^{10} \simeq 0.028$$

corresponding to a probability of 2.8%.

8.7 CONTINUOUS PROBABILITY DENSITIES

8.7.1 Gaussian Distribution

We begin by considering the normal, or Gaussian, probability density function, which can be derived as a limit for large n of the binomial distribution, which derivation we now sketch. Beginning with (8.63) we have

$$p(m) = \frac{n!}{m!(n-m)!}p^m(1-p)^{n-m}$$

$$\simeq \frac{n^n e^{-n}\sqrt{2\pi n}}{m^m e^{-m}\sqrt{2\pi m}(n-m)^{n-m}e^{-n+m}\sqrt{2\pi(n-m)}}p^m(1-p)^{n-m} \qquad (8.78)$$

in which Stirling's approximation for each factorial has been used and embodies the assumption of large n. Rearranging (8.78) yields

$$p(m) \simeq \frac{1}{\sqrt{2\pi n}}\exp\left[\left(-m-\frac{1}{2}\right)\ln\left(\frac{m}{n}\right) - \left(n-m+\frac{1}{2}\right)\right]$$

$$\times \exp\left[\ln\left(\frac{n-m}{n}\right) + n\ln(p) + (n-m)\ln(1-p)\right] \qquad (8.79)$$

We now set $m = np + \zeta$, where $\zeta \ll np$, which reflects the fact that for large np, the average value of m of the binomial distribution, given in Eq. (8.66), is expected to be the maximum value of $p(m)$. Substituting this value of m into (8.78) yields

$$p(m) = \frac{1}{\sqrt{2\pi n}} \exp\left[\left(-np - \zeta - \frac{1}{2}\right)\ln\left(\frac{mp+\zeta}{n}\right)\right]$$

$$\times \exp\left[-\left(n - np - \zeta + \frac{1}{2}\right)\ln\left(\frac{n - np - \zeta}{n}\right)\right]$$

$$\times \exp\left[(np + \zeta)\ln(p) + (n - np - \zeta)\ln(1 - p)\right] \tag{8.80}$$

Expanding $\ln(p + \zeta/n)$ and $\ln(1 - p - \zeta/n)$ in series using the result

$$\ln(x + a) = \ln(x) + 2\left(\frac{a}{2x + a} + \cdots\right) \tag{8.81}$$

yields

$$\ln\left(p + \frac{\zeta}{n}\right) = \ln(p) + \frac{2\zeta/n}{2p + \zeta/n} + \cdots \tag{8.82a}$$

and

$$\ln\left(1 - p - \frac{\zeta}{n}\right) = \ln(1 - p) - \frac{2\zeta/n}{2(1 - p) - \zeta/n} + \cdots \tag{8.82b}$$

Substituting (8.82a) and (8.82b) into (8.79) and rearranging yields, upon noting that $p(m)$ becomes $p(\zeta)$,

$$p(\zeta) \simeq \frac{1}{\sqrt{2\pi np(1 - p)}} e^{-2(np+\zeta+1/2)(\frac{\zeta/n}{2p+\zeta/n})}$$

$$\times e^{[2(n-np-\zeta+\frac{1}{2})(\frac{\zeta/n}{2(1-p)-\zeta/n})]} \tag{8.83}$$

where all the $\ln()$ terms have cancelled. Rearranging (8.83) and dropping higher-order terms yields eventually

$$p(\zeta) = \frac{1}{\sqrt{2\pi np(1 - p)}} e^{-\zeta^2/[2np(1-p)]} \tag{8.84}$$

In Eq (8.84) we identify $\sqrt{np(1 - p)} = \sigma$ with the standard deviation of the binomial distribution given in Eq. (8.70). Thus, $p(\zeta)$ can be written as

$$p(\zeta) = \frac{1}{\sqrt{2\pi}\sigma} e^{-\zeta^2/(2\sigma^2)} \tag{8.85}$$

This point will be revisited shortly in applications of the Gaussian distribution as well as in some further applications of the binomial distribution.

In order to consider the Gaussian distribution from a more usual point of view, we begin by assuming the existence of a continuous, real, frequency distribution, $f(x)$, with x in the range $-\infty \leq x \leq \infty$ and $f(x) > 0$. The statement that $f(x)$ is normalized to one is equivalent to the relation

$$\int_{-\infty}^{\infty} f(x)\, dx = 1 \tag{8.86}$$

For the case in which $f(x)$ is normalized to 1 the average, $< g >$, of a function $g(x)$ over the probability density function $f(x)$ is given by

$$< g(x) > = \int_{-\infty}^{\infty} g(r)f(x)\, dx \tag{8.87}$$

In the case in which $f(x)$ is not normalized, $< g(x) >$ would be computed using

$$< g(x) > = \frac{\int_{-\infty}^{\infty} g(x)f(x)\, dx}{\int_{-\infty}^{\infty} f(x)\, dx} \tag{8.88}$$

Examples of $< g(x) >$ are the average of the mth moment of $f(x)$ given by

$$\alpha_m = < x^m > = \int_{-\infty}^{\infty} x^m f(x)\, dx \tag{8.89}$$

and the mth central moment, $< (x- < x >)^m >$, the average of $(x- < x >)^m$ over $f(x)$, which is

$$\beta_m = < (x- < x >)^m > = \int_{-\infty}^{\infty} (x- < x >)^m f(x)\, dx \tag{8.90}$$

The second central moment, the central moment for $m = 2$, is the variance of the distribution as given previously. The median, x_a, which is the value of x that divides a plot of $f(x)$ versus x into two equal areas, is given by

$$\frac{1}{2} = \int_{-\infty}^{x_a} f(x)\, dx \tag{8.91}$$

while, for a unimodal distribution, that is, one having a single maximum, f_m, the value of x for which the probability density function is a maximum, is the root of the equation

$$\frac{df(x)}{dx} = 0 \tag{8.92}$$

For the case of a bimodal distribution, namely one having two maxima, Eq. (8.92) would have three roots, namely the three values of x corresponding to the two maxima and the relative minimum between them.

By way of example, we choose the following unnormalized probability density function, $f_u(x)$, namely

$$f_u(x) = ae^{-b(x-c)^2} \tag{8.93}$$

and seek to determine the constants a, b, and c, in terms of some of the statistical quantities discussed. We begin by normalizing $f_u(x)$ to one to yield $f(x)$ as follows.

$$1 = \int_{-\infty}^{\infty} f(x)\, dx = a \int_{-\infty}^{\infty} e^{-b(x-c)^2}\, dx = a\sqrt{\frac{\pi}{b}} \tag{8.94}$$

Solving (8.94) for a yields for $f(x)$, so far,

$$f(x) = \sqrt{\frac{b}{\pi}} e^{-b(x-c)^2} \tag{8.95}$$

We now compute the average of x, $< x >$, the first moment of the distribution, over the distribution to find

$$< x > = \sqrt{\frac{b}{\pi}} \int_{-\infty}^{\infty} xe^{-b(x-c)^2}\, dx = c \tag{8.96}$$

Thus, $f(x)$ now becomes

$$f(x) = \sqrt{\frac{b}{\pi}} e^{-b(x-<x>)^2} \tag{8.97}$$

We now compute the variance, σ^2, for this theoretical distribution as follows [compare Eq. (8.68), which was written for the variance of a discrete distribution also normalized to 1].

$$\sigma^2 = < x^2 - < x >^2 > = < x^2 > - < x >^2$$

$$= \sqrt{\frac{b}{\pi}} \int_{-\infty}^{\infty} (x^2 - < x >^2) e^{-b(x-<x>)^2}\, dx$$

$$= \sqrt{\frac{b}{\pi}} \frac{\sqrt{\pi}}{2b^{3/2}} = \frac{1}{2b} \tag{8.98}$$

Thus,

$$b = \frac{1}{2\sigma^2} \tag{8.99}$$

and $f(x)$ becomes

$$f(x) = \frac{1}{\sqrt{2\pi}\sigma} e^{-(x-<x>)^2/2\sigma^2} \tag{8.100}$$

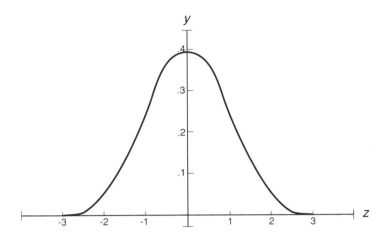

Figure 8.4: A plot of the probability density for the Gaussian distribution for the case in which the average equals zero and the standard deviation equals one.

The probability density function of (8.98) will be recognized as the Gaussian or normal distribution. It can be shown that the median of this distribution is $< x >$ and that the mode, the value of x for which $f(x)$ is a maximum, is $< x >$ as well. From Eq. (8.100) it is seen that Eq. (8.85), containing the form of the Gaussian distribution derived from the binomial distribution, is written for the case in which $< x > = 0$. A plot of $f(x)$ versus x for values of $< x > = 0$ and $\sigma = 1$ is given in Figure 8.4. The cases in which experimental data can be fitted to assumed continuous probability density functions are often more important theoretically and practically than the cases in which discrete distributions would be used. Examples of both will now be given.

The probability of occurrence of any particular range of values of x lying between the values a and b, say, which represents a particular physical quantity, depending on the situation being considered, is equal to the area under the curve of $f(x)$ versus x delimited by the endpoints of the range of x. Since the area under the curve over a single point is zero, the probability of occurrence of a particular value of x is equal to zero. Thus, while it is usually important to state clearly whether or not the endpoints of an interval are included within the interval, the range of x in this instance, here due to the fact of the area over a single point being equal to zero, the same results are obtained whether or not the endpoints of the interval over which x is being considered are included or not, and it is only meaningful to consider the probabilities associated with ranges of x. Thus, the probability of occurrence of x in the range from a to b, $\Pr(a \leq x \leq b)$, for the Gaussian distribution is

$$\Pr(a \leq x \leq b) = \frac{1}{\sqrt{2\pi}\sigma} \int_a^b e^{-(x-\mu)^2/2\sigma^2} \, dx \qquad (8.101)$$

The use of the Gaussian distribution is facilitated if differences of values of x from the mean, which we have denoted by $x = \mu$ in Eq. (8.101), divided by the standard deviation are considered. The relevant quantity, denoted by z, is called the *standard normal random variable*, and is

$$z = \frac{x - \mu}{\sigma} \qquad (8.102)$$

The variable z has a Gaussian distribution (is said to be normally distributed), with a mean of zero and a standard deviation equal to one. For calculational convenience a table of areas under the right half of the curve of Figure 8.4, which is symmetric about the mean, has been calculated as a function of z using the series expansion

$$\Pr(0 \leq z) = \Phi(z) = \frac{1}{\sqrt{2\pi}} \int_0^z e^{-x^2/2} \, dx = \frac{1}{\sqrt{2\pi}} \sum_{k=o}^{\infty} \frac{(-\frac{1}{2})^k z^{2k+1}}{(2k+1)k!} \qquad (8.103)$$

and is given in Table D.3b while values of z as a function of area, critical values of the Gaussian distribution, are given in Table D.3a. Computations using the table are aided by noting that the area under the curve of Figure 8.4 equals 1 for $-\infty \leq x \leq \infty$, so that in the limit as $z \to \infty$, as could be shown directly by evaluating the improper integral using the gamma function,

$$\Pr(0 \leq \infty) = \frac{1}{\sqrt{2\pi}} \int_0^{\infty} e^{-x^2/2} \, dx = \frac{1}{2} \qquad (8.104)$$

We now consider an example as follows. A set of soil samples is analyzed for the presence of transuranic radionuclides. If the average, m, of the values found in 100 μCi/ℓ, and the standard deviation s of the readings is 25 μCi/ℓ, and it is assumed that the observed values are normally distributed, what is the probability of the soil samples having levels of activity in the range of 100 to 250 μCi/ℓ? We begin by calculating the value of the z statistic for this situation, which is

$$z = \frac{200 - 150}{25}$$

where the units have been divided out. The value of $z = 2$ corresponds to an area between the values of $z = 0$ and 2 of 0.4772, as seen from Table D.3b, which represents a 47.72 % probability of any one soil sample having a level of activity of transuranics in the 100 to 200 $\mu Ci/\ell$ range. The probability of encountering a soil sample having a level of activity greater than 200 $\mu Ci/\ell$ is given by the area under the normal curve between $z = 2$ and ∞, where the area between $z = 0$ and ∞ corresponds to an area of 0.5, as noted in connection with Eq. (8.104). Thus the required probability is given by $[\Pr(0 \leq z \leq \infty) - \Pr(0 \leq z \leq 2)] \times 100\% = (0.5 - 0.4772) \times 100\% \simeq$ 2.3%. Note that if the mean, standard deviation, and z statistic, this last possibly corresponding to a known or chosen percentile, are known, the corresponding value of x is found by solving Eq. (8.102) for x to yield

$$x = z\sigma + \mu \qquad (8.105)$$

We now note that when certain conditions are satisfied, the Gaussian distribution is a good approximation to the binomial distribution. The general condition to be satisfied is that the interval of three standard deviations above and below the mean of the Gaussian distribution be contained in the range of 0 to n of the binomial distribution. Thus, if the mean and standard deviation are computed for the binomial distribution, this condition may be written as

$$0 < \mu \pm 3\sigma < n \qquad (8.106a)$$

or

$$0 < np \pm \sqrt{p(1 - p)} \qquad (8.106b)$$

The z statistic is computed as usual, with the caveat that a correction factor can be included to account for the fact that the value of the random variable for the case in which the binomial probability is being computed, m, in Eq. (8.64), for example, proceeds in integer steps, whereas the Gaussian independent variable is continuous. For x in the range between a and b, where $a < b$,

$$z = \frac{b \pm 0.5 - \mu}{\sigma} = \frac{b \pm 0.5 - np}{\sqrt{p(1 - p)}} \qquad (8.107a)$$

where the upper sign is used if $\Pr(x > b)$ and the lower sign is used if $\Pr(x \leq b)$. Similarly, for the lower limit of the range of x, we have

$$z = \frac{a \pm 0.5 - \mu}{\sigma} = \frac{a \pm 0.5 - np}{\sqrt{p(1 - p)}} \qquad (8.107b)$$

where the upper sign is chosen if $\Pr(x \leq a)$ and the lower sign is chosen if $\Pr(x > a)$. The practical utility of this approximation lies in the fact than tables of the Gaussian distribution are generally more readily available and cover a wider range than do tables of the binomial distribution.

We now consider three other continuous distributions, the χ^2 distribution, Student's t distribution, and the F distribution. All of these distributions have the common characteristic of being functions of normally distributed independent variables. In general, for a particular distribution, $f(x)$, say, of a number of independent variables, $x_1, x_2, \ldots, x_n$, say, each having the same distribution is being considered, then the distribution of the combination of of the distributions of the n independent variables, $g(x_1, \ldots, x_n)$, say, is given by

$$g(x_1, \ldots, x_n) = \prod_{i=1}^{n} f(x_i) \tag{8.108}$$

We now apply this result to consideration of the χ^2 distribution as follows.

8.7.2 χ^2 Distribution

We consider ν independent random variables, $x_1, \ldots, x_\nu$, and assume that each variable is normally distributed with a mean of zero and a standard deviation of one. This is the distribution of the z statistic just discussed and plotted in Figure 8.4. From Eqs. (8.85) and (8.108) we have for $g(x_1, \ldots, x_\nu)$ in this case

$$g(x_1, \ldots, x_\nu) = \frac{1}{(2\pi)^{\nu/2}} e^{-(x_1^2 \cdots x_\nu^2)/2} \tag{8.109}$$

We now consider $\chi^2 = x_1^2 \cdots x_\nu^2$ to be a random variable and seek its probability density function and the properties and uses of that function. The probability that χ^2 is less than or equal to some particular value, χ_0^2, say, is written as, after some development that involves integrating $g(x_1, \ldots, x_\nu)$ over a volume in the hyperspace of the $x_1 \cdots x_\nu$ such that χ_0^2 is the boundary of the volume (a many-to-one mapping) and normalizing the result to one, and observing that since the x_i are squared, the quantities comprising χ^2 are all positive, so that $0 \leq \chi^2 \leq \infty$ and

$$\Pr(\chi^2 \leq \chi_0^2) = \frac{1}{2^{\frac{\nu}{2}} \Gamma\left(\frac{\nu}{2}\right)} \int_0^{\chi_0^2} x^{\frac{\nu}{2}-1} e^{-x/2} \, dx \tag{8.110}$$

where $\Gamma()$ is the usual gamma function, while the probability density function of Eq. (8.110), $d(x, \nu)$, is

$$d(x, \nu) = \frac{1}{2^{\nu/2} \Gamma\left(\frac{\nu}{2}\right)} x^{\frac{\nu}{2}-1} e^{-x/2} \tag{8.111}$$

where the probability density function $d(x,\nu)$ has been normalized to one so that

$$\int_0^\infty d(x,\nu)\,dx = 1 \qquad (8.112)$$

The number of independent random variables, ν, is called the *number of degrees of freedom* of the χ^2 distribution. One of the uses to which the χ^2 distribution can be put is to test the goodness of fit of an assumed distribution to data and will be discussed shortly. We now discuss the t distribution.

8.7.3 Student's t Distribution

We now consider Student's ("Student" was a pseudonym used by Gosset, a British mathematician employed by a brewery that frowned on publications by its employees-hence the necessary anonymity.) t distribution as follows. The variable t of the distribution is composed of two independent variables, one, ζ, say, having a Gaussian distribution with a mean of zero and a standard deviation of one and a variable, ξ, say, that has a χ^2 distribution as given in eq.(8.110) such that

$$t = \sqrt{\nu}\frac{\zeta}{\xi} \qquad (8.113)$$

where ν is the number of degrees of freedom of the χ^2 distribution. The distribution of the variable t is found by integrating the joint frequency function, $g(\zeta, y)$, over a volume in ζ–y space. Thus, using (8.108) along with (8.85) with $\sigma = 1$, yields

$$g(\zeta, y) = \frac{1}{\sqrt{2\pi}}e^{-\zeta^2/2}\frac{1}{2^{\nu/2}\,\Gamma\left(\frac{\nu}{2}\right)}\zeta^{\frac{\nu}{2}-1}e^{-\zeta/2} \qquad (8.114)$$

The probability of t lying between $-\infty$ and x, $\Pr(t \le x)$, where

$$x \ge \sqrt{\nu}\frac{\zeta}{\xi} \qquad (8.115)$$

is given by

$$\Pr(t \le x) = \frac{k(\nu)}{\sqrt{2\pi}2^{\nu/2}\,\Gamma\left(\frac{\nu}{2}\right)}\int_0^\infty y^{\frac{\nu}{2}-1}e^{-y/2}\,dy\int_{-\infty}^{x\sqrt{\zeta}/\sqrt{\nu}}e^{-\zeta^2/2}\,d\zeta \qquad (8.116)$$

where the upper limit of the second integral on the right-hand side of (8.116) is obtained by solving (8.115) for ζ, and $k(\nu)$ is found by applying the condition that the probability density function for t, denoted here by $s(x, \nu)$, is normalized to one over the interval $-\infty \le x \le \infty$. Carrying out the indicated integration, facilitated by the change of variable in the second integral of $y = \zeta \sqrt{\nu}/\xi$,

$$\Pr(t \le x) = T(x, \nu) = \frac{\Gamma\left(\frac{\nu+1}{2}\right)}{\sqrt{\nu\pi}\,\Gamma\left(\frac{\nu}{2}\right)} \int_{-\infty}^{x} \frac{1}{\left(1 + \frac{y^2}{\nu}\right)^{\frac{\nu+1}{2}}}\, dy \tag{8.117}$$

where $s(x, \nu)$, the probability density function for the t distribution is now seen to be

$$s(x, \nu) = \frac{\Gamma\left(\frac{\nu+1}{2}\right)}{\sqrt{\nu\pi}\,\Gamma\left(\frac{\nu}{2}\right)} \frac{1}{\left(1 + \frac{x^2}{\nu}\right)^{\frac{\nu+1}{2}}} \tag{8.118}$$

where ν is called the *number of degrees of freedom* of the distribution. The t distribution is used in estimating population parameters for small numbers of measurements of population properties and will be discussed in more detail shortly.

We now consider the F distribution, which is composed of two χ^2 distributions, as follows.

8.7.4 F Distribution

The function $g(\zeta, \xi, \nu_1, \nu_2)$ composed of two χ^2 distributions having independent variables ζ and ξ, having ν_1 and ν_2 degrees of freedom, respectively, is given by

$$g(\zeta, \xi, \nu_1, \nu_2) = \frac{\zeta^{\frac{\nu_1}{2}-1} e^{-\zeta/2} \xi^{\frac{\nu_2}{2}-1} e^{-\xi/2}}{2^{\nu_1/2}\,\Gamma\left(\frac{\nu_1}{2}\right)\, 2^{\nu_2/2}\,\Gamma\left(\frac{\nu_2}{2}\right)} \tag{8.119}$$

The variable F, the distribution of which we require, is given by

$$F = \frac{\zeta/\nu_1}{\xi/\nu_2} = \frac{\zeta\nu_2}{\xi\nu_1} \tag{8.120}$$

Introducing a normalization parameter, $k(\nu_1, \nu_2)$, and integrating to find the probability that F is less than or equal to a particular value, x, $\Pr(F \le x)$, leads to

$$\Pr(F \le x) = \frac{k(\nu_1, \nu_2)}{2^{\frac{\nu_1+\nu_2}{2}}\,\Gamma\left(\frac{\nu_1}{2}\right)\,\Gamma\left(\frac{\nu_2}{2}\right)}$$

$$\times \int_0^\infty \int_0^{x\xi\nu_1/\nu_2} \zeta^{\frac{\nu_1}{2}-1} \xi^{\frac{\nu_2}{2}-1} e^{-(\zeta+\xi)/2}\, d\zeta\, d\xi \tag{8.121}$$

Denoting the probability density function by $f(x, \nu_1, \nu_2)$, making the change of variable $y = \zeta \nu_2/\nu_1 \xi$, evaluating $k(\nu_1, \nu_2)$ from the normalization condition

$$\int_0^\infty k(x, \nu_1, \nu_2)dx = 1 \tag{8.122}$$

and carrying out the integration over ζ yields

$$\Pr(F \le x) = F(x, \nu_1, \nu_2) = \frac{\Gamma\left(\frac{\nu_1+\nu_2}{2}\right)}{\Gamma\left(\frac{\nu_1}{2}\right)\Gamma\left(\frac{\nu_2}{2}\right)} \int_0^x \frac{y^{\frac{\nu_2}{2}-1}}{(y+1)^{\frac{\nu_1+\nu_2}{2}}} \, dy \tag{8.123}$$

and we make the identification

$$f(x, \nu_1, \nu_2) = \frac{\Gamma\left(\frac{\nu_1+\nu_2}{2}\right) x^{\frac{\nu_2}{2}-1}}{\Gamma\left(\frac{\nu_1}{2}\right)\Gamma\left(\frac{\nu_2}{2}\right)(x+1)^{\frac{\nu_1+\nu_2}{2}}} \tag{8.124}$$

The F distribution is used in the analysis of variance, which will be discussed shortly. We will consider applications of the distributions given so far in statistical inferences about characteristics of members of populations.

8.8 PARAMETER ESTIMATION

8.8.1 Central Limit Theorem

The central limit theorem of probability, which we shall give without proof, states the following. If $x_1 \cdots x_n$ are independent random variables that may or may not be normally distributed, then as n becomes large the sum $x = \sum_{i=1}^n x_i$ is asymptotically normally distributed with a mean μ given by

$$\mu = \sum_{i=1}^n \mu_i \tag{8.125}$$

where the μ_i are the standard deviations of the individual distributions and a variance σ^2 given by

$$\sigma^2 = \sum_{i=1}^n \sigma_i^2 \tag{8.126}$$

where the σ_i^2 are the variances of the individual distributions. Further, in the case that the x_i each have the same underlying probability distribution, normal or not, and all distributions have the same mean, μ_d, and standard deviation, σ_d, the average of the individual values of the x_i, $\bar{x}$, is given by

$$\bar{x} = \frac{1}{n} \sum_{i=1}^n x_i \tag{8.127}$$

and its probability distribution tends asymptotically to a normal distribution for large n with mean $\mu = \mu_d$ and standard deviation $\sigma_s = \sigma_d/\sqrt{n}$. The quantity σ_s is called the standard deviation of the mean μ_s.

If we now consider the x_i to be individual observations of a variable X of a single population and the variable has a mean μ and standard deviation σ, the mean of the distribution of the sum, x, of the x_i, the sampling distribution in this case, is given by $n\mu$, and the standard deviation of the distribution of the sum of the x_i is given by $n\sigma$, as can be seen from Eqs. (8.125) and (8.126) if the μ_i and σ_i, respectively, are equal. In practice we can estimate the value of a population property by the average of its measured values. If we apply the results of the central limit theorem, we can construct an interval estimator of the population parameter of the form

$$I = \bar{x} \pm z\sigma_s = \bar{x} \pm z\frac{\sigma_d}{\sqrt{n}} \tag{8.128}$$

where the x_i of Eq. (8.127) are taken to be individual measurements of the population attribute, n is the number of measurements of the attribute, σ_d is the standard deviation of the distribution of the population attribute, while σ_s is the standard deviation of the sampling distribution of $\bar{x}$, z is the (possibly fractional) number of standard deviations above and below $\bar{x}$ comprising the confidence interval I, and it is recalled that the sampling distribution of $\bar{x}$ is asymptotically normal so that z here has the same meaning as in Eq. (8.102). Computationally, evaluating I is hampered by the fact that σ_d is a population parameter and hence unknown, so that σ_s is unknown as well. For large values of n, however (in practice for $n \geq 30$), we may use the standard deviation, s, of the measured values as a point estimator of σ_d, where we recall that s, an unbiased estimator for σ, the unknown standard deviation of the population, is given by

$$s = \left(\frac{\sum_{i=1}^{n} (x_i - \mu)^2}{n} \right)^{\frac{1}{2}} \tag{8.129}$$

where μ is the unknown population mean. We note that since s is a function of a random variable, the set of the x_i, it is itself a random variable and as a point estimator of σ_d has its own sampling distribution, average value, and variance as well. It can be shown that the average of the sample variance, $\overline{s^2}$, is related to σ_d for situations in which the central limit theorem applies via

$$\overline{s^2} = \frac{n-1}{n}\sigma_d^2 \tag{8.130}$$

where n is the number of samples and in the large-sample limit we have

$$s = \left(\frac{\sum_{i=1}^{n} (x_i - \bar{x})^2}{n-1} \right)^{\frac{1}{2}} \qquad (8.131)$$

where $\bar{x}$ is the sample mean given by

$$\bar{x} = \frac{1}{n} \sum_{i=1}^{n} x_i \qquad (8.132)$$

Further, it can be shown that the standard deviation, s_s, of the sample variance is

$$s_s = \left(\frac{\beta_4 - \beta_2^2}{n} - \frac{2(\beta_4 - 2\beta_2^2)}{n^2} + \frac{\beta_4 - 3\beta_2^2}{n^3} \right)^{\frac{1}{2}} \qquad (8.133)$$

where β_2 and β_4 are the second and fourth central moments of the distribution as given in Eq. (8.59) for a discrete distribution or Eq. (8.90) for a continuous distribution.

We now consider examples of the uses of these distributions beginning with an application of the central limit theorem. Thirty measurements of the levels of cadmium in samples of hazardous waste extract are made. The average of the measurements is 0.6 μg of cadmium per liter of extract (0.6 μg/ℓ) and s as computed using Eq. (8.131) is found to be 0.2 μg/ℓ. The underlying distribution of the cadmium in the extract is not known, but yet it is desired to compute a confidence interval at the 95% level, within which the value of the population average is represented by the sample average. We thus employ Eq. (8.128) with σ_d estimated by s and with $z = 1.96$, which is the z value corresponding to 0.475 (0.95/2) of the area under a Gaussian distribution having a mean of zero and a standard deviation of one from Table D.3a. Thus, from Eq. (8.128)

$$I = 0.6 \, \mu\text{g}/\ell \pm 0.072 \qquad (8.134)$$

or, at the 95% confidence level, the value of the population mean is found to lie in the interval $0.528 \leq \mu \leq 0.672 \, \mu\text{g}/\ell$.

8.8.2 χ^2 Distribution

We next consider applications of the χ^2 distribution and begin with a short discussion of the computation and arrangement of Table D.5. The values in the table were computed using a combination of Newton–Raphson iteration (Scarborough, 1955, pp. 192–198) and bisection (Press et al., 1986, pp.

246–247) with the starting values of the computations taken from Pearson and Hartley (1954, pp. 130–131). It will be noted that more significant figures are displayed in Table D.5 than is customary and that in some cases the accuracy of the entries has been improved as well. The values of χ^2 in the tables are those of the probability integral rather than the value of the integral subtracted from 1, as is customary in many tables. It is the latter form that is used in evaluating the implications of particular values of computed statistics and corresponds to values of the integral that would exceed the values of χ^2 computed from data in particular situations at a particular level of significance–this term to be made precise shortly–for the test.

The first application is to test hypotheses regarding population variances estimated from measurements on small numbers of samples. The statistic we shall use is

$$\chi_i^2 = \frac{(n-1)s^2}{\sigma_i^2} \tag{8.135}$$

where n is the number of sample points, $n-1$ is the number of degrees of freedom, ν for short, for this application, s^2 is the variance of the sample values, the x_i, and is given by

$$s^2 = \frac{\sum_{i=1}^n (x_i - \bar{x})^2}{n-1} \tag{8.136}$$

and σ_i^2 is the assumed/inferred value of population variance, the null hypothesis, that is to be tested against the alternative hypothesis that the population variance has a different value. Either one- or two-tailed tests are possible, and we shall consider an example of each. From the mathematical form of the test statistic, specifically the fact that it is inversely proportional to σ_i^2, we see that for a one-tailed test the probability that the test statistic, χ^2, for particular circumstances is larger than the value of $\chi_{1-\alpha}^2$, where α is the level of significance of the test, is the area under a curve of χ^2 between the value $\chi_{1-\alpha}^2$ and infinity where $1-\alpha$ is the confidence coefficient associated with the test and $(1 - \alpha) \times 100\%$ is the confidence level of the test. Thus, we write for the probability that χ^2 exceeds $\chi_{1-\alpha}^2$

$$\Pr\left(\chi^2 > \chi_{1-\alpha}^2\right) = \alpha \tag{8.137}$$

where χ^2 and $\chi_{1-\alpha}^2$ each depend on $\nu = n - 1$.

Six samples are taken from a hazardous waste stream and found to contain cadmium at the levels 0.5, 0.7, 0.4, 0.8, 0.6, and 0.9 micrograms per liter ($\mu g/\ell$) of extract. The average of these measurements is .65 $\mu g/\ell$, and it is desired to test the hypothesis that at the 95% confidence level the standard deviation of the population is 0.35 or less since the regulatory action level for cadmium in waste extract is 1 $\mu g/\ell$. The null hypothesis to be tested is thus that $\sigma = \sigma_t$, where σ_t, the test value of standard deviation for the population, is 0.35. Since the hypothesis would be rejected if, statistically speaking, σ were found to be greater than σ_t, a one-tailed test will suffice. From the mathematical form of the test statistic given in Eq. (8.135), we see that if its value is greater than the value, $\chi^2_{1-\alpha}$, found from Table D.5, for particular values of α and $n - 1$, then the hypothesis is to be rejected. From the data given we have for the test statistic

$$\chi^2_i = \frac{5(0.837)}{0.1225} \simeq 34.2$$

From Table D.5b, for $\nu = n - 1 = 5$, and $1 - \alpha = 0.90$, so that $\alpha = 0.1$, we find $\chi^2_{0.9} \simeq 9.24$, and since $\chi^2_i > \chi^2_{1-\alpha}$, in this case, we reject the hypothesis that σ for the waste stream population is 0.35 or less.

For a two-tailed test, the hypothesis being tested is that the variance of the population has a particular value and is tested against the alternative hypothesis that the variance may have a larger or smaller value than the test value. Since the rejection regions are at the upper and lower ends of a plot of χ^2, the area of each rejection region is $1 - \alpha/2$. Thus, for the lower tail of the distribution, the value of χ^2 bounding the area between zero and $1 - \alpha/2$ is denoted by $\chi^2_{\alpha/2}$, with the understanding that the subscript denotes the area under the χ^2 curve between zero and $1 - \alpha/2$, while $\chi^2_{1-\alpha/2}$ denotes the value of χ^2 for which the area $1 - \alpha/2$ is contained under the upper end of the curve. Since the χ^2 distribution is not symmetric, $\chi^2_{\alpha/2} \neq \chi^2_{1-\alpha/2}$. Finally, for a two-tailed test the rejection values of the null hypothesis for the test statistic of Eq. (8.135) are $\chi^2_i < \chi^2_{\alpha/2}$ and $\chi^2_i > \chi^2_{1-\alpha/2}$.

Extreme variability in measured values of hazardous waste is often observed. In such cases it is of value to test the hypothesis that the standard deviation of the population has a particular value, the null hypothesis, against the alternative hypothesis that the standard deviation of the population is either larger or smaller than the the test value. Samples from a hazardous waste stream are analyzed for acetone. The measured values in six of the samples are 180, 84, 73, 37, 35, and 30, $\mu g/kg$. It is hypothesized that the standard deviation of the measurements is 10 $\mu g/kg$. At the 90% confidence level, the rejection regions would be $\chi^2_{0.05}$ for the lower tail, and $\chi^2_{0.95}$ for the upper tail of the distribution. For $n - 1 = 5$ degrees of freedom in this case, we see from Table D.5a that $\chi^2_{0.05} \simeq 1.145$ and from Table D.5c that $\chi^2_{0.95} \simeq 11.07$. Computing the value of the test statistic, χ^2_i, yields χ^2_i, yields $\chi^2_i \simeq 161.6$, which is in the rejection region for the upper tail and further shows that the standard deviation of the population is larger than the test value.

We can identify a confidence interval for the population variance based on the two-tailed test using the χ^2 statistic as

$$\frac{(n-1)s^2}{\chi^2_{1-\alpha/2}} < \sigma^2 < \frac{(n-1)s^2}{\chi^2_{\alpha/2}} \qquad (8.138)$$

which will be recognized as being derived from rearrangement of Eq. (8.135) for both the upper and lower rejection regions. Computing the confidence limits for the example just given leads to $7.3 \times 10^3 < \sigma^2 < 7.06 \times 10^4$, or, upon taking the square root of each term (compare Kazarinoff, 1961, p. 4) $85.4 < \sigma < 265.6$, which does not even contain the hypothesized test value within its range.

We now consider the use of the χ^2 distribution for testing the "goodness of fit" in the statistical sense of an assumed distribution to measured data. We note in this connection that the theoretical distributions to which data are fitted are not necessarily unique and that a good fit is a necessary but not a sufficient condition for the assumed theoretical distribution to be correct.

We begin by grouping the experimental data into k groups with the number of measurements in the ith group given by n_i and the total number of experimental measurements, n, given by

$$n = \sum_{i=1}^{k} n_i \qquad (8.139)$$

The corresponding probability values taken from the frequency distribution being tested for goodness of fit to the data are denoted by p_i, where

$$1 = \sum_{i=1}^{k} p_i \qquad (8.140)$$

The test statistic, χ^2_f (the subscript "f" stands for "fit") is given by

$$\chi^2_f = \sum_{i=1}^{k} \frac{(o_i - e_i)^2}{e_i} = \sum_{i=1}^{k} \frac{(n_i - np_i)^2}{np_i} \qquad (8.141)$$

where the observed values, the o_i, equal the n_i in this case, while the expected values, the e_i, equal the np_i. This statistic is used in a one-tailed test of the null hypothesis that the n_i are equal to the np_i. The rejection region for the test is $\chi^2 > \chi^2_f$ and the number of degrees of freedom for this test is $\nu = k - 1$, where k is the number of class intervals, unless the data are used to estimate parameters in the distribution being fitted. In this case, if r is the number of parameters being estimated, then the number of degrees of freedom is $\nu = k - r - 1$.

We now discuss the application of this test to fits to the normal distribution. In particular we shall see how the p_j are determined in this case. From Eq. (8.101) we have, on substituting $\bar{x}$ for μ, s for σ, and carrying out the change of variable $y = (x - \bar{x})/s$

$$\Pr(a \leq x \leq b) = \frac{1}{\sqrt{2\pi}s} \int_a^b e^{-(x-\bar{x})/2s^2} \, dx = \frac{1}{\sqrt{2\pi}} \int_{(a-\bar{x})/s}^{(b-\bar{x})/s} e^{-y^2/2} \, dy$$

$$= \frac{1}{\sqrt{2\pi}} \int_0^{(b-\bar{x})/s} e^{-y^2/2} \, dy - \frac{1}{\sqrt{2\pi}} \int_0^{(a-\bar{x})/s} e^{-y^2/2} \, dy \quad (8.142)$$

Setting $a = a_{i-1}$ and $b = a_i$, where the a_i are the ends of the class interval, $1 \leq i \leq k$, a_0 is the lower bound of the first class interval and a_k is the upper bound of the last class interval. Denoting the area under a Gaussian distribution having a mean of zero and a standard deviation equal to one between zero and $(a_i - \bar{x})/s$ by $\Phi((a_i - \bar{x})/s)$, we have

$$\Pr\left(0 \leq \frac{a_i - \bar{x}}{s}\right) = \Phi\left(\frac{a_i - \bar{x}}{s}\right) \quad (8.143)$$

and for the probability p_i appearing in Eq. (8.140) corresponding to the interval $a_i - a_{i-1}$

$$p_i = \Phi\left(\frac{a_i - \bar{x}}{s}\right) - \Phi\left(\frac{a_{i-1} - \bar{x}}{s}\right) \quad (8.144)$$

We note further that by making the change of variable of integration $y = -y$ in the expression for $\Phi(-z)$ [compare Eq. (8.104)]

$$\Phi(-z) = \frac{1}{\sqrt{2\pi}} \int_0^{-z} e^{-y^2/2} \, dy = -\frac{1}{\sqrt{2\pi}} \int_0^z e^{-y^2/2} \, dy = -\Phi(z) \quad (8.145)$$

Thus, values of $\Phi((a_i - \bar{x})/s)$ for which $(a_i - \bar{x})/s < 0$ can be found from Table D.3.

We now consider an example of testing the fit of data to a Gaussian distribution. Values of acetone are measured in samples from a hazardous waste stream. The numbers of measurements falling into class intervals having upper and lower endpoints a_i and a_{i-1}, respectively, which are not of equal width, are given in Table 8.7, as are the numbers of measurements falling into each interval, the probabilities of the measurements in each interval fitting a Gaussian curve, as well as the values of np_i, the probable numbers of measurements falling into the intervals previously chosen that would be predicted if the data followed a Gaussian distribution.

Table 8.7: Class intervals of measured data

$a_i - a_{i-1}$	n_i	p_i	np_i
25.0–25.1	10	0.0069	3.45
25.1–25.5	60	0.1718	85.90
25.5–26.0	360	0.6114	305.70
26.0–26.5	60	0.2003	100.15
26.5–27.0	10	0.0054	2.70

The average of the measured values is computed as indicated in Eq. (8.132) and is $\bar{x} \simeq 25.76$, while the standard deviation is computed as in Eq. (8.131) and is $s \simeq 0.2898$. Values of the p_i are computed as indicated in Eqs. (8.142)–(8.145). The test statistic is computed using the values in Table 8.7 in Eq. (8.141) and is approximately 65.7. Since estimates of two parameters of the assumed distribution are computed using the data being tested, and the number of class intervals in this case is $k = 5$, we have for the number of degrees of freedom $\nu = 5 - 2 - 1 = 3$. From Table D.5c we have $\chi^2_{0.95} \simeq 7.81$. Since $\chi^2_t > \chi^2_{0.95}$, we reject, at the 95% confidence level, the null hypothesis that the data are described by a Gaussian distribution.

We now consider another application of the χ^2 distribution, namely to test the independence of two or more attributes of members of a population. Observed values of these attributes are often displayed in arrays called *contingency* (or classification) *tables*. For a two-dimensional array of h rows and k columns, we compute the average, or expected, values of the elements of the array from the observed values of the attributes comprising the elements of the array as follows. Denoting the elements of the array containing the numbers of original observations of each attribute by n_{ij} with $1 \le i \le h$ and $1 \le j \le k$, we note that the total number of observations of each attribute is given by

$$n = \sum_{i=1}^{h} \sum_{j=1}^{k} n_{ij} = \sum_{j=1}^{h} \sum_{i=1}^{k} n_{ij} \qquad (8.146)$$

We denote the sum of the numbers of attributes displayed in the ith of the two-way classification array by $_i n$ and the sum of each of the j columns by n_j. Thus,

$$_i n = \sum_{j=1}^{k} n_{ij} \qquad (8.147a)$$

$$n_j = \sum_{i=1}^{h} n_{ij} \qquad (8.147b)$$

The probabilities, p_{ij}, to be expected under the assumption that the attributes are independent are

$$p_{ij} = \frac{_in}{n}\frac{n_j}{n} \tag{8.148}$$

while the numbers, N_{ij}, of each attribute to be expected under this assumption are

$$N_{ij} = np_{ij} = \frac{_inn_j}{n} \tag{8.149}$$

The statistic to be computed to test the assumption that the different attributes of the population are independent is

$$X^2 = \sum_{i=1}^{h}\sum_{j=1}^{k} \frac{(n_{ij} - N_{ij})^2}{N_{ij}} \tag{8.150}$$

where we identify X^2 with the variable χ^2 distribution for sufficiently large values of n_{ij} and N_{ij}. The number of degrees of freedom, ν, associated with this statistic is given by

$$\nu = (h - 1)(k - 1) - m \tag{8.151}$$

where m is the number of population parameters, if any, estimated from the observed data. We now consider an example of this use of the χ^2 distribution as follows. Samples of material from four waste ponds are found to contain chromium and acetone. The measured values of chromium are grouped in ranges below and above the regulatory action level. The observed data, the n_{ij}, are arranged in an array such that the index j corresponds to the number of the pond so that $k = 4$ and $i = 1$ correspond to the data falling below the regulatory action level, while $i = 2$ denotes values of chromium above the regulatory limit. The numbers, n_{ij}, of data points are given in the array of Eq. (8.152).

$$n_{ij} = \begin{bmatrix} 10 & 15 & 20 & 12 \\ 40 & 25 & 30 & 23 \end{bmatrix} \tag{8.152}$$

The $_in$ as computed from Eq. (8.147a) are $_1n = 57$ and $_2n = 118$, while the n_j as computed from Eq. (8.147b) are $n_1 = 50$, $n_2 = 40$, $n_3 = 50$, and $n_4 = 35$. The N_{ij} computed from Eqs. (8.148) and (8.149) are given in the array

$$N_{ij} = \begin{bmatrix} 16.29 & 13.03 & 16.29 & 11.40 \\ 33.71 & 26.97 & 33.71 & 23.60 \end{bmatrix} \tag{8.153}$$

Computing the statistic X^2 from Eq. (8.150) yields $X^2 \simeq 5.344$. From Table D.5c we note that the value of χ^2 at the 95% confidence level for

$(2-1)(4-1) = 3$ degrees of freedom is $\chi^2_{0.95} = 7.8$, which is larger than the computed value of the test statistic, and hence in the rejection region, so that the hypothesis of independence of the population attributes of pond numbers and chromium contents can be rejected at the 95% confidence level.

8.8.3 t Distribution

We now consider the use of the next distribution function derived, the t distribution. The features of the t distribution that are of interest in this connection are that it is symmetric and is applicable to small-sample testing in that fewer than thirty samples will suffice. When more than thirty samples are used, the t distribution closely approximates the normal distribution. Further, the population parameters do not appear in the distribution function–only parameters derived from sample values are required. The number of degrees of freedom, ν, is

$$\nu = n - 1 \tag{8.154}$$

where n is the number of data points in the sample.

The first use of the t distribution we will consider is to test hypotheses regarding the mean. As previously, the null hypothesis is that the sample mean is, at some confidence level, equal to the population mean. The test statistic is

$$t = \frac{\bar{x} - \mu}{s/\sqrt{n}} \tag{8.155}$$

where μ is the hypothesized value of the population mean, s is the standard deviation of the observed data values as given by Eq. (8.131), and $\bar{x}$ is the arithmetic mean of the sample values. For a two-tailed test the critical values of t are $t < -|t_{1-\alpha}|$ and $t > t_{1-\alpha}$, where $(1 - 2\alpha) \times 100\%$ is the confidence level of the test, and $1 - 2\alpha$ is the area under the t-distribution plot between the critical values shown in Figure 8.5. The area 2α under the curve in Figure 8.5 is given, from Eq. (8.117), by

$$\alpha = \frac{\Gamma\left(\frac{\nu+1}{2}\right)}{\sqrt{\nu\pi}\,\Gamma\left(\frac{\nu}{2}\right)} \int_{-|t_\alpha|}^{t_\alpha} \frac{1}{\left(1 + \frac{y^2}{\nu}\right)^{\frac{\nu+1}{2}}} \, dy \tag{8.156}$$

For a one-tailed test in which the hypothesis that the population mean has a particular value at a given confidence level is tested against the alternative hypothesis that the true population takes on a larger or smaller value the statistic of Eq. (8.155) is calculated and the rejection regions have the critical values $t < -|t_{2\alpha}|$ or $t > t_{2\alpha}$. In these cases the area under the upper tail of the distribution curve $1 - 2\alpha$, is given by

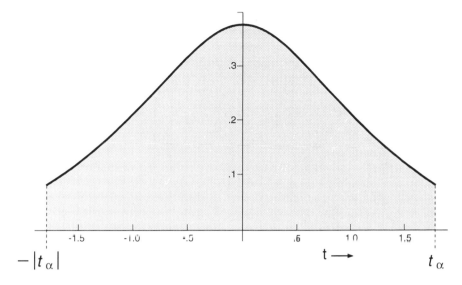

Figure 8.5: A plot of the t-distribution showing its symmetric nature and critical values in terms of $(1 - 2\alpha)$, the area under each tail of the distribution.

$$1 - 2\alpha = \frac{2\Gamma\left(\frac{\nu+1}{2}\right)}{\sqrt{\nu\pi}\,\Gamma\left(\frac{\nu}{2}\right)} \int_{t_{2\alpha}}^{\infty} \frac{1}{\left(1 + \frac{y^2}{\nu}\right)^{\frac{\nu+1}{2}}} \, dy \qquad (8.157)$$

Values of t_α for values of 2α were computed using a combination of Newton Raphson and bisection methods with the starting values of the calculation taken from Pearson and Hartley (1956, p. 138). We now consider an example of the use of the t distribution as follows. Values of tritium emissions from a generating facility are measured. The generating facility states that average values of 10 pCi/m³ (1 pCi $= 10^{-12}$ Ci) are normally observed. The data are, in units of pCi/m³, 8.0, 9.5, 10.2, 10.6, and 10.8. In this case, $n = 5$, $\nu = 4$, and for a one-tailed test, the question is to judge whether or not the null hypothesis of an average value of $\mu = 10$ is to be rejected as being less, statistically speaking, than the data actually show. The critical value of the t statistic for the 90% confidence level ($2\alpha = 0.9$ in Table D.4)for rejection is approximately 2.132. Thus, values of the t statistic computed from data less than this value indicate that insufficient data exist to reject the null hypothesis, while computed values of the statistic larger than this value indicate that the average value of tritium emissions is actually larger than the stated value. Using the given data to compute the t statistic from Eq. (8.155) yields -0.355, which lies in the $-\infty \leq t \leq 2.132$ acceptance region. Thus, at the 90% confidence level the hypothesis is rejected. We now consider an example of a two-tailed test using the t distribution as follows. Measurements of gross beta particle activity in water at a facility are made. The facility operator states that the most recently observed values of 25 pCi/ℓ (10^{-12} Curies per liter of water) are typical and further that values significantly below this are not to be expected due to the fact that the beta particles are of natural geologic origin and not due to external contamination. Thus, a two-tailed test is appropriate in this case. The lower tail component is used to check the existence of an irreducible minimum contamination, while the upper tail component of the test is used to check the assertion of a natural maximum value. The observed values, in units of pCi/ℓ, are 19.0, 21.5, 23.0, 24.0, 25.0, 26.0, and 28.0. The area of the rejection region under each tail of the t distribution curve is given by $[1 - 2\alpha(\nu)]$, and for $n - 1 = 7 - 1 = \nu = 6$ we have for the corresponding critical values of t, namely $\pm t_{0.025}$, $\pm t_{0.025} = \pm 2.447$. Computing the t statistic yields $t = -1.082$, which falls into the acceptance region $-2.447 \leq t \leq 2.447$. Thus, at the 95% confidence level, corresponding to $2\alpha = 0.95$ in Table D.4, the null hypothesis cannot be rejected.

8.8.4 F Distribution

We now consider the use of the F-distribution in the analysis of variance. We consider m sets of data, which may result from successive sampling of a population following treatments of some kind, or from different populations whose means we wish to compare. The null hypothesis that is tested using the F distribution is that the mean values of each of the m data sets are equal. Comparisons are made between the variability of the means of each set using a statistic normalized to a measure of the internal variability of each set, each measure of variability normalized to a common scale, so to speak, by dividing it by its own number of degrees of freedom. Following established historical convention, we term a measure of the variability between data sets, the sum of the squared deviations between the mean of each data set and the mean of the values of all sets as the sum of squares of treatments, or SST, which is defined as

$$\text{SST} = \sum_{i=1}^{m} n_i (\bar{x}_i - <x>)^2 \tag{8.158}$$

where n_i is the number of points in the ith data set, x_i is the average of the values of the ith data set, $<x>$ is the arithmetic mean of the data values in all the data sets combined, and the sum is over all data sets. The measure of variability within a data set is called the sum of squares of error (SSE) and is defined as

$$SSE = \sum_{i=1}^{m} \sum_{j=1}^{n_i} (x_{ij} - \bar{x}_i)^2 \tag{8.159}$$

where x_{ij} is the jth member of the ith data set, and the rest of the symbols are as previously defined. In order to normalize these measures of variability to a common scale, we divide each of them by their respective numbers of degrees of freedom. This results in the mean square for treatments, or MST, defined as

$$MST = \frac{SST}{m - 1} \tag{8.160}$$

and the mean square for error, MSE, defined as

$$MSE = \frac{SSE}{n - m} \tag{8.161}$$

where

$$n = \sum_{i=1}^{m} n_i \tag{8.162}$$

and the rest of the quantities are as previously defined. The test F statistic, F_t, is now written as

$$F_t = \frac{MST}{MSE} \tag{8.163}$$

Comparing (8.103) with the definition of the F statistic in (8.120) leads to the identifications

$$\zeta = SST \tag{8.164a}$$

$$\nu_1 = m - 1 \tag{8.164b}$$

$$\xi = SSE \tag{8.165a}$$

$$\nu_2 = n - m \tag{8.165b}$$

We now consider an example of the use of the F distribution as follows. Samples of acetone are found in wastes that are planned to be disposed of by burial. The acetone levels are of potential regulatory concern, however, and it is desired to make judgments as to the adequacy of the sampling and analyses that have been carried out by the waste generator. The data from two sets of samples are analyzed to determine whether or not, statistically, they represent values from the same population. Thus, the hypothesis to be tested is that, statistically speaking, the means of the two data sets are equal. The data are shown in the array:

$$n_{ij} = \begin{bmatrix} 17.5 & 18.0 & 19.0 & 22.0 & 24.0 & \\ 90.0 & 120.0 & 123.0 & 125.0 & 132.0 & 134.0 \end{bmatrix}$$

From (8.158) we have, using $n_1 = 5$, $n_2 = 6$, $m = 2$, $\nu_1 = 4$, $\nu_2 = 9$, and the computed values $\bar{x}_1 = 20.1$, $\bar{x}_2 = 120.7$,

$$\text{SST} = \sum_{i=1}^{2} n_i(\bar{x}_i - <x>)^2 = 5(20.1-74.95)^2 + 6(120.7-74.95)^2 = 2.76 \times 10^4$$

$$\text{SSE} = \sum_{i=1}^{2}\sum_{j=1}^{n_i}(x_{ij} - \bar{x}_i)^2 = \sum_{j=1}^{n_i}(x_{1j} - \bar{x}_1)^2 + \sum_{j=1}^{n_2}(x_{2j} - \bar{x}_2)^2$$

$$= (17.5 - 20.1)^2 + (18.0 - 20.1)^2 + (19.0 - 20.1)^2 + (24.0 - 20.1)^2$$
$$+ (90.0 - 120.7)^2 + (120.0 - 120.7)^2 + (123.0 - 120.7)^2$$
$$+ (132.0 - 120.7)^2 + (134.0 - 120.7)^2 + (125.0 - 120.7)^2$$

$$= 31.2 + 1271.3 = 1302.5$$

$$\text{MST} = \frac{2.76 \times 10^4}{1} = 2.76 \times 10^4$$

$$\text{MSE} = \frac{1302.5}{9} = 190.7$$

Thus,

$$F_t = \frac{\text{MST}}{\text{MSE}} \simeq 47.68$$

At the 95% confidence level, with $\nu_1 = 4$ and $\nu_2 = 9$, we find from Table D.6, computed for this work using the same methodology as for computing the values of the t distribution, $F = 3.64$. Since F_t is larger than F, we reject the hypothesis at the 95% confidence level that the acetone samples being measured, and by extension the waste samples being studied, come from the same population.

8.9   EXTREME VALUES

The interest in extreme values of data stems from two related points of view. The first is the statistically based identification of true outlier data values that do not belong to the population being sampled. The second, which we discuss here, proceeds from the assumption that the largest and/or smallest data values are members of the population being sampled. The extreme values are random variables having their own statistical distribution depending in general on the distribution describing the population from which the extreme values are drawn. Further, the probability of encountering a value larger than any previously observed increases with the number of data points collected (trials). If $\Phi(x)$ is the cumulative probability distribution, normalized to one, of a given attribute of the population, assumed to be continuous, and $\phi(x) = \Phi'(x)$ is the probability density of the attribute, we have

$$\Phi(x) = \int_{-\infty}^{x} \phi(x') \, dx' \qquad (8.166)$$

We observe that $\Phi(x)$ is the probability that observed values of the population attribute will be less than x and that

$$1 - \Phi(x) = \int_{x}^{\infty} \phi(x') \, dx' \qquad (8.167)$$

is the probability of observing values of the population attribute larger than x. The *return period*, $T(x)$, of the largest value being considered, x_0, is

$$T(x_0) = \frac{1}{1 - \Phi(x_0)} \qquad (8.168)$$

and if the observations were taken at fixed intervals of time, as in the case of yearly streamflow records, for example, $T(x_0)$ is the length of time that, on average, may elapse before x_0 is recorded. It can be shown that if the extreme-value data are arranged in order from the largest to the smallest values, then the frequency distribution, $\phi(x, m)$, of the mth value from the largest for which m would equal one, is (Cramér, 1946, p. 370)

$$f(x, m) = m \binom{n}{m} \Phi(x)^{n-m} [1 - \Phi(x)]^{m-1} \phi(x) \qquad (8.169)$$

Alternatively, in the case of n independent observations the probability, $\zeta(x, n)$, of observing values of the population attribute less than x is

$$\zeta(x, n) = \Phi(x)^n \qquad (8.170a)$$

The derivative of $\zeta(x, n)$, $\xi(x, n)$

$$\xi(x, n) = \zeta'(x, n) = n\Phi(x)^{n-1} \phi(x) \qquad (8.170b)$$

is the probability density of the largest value of the population attribute seen in n repetitions of the experiment and is seen to be a special case of (8.169) for $m = 1$. For distributions of the exponential type, which include the normal, log-normal (in which the logarithm of the the variate is normally distributed), and the Poisson, asymptotic representations of the distribution function for the largest and smallest observed values are available. The asymptotic distribution, $D(x, m)$, of the mth extreme value from the largest one observed is (Cramér, 1946, p. 375)

$$D(x, m) = \frac{1}{\Gamma(m)} e^{-mye^{-y}} \tag{8.171a}$$

where

$$y = \frac{1}{\beta}(x - \alpha) \tag{8.171b}$$

where $m = 1$ is the case usually considered, y is the standardized variate, α is the mode of the distribution, and, along with β, is to be evaluated from the original data, the x_i, as will be discussed. The asymptotic distribution for the minimum extreme value, $E(x)$, is (Hastings and Peacock, 1974, p. 60)

$$E(x) = 1 - e^{-e^{-y}} \tag{8.172}$$

where y is as just given. We now consider the evaluation of α and β. It can be shown (Gumbel, 1941, pp. 175–176) that the mean of the theoretical distribution, $< x >$, is given by

$$< x >= \alpha + \gamma\beta = \frac{1}{n} \sum_{i=1}^{n} x_i \tag{8.173}$$

where $\gamma = 0.5772157 \cdots$ is Euler's constant. It can further be shown (Gumbel, loc. cit.) that the standard deviation of the theoretical distribution $\sigma(n)$, of largest extreme values is

$$\sigma(n) = \frac{\pi}{\alpha\sqrt{6}} = \left(\frac{n}{n-1} (< x^2 > - < x >^2) \right)^{\frac{1}{2}} \tag{8.174}$$

where

$$< x^2 >= \frac{1}{n} \sum_{i=1}^{n} x_i^2 \tag{8.175}$$

and $< x >$ is given by (8.173). Solving (8.175) for α yields

$$\alpha = \frac{\pi}{\sqrt{6}\sigma(n)} \tag{8.176}$$

while from (8.173) we have

$$\beta = \frac{1}{\gamma}\left(<x> - \frac{\pi}{\sqrt{6}\sigma(n)}\right) \qquad (8.177)$$

The values of y that are computed from the x_i after α and β have been determined can be plotted on extreme-value probability paper to interpolate and extrapolate the observed data. The values on the abscissa of extreme-value probability paper, $D(y, 1)$, are computed from (8.171a) for $\alpha = 0$ and $\beta = 1$.

8.10 REGRESSION AND CORRELATION

8.10.1 Regression

Up to this point we have considered descriptions of single attributes of members of populations. We now consider relationships between more than one attribute that members of the population may have. We begin with possible linear relationships between two population attributes. A hypothesized linear relationship between two attributes, x and y, say, can be expressed as

$$\underline{y} = b_1 + b_2 x \qquad (8.178)$$

where x_i and y_i are paired values of the population attributes, x is a value of x that is not necessarily one of the data points, $\underline{y}$ is the value of the attribute computed from a given value of x, b_2 is the slope of a plot of y_i versus x_i, while b_1 is the y intercept of the straight-line fit to the plotted points. In order to determine b_1 and b_2 from the given data points in the sense of least squares, we write

$$\sum_{i-1}^{n} y_i = nb_1 + b_2 \sum_{i=1}^{n} x_i \qquad (8.179a)$$

$$\sum_{i=1}^{n} x_i y_i = b_1 \sum_{i=1}^{n} x_i + b_2 \sum_{i=1}^{n} x_i^2 \qquad (8.179b)$$

Arranging Eqs. (8.179a) and (8.179b) in matrix form yields

$$\begin{bmatrix} n & \sum_{i=1}^{n} x_i \\ \sum_{i=1}^{n} x_i & \sum_{i=1}^{n} x_i^2 \end{bmatrix} \begin{bmatrix} b_1 \\ b_2 \end{bmatrix} = \begin{bmatrix} \sum_{i=1}^{n} y_i \\ \sum_{i=1}^{n} x_i y_i \end{bmatrix} \qquad (8.180)$$

Indicating the inverse of the first matrix on the left and multiplying yields

$$\begin{bmatrix} b_1 \\ b_2 \end{bmatrix} = \begin{bmatrix} a_{11} & a_{12} \\ a_{21} & a_{22} \end{bmatrix}^{-1} \begin{bmatrix} c_1 \\ c_2 \end{bmatrix} \qquad (8.181)$$

where

$$a_{11} = n \tag{8.182a}$$

$$a_{12} = a_{21} = \sum_{i=1}^{n} x_i \tag{8.182b}$$

$$a_{22} = \sum_{i=1}^{n} x_i^2 \tag{8.182c}$$

$$c_1 = \sum_{i=1}^{n} y_i \tag{8.182d}$$

$$c_2 = \sum_{i=1}^{n} x_i y_i \tag{8.182e}$$

$$\begin{bmatrix} a_{11} & a_{12} \\ a_{21} & a_{22} \end{bmatrix}^{-1} = \begin{bmatrix} A_{11} & A_{12} \\ A_{21} & A_{22} \end{bmatrix} \tag{8.182f}$$

where the A_{ij} are the elements of the matrix that is the inverse of the a matrix and are found by taking the transpose of the adjoint of a matrix divided by the determinant of the a matrix. Thus,

$$a^{-1} = A = \mathrm{Tr} \begin{bmatrix} a_{22} & -a_{21} \\ -a_{12} & a_{11} \end{bmatrix} \div \det \begin{bmatrix} a_{11} & a_{12} \\ a_{21} & a_{22} \end{bmatrix} \tag{8.183}$$

where the elements in the matrix to be transposed are found by replacing each element in the original matrix by its signed cofactor, and the cofactor of a given array element is the determinant of the elements remaining after the row and column containing the element whose cofactor is to be found have been removed. Thus,

$$A = \begin{bmatrix} a_{22} & -a_{12} \\ -a_{21} & a_{11} \end{bmatrix} \div (a_{11}a_{22} - a_{12}a_{21}) \tag{8.184}$$

Thus, substituting (8.183a)–(8.183d) into (8.181) and carrying out the indicated multiplication on the right-hand side yields

$$b_1 = \frac{a_{22}c_1 - a_{12}c_2}{a_{11}a_{22} - a_{12}a_{21}} \tag{8.185a}$$

$$b_2 = \frac{-a_{21}c_1 + a_{11}c_2}{a_{11}a_{22} - a_{12}a_{21}} \tag{8.185b}$$

We can test the hypothesis that a relationship between x and y exists by constructing a t statistic of the form

$$t_s = \frac{b_2}{s/\sqrt{\sum_{i=1}^{n}(x_i - <x>)^2}} \qquad (8.186)$$

where

$$<x> = \frac{1}{n}\sum_{i=1}^{n} x_i \qquad (8.187)$$

and s is the standard deviation estimated from data, namely,

$$s = \frac{\sqrt{\sum_{i=1}^{n}(y_i - \underline{y}_i)^2}}{n-2} \qquad (8.188)$$

where $\underline{y}_i$ is the value predicted from Eq. (8.179) using x_i and the values of b_1 and b_2 from Eqs.(8.185a) and (8.185b) and $n-2$ is the number of degrees of freedom used with the t distribution. It can be shown that

$$\sum_{i=1}^{n}(x_i - <x>)^2 = \frac{1}{n}(a_{11}a_{22} - a_{12}a_{21}) \qquad (8.189)$$

so that the test statistic becomes

$$t_s = \frac{a_{11}c_2 - a_{21}c_1}{ns} = \frac{(n-2)(a_{11}c_2 - a_{21}c_1)}{n\left[\sum_{i=1}^{n}(y_i - \underline{y}_i)^2\right]^{1/2}} \qquad (8.190)$$

For a one-tailed test of the hypothesis that $b_2 = 0$ at the $(1-2\alpha) \times 100\%$ confidence level, the rejection region is $t_s > t_{2\alpha}$ for values of $b_2 > 0$, while for $b_2 < 0$ the rejection region is $-t_{2\alpha} < t_s$. For a two-tailed test of the hypothesis that $b_2 = 0$, the rejection regions are $-t_\alpha < t_s$ and $t_s > t_\alpha$, respectively. We now consider the following example. Values of toluene and xylene are measured in waste extract. The question as to whether or not they are related in some way and knowledge of values of one would allow predictions of values of the other. The data for pairs of measurements are given in normalized units with x values for xylene and y values for toluene in Table 8.8. We hypothesize a linear relationship between y and x and compute b_1 and b_2 using Eqs. (8.182a)–(8.182e) and (8.185a) as follows. Using $a_{11} = 8$, $a_{12} = a_{21} = 122$, $a_{22} = 2016$, $c_1 = 252$, and $c_2 = 4121$ as computed from the given data, we have $b_1 \simeq 4.23633$ and $b_2 \simeq 1.78778$. The test statistic t_s is found to be equal to 21.13, and from Table D.4 we find for a one-tailed test with six degrees of freedom and $1 - 2\alpha = 0.05$ $t_{0.95} \simeq 2.447$. Since $t_s > t_{2\alpha}$ in this case, b_2 as computed

Table 8.8: Values of toluene and xylene in normalized units

i	y_i	x_i	y_i	$y_i - y_i$
1	20	10	22.11	-2.11
2	23	12	25.69	-2.69
3	25	11	23.90	1.10
4	29	13	27.48	1.52
5	31	16	32.84	-1.84
6	37	19	38.20	-1.20
7	42	17	34.63	7.37
8	45	24	47.14	-2.14

from data is > 0, and we reject the hypothesis that b_2 is zero and that no relation exists between y and x. The fact that a relation does exist, however, does not demonstrate causality, and in fact in the present case the underlying processes that produced the waste are the more likely causes of a relationship between the two.

We note finally that higher-order fits to data than linear are possible and often necessary. For example, the equations to be solved for the regression coefficients for a quadratic fit to data are

$$\sum_{i=1}^{n} y_i = a_0 n + a_1 \sum_{i=1}^{n} x_i + a_2 \sum_{i=1}^{n} x_i^2 \tag{8.191a}$$

$$\sum_{i=1}^{n} x_i y_i = a_0 \sum_{i=1}^{n} x_i + a_1 \sum_{i=1}^{n} x_i^2 + a_2 \sum_{i=1}^{n} x_i^3 \tag{8.191b}$$

$$\sum_{i=1}^{n} x_i^2 y_i = a_0 \sum_{i=1}^{n} x_i^2 + a_1 \sum_{i=1}^{n} x_i^3 + a_2 \sum_{i=1}^{n} x_i^4 \tag{8.191c}$$

which can be arranged into matrix form and solved for the a_i as was done previously. Regression can also be carried out using powers of more than one independent variable and partial regressions can be carried out for each independent variable separately.

8.10.2   Correlation

We now consider the related topic of the correlation and consider two variables only. For two variables, x and y, say, the correlation coefficient, r_c, is given by

$$r_c = \frac{\sum_{i=1}^{n}(x_i - <x>)(y_i - <y>)}{\{[\sum_{i=1}^{n}(x_i - <x>)^2][\sum_{i=1}^{n}(y_i - <y>)^2]\}^{1/2}} \tag{8.192}$$

Recalling Eq. (8.189) and noting the relation

$$\sum_{i=1}^{n}(x_i - <x>)(y_i - <y>) = \frac{1}{n}(a_{11}c_2 - a_{12}c_1) \tag{8.193}$$

allows us to write r_c in the form

$$r_c = \frac{a_{11}c_2 - a_{12}c_1}{[n(a_{11}a_{22} - a_{12}a_{21})(\sum_{i=1}^{n}(y_i - <y>)^2)]^{1/2}} \tag{8.194}$$

where the numerator is the numerator in the expression for b_2 and a nonzero value implies correlation between the variables. The correlation coefficient varies between -1 and 1. For $r_c = 1$ we have perfectly correlated variables, while for $r_c = -1$ the variables are anticorrelated in the sense that large values of x imply small values of y and vice versa. The correlation coefficient measures the amount of variation explained by the quantity in the numerator of Eq. (8.192) relative to the total variations of x and y in the denominator. We now compute r_c for the data in the example just given and find $r_c = 0.929$, which shows high correlation but not necessarily cause and effect. For a continuous distribution we have for the correlation coefficient

$$r_c = \frac{<(x - <x>)(y - <y>)>}{(<x - <x>>^2 <y - <y>>^2)^{1/2}} \tag{8.195}$$

where

$$<x - <x>> = \int_{-\infty}^{x}\int_{-\infty}^{\infty}(x' - <x>)f(x',y')\,dy'\,dx' \tag{8.196a}$$

$$<y - <y>> = \int_{-\infty}^{y}\int_{-\infty}^{\infty}(y' - <y>)f(x',y')\,dx'\,dy' \tag{8.196b}$$

$$<(x - <x>)(y - <y>)>$$
$$= \int_{-\infty}^{x}\int_{-\infty}^{y}(x' - <x>)(y' - <y>)f(x',y')\,dy'\,dx' \tag{8.196c}$$

$$<x> = \int_{-\infty}^{x}\int_{-\infty}^{\infty}x'f(x',y')\,dy'\,dx' \tag{8.196d}$$

$$<y> = \int_{-\infty}^{y}\int_{-\infty}^{\infty}y'f(x',y')\,dx'\,dy' \tag{8.196e}$$

$$1 = \int_{-\infty}^{\infty}\int_{-\infty}^{\infty}f(x',y')\,dx'\,dy' \tag{8.196f}$$

where $f(x, y)$ is the joint frequency function of x and y. The reduced frequency distributions, which are functions of x and y, are

$$f_1(x) = \int_{-\infty}^{\infty} f(x, y') \, dy' \tag{8.197}$$

$$f_2(y) = \int_{-\infty}^{\infty} f(x', y) \, dx' \tag{8.198}$$

The conditional probability distribution for x, $f_{xc}(x|y)$, given the occurrence of y, is

$$f_{xc}(x|y) = \frac{f(x, y)}{f_2(y)} \tag{8.199a}$$

and similarly for $f_{yc}(y|x)$, the probability distribution for y given the fact that x has occurred

$$f_{yc}(y|x) = \frac{f(x, y)}{f_1(x)} \tag{8.199b}$$

Eliminating $f(x, y)$ between (8.199a) and (8.199b) yields

$$f_{xc}(x|y) = \frac{f_1(x) f_{yc}(y|x)}{f_2(y)} \tag{8.200}$$

which is the analogue of Bayes's Theorem of Eq. (8.49). The conditional mean of of x, $< x >_c$, is

$$< x >_c = \frac{\int_{-\infty}^{\infty} x' f(x', y) \, dx'}{f_2(y)} \tag{8.201}$$

and similarly for the conditional mean of y, $< y >_c$,

$$< y >_c = \frac{\int_{-\infty}^{\infty} y' f(x, y') \, dy'}{f_1(x)} \tag{8.202}$$

The conditional variance of x, σ_{xc}, is

$$\sigma_{xc} = \frac{\int_{-\infty}^{\infty} (x' - < x >_c)^2 f(x', y) \, dx'}{f_2(y)} \tag{8.203}$$

and similarly for σ_{yc}.

An example of $f(x, y)$ is the two-dimensional Gaussian distribution, $f_g(x, y)$, given by

$$f_g(x, y) = \frac{1}{2\pi \sigma_{xg} \sigma_{yg} \sqrt{1 - r_g^2}} e^{-e_g(x, y)} \tag{8.204a}$$

where

$$e_g(x,y) = \frac{1}{(1-r_g)^2} \times \left(\frac{(x- <x>_g)^2}{\sigma_{xg}^2} - \frac{2r_g(x- <x>_g)(y- <y>_g)}{\sigma_{xg}\sigma_{yg}} \right.$$

$$\left. + \frac{(y- <y>_g)^2}{\sigma_{yg}^2} \right) \qquad (8.204b)$$

where

$$<x>_g = \int_{-\infty}^{\infty} \int_{-\infty}^{\infty} x' f_g(x',y') \, dx' \, dy' \qquad (8.205)$$

and similarly for $<y>_g$. The Gaussian variance on x, σ_{xg}^2, is given by

$$\sigma_{xg}^2 = \int_{-\infty}^{\infty} \int_{-\infty}^{\infty} (x'- <x>_g)^2 f_g(x',y') \, dx' \, dy' \qquad (8.206)$$

and similarly for σ_{yg}^2. The Gaussian correlation coefficient, r_g, is given as an application of Eq. (8.194). For the case in which $r_g = 0$, $f_g(x,y)$ reduces to $f_{gr}(x,y)$, given by

$$f_{gr}(x,y) = \frac{1}{2\pi\sigma_{xg}\sigma_{yg}} \times \exp - \left(\frac{(x- <x>)^2}{\sigma_{xg}^2} + \frac{(y- <y>_g)^2}{\sigma_{yg}^2} \right) \quad (8.207)$$

and is an example of the form taken by $f(x,y)$ for the case in which x and y are completely independent, which is, in general,

$$f(x,y) = f_1(x)f_2(y) \qquad (8.208)$$

Partial correlations can be defined in a manner analogous to the partial regressions mentioned previously.

SELECTED REFERENCES

Cramér, Harald, *Mathematical Methods of Statistics*, Princeton University Press, Princeton, NJ, 1946.

Fix, Evelyn, and J. L. Hodges, Jr., "Significance Probabilities of the Wilcoxon Test," *Ann. Math. Statistics*, Vol. 26, pp. 301–312, 1955.

Gumbel, E. J., "The Return Period of Flood Flows," *Ann. Math. Statistics*, Vol. 12, pp. 163–190, 1941.

Gupta, Hansraj, "Tables of Partitions," *Royal Society Mathematical Tables*, Volume 4, Cambridge University Press, Cambridge, 1958.

Hastings, N. A., and J. B. Peacock, *Statistical Distributions*, Butterworth & Co., London, 1974.

Kazarinoff, N. D., *Analytic Inequalities*, Holt, Rinehart and Winston, New York, 1961.

Mann, H. B., and D. R. Whitney, "On a Test of Whether One of Two Random Variables is Stochastically Larger Than the Other," *Ann. Math. Statistics*, Vol. 18, pp. 50–60, 1947.

Pearson, E. S., and H. O. Hartley, *Biometrika Tables for Statisticians*, Vol. I, Cambridge University Press, Cambridge, 1954.

Press, William H., et al., *Numerical Recipes*, Cambridge University Press, New York, 1986.

Scarborough, J. B., *Numerical Mathematical Analysis*, The Johns Hopkins Press, Baltimore, 1955.

Wilcoxen, Frank, "Individual Comparisons by Ranking Methods," *Biometrics*, Vol. 1, pp. 80–83, 1945.

PROBLEMS

8.1 (a) Show that

$$\int_{-\infty}^{\infty} e^{-b(x'-c)^2} \, dx' = \sqrt{\frac{\pi}{b}}$$

Hint: Make use of the known definite integral

$$\int_{0}^{\infty} y^p e^{-bx^2} \, dx = \frac{1}{2} b^{-(p+1)/2} \Gamma\left(\frac{p+1}{2}\right)$$

where $\Gamma()$ is the usual gamma function. (b) Using the result of (a), show that the value of a in the relation

$$a \int_{-\infty}^{\infty} e^{-b(x-c)^2} \, dx = 1$$

is $\sqrt{b/\pi}$.

8.2 Show that

$$<x> = \sqrt{\frac{b}{\pi}} \int_{-\infty}^{\infty} x e^{-b(x-c)^2} \, dx = c$$

by first changing the dummy variable of integration from x to $y = x - c$, breaking up the range of integration from $-\infty$ to ∞ to the sum of two integrals having ranges of integration $-\infty$ to 0 and 0 to ∞, respectively, changing the dummy variable of integration from y to $z = -y$ in the first of the two integrals just mentioned, and then using a special case of the known definite integral given in Problem 8.1.

8.3 Solve the equation $df(x)/dx = 0$ for the value of x that makes $f(x)$ a maximum for the case in which

$$f(x) = e^{-(x-<x>)^2/2\sigma^2}$$

8.4 Evaluate the integral

$$\frac{1}{\sqrt{2\pi}\sigma} \int_{-\infty}^{<x>} e^{(x-<x>)^2/2\sigma^2} \, dx$$

to verify that $<x>$ is the value of x dividing the area under a plot of the Gaussian probability frequency function into two equal areas.

8.5 Beginning with Eq. (8.29) and recalling the binomial expansion

$$(a + b)^n = \sum_{k=o}^{n} a^k b^{n-k} \binom{n}{k}$$

show that $_TC_n^n = 2^n - 1$.

8.6 By substituting $E_j \cup E_k$ for E_j in Eq. (8.32) and arranging the result appropriately, arrive at Eq. (8.34).

8.7 The probability density function, $p(x)$, for a Poisson distribution has the form $p(x) = [\lambda^x e^{-x}]/\Gamma(x + 1)$ where λ can be estimated as the average of x over an experimental data distribution, $f_i(x_i$ where x_i are experimental data, and the f_i are the experimental weights associated with the x_i. Thus, $\lambda \simeq (\sum_{i=1}^{n} f_i x_i) / \sum_i^n f_i$ where n is the number of data points. If values of x_i represent successive values of time, and the f_i represent absolute values of measurements made at successive times short to the length of time changes in field/laboratory situation would be expected to arrive at a "known" absolute value, fit a Poisson distribution by finding the value of λ and use the relation given for a Poisson distribution in terms of λ and x for the data given.

Measurements versus time

x_i	0	1	2	3	4	5
f_i	4	7	11	8	6	2

APPENDIX A

Review of Elements of Tensor Analysis

We begin with vector notation as being more familiar than the superscript-subscript notation of tensor analysis and consider unitary base vectors which from the components of the vectors and tensors we shall consider. Figure A-1 shows three intersecting coordinate surfaces embedded in a Cartesian coordinate system and having equations of the form

$$x^i(x, y, z) = c_i \tag{A.1}$$

where $1 \leq i \leq 3$ and the c_i are constants. Two sets of unitary base vectors may be computed for this system. The first set, shown in Figure A.1 (a), and denoted by $\vec{a}_i$, lies along the lines of intersection of the surfaces, and its components are given by

$$\vec{a}_i = \frac{\partial \vec{r}}{\partial x^i} \tag{A.2}$$

where $\vec{r}$ is the position vector in the x^i space. The second possible set of unitary base vectors, denoted by $\vec{a}^i$, and shown in Figure A.1 (b) above, consists of the gradients of the x^i (derivatives with respect to x, y, and z) and as such are vectors perpendicular to the coordinate surfaces expressing the maximum rate of change at the point at which the set is evaluated. Thus,

$$\vec{a}^i = \vec{\nabla} x^i \tag{A.3}$$

The two possible basis vector sets are inverses of each other in the sense that

$$\vec{a}^i = \vec{a}^i \cdot \vec{a}_j = \delta^i_j \tag{A.4}$$

where δ^i_j is the Kronecker delta, a second-rank tensor with the properties

$$\delta^i_j = 1, \qquad \text{for } i = j$$

$$= 0, \qquad \text{for } i \neq j \tag{A.5}$$

and $\cdot$ indicates the usual inner product operation. This "inverse" property can be demonstrated as follows. We note that

$$\vec{a}^1 \cdot d\vec{r} = \vec{\nabla} x^1 \cdot d\vec{r} = \frac{\partial x^1}{\partial x} dx + \frac{\partial x^1}{\partial y} dy + \frac{\partial x^1}{\partial z} dz = dx^1 \tag{A.6a}$$

and similarly

$$\vec{a}^2 \cdot d\vec{r} = \vec{\nabla} x^2 \cdot d\vec{r} = \frac{\partial x^2}{\partial x} dx + \frac{\partial x^2}{\partial y} dy + \frac{\partial x^2}{\partial z} dz = dx^2 \tag{A.6b}$$

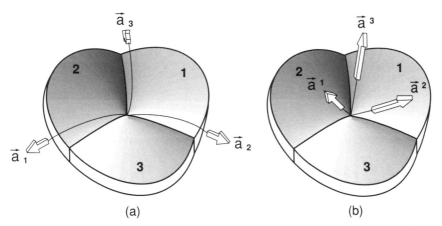

Figure A.1: (a) Unitary base vectors parallel to the intersections of the coordinate surfaces of a curvilinear coordinate system; (b) Unitary base vectors perpendicular to the intersecting coordinate surfaces of a curvilinear coordinate system.

$$\vec{a}^3 \cdot d\vec{r} = \vec{\nabla} x^3 \cdot d\vec{r} = \frac{\partial x^3}{\partial x} dx + \frac{\partial x^3}{\partial y} dy + \frac{\partial x^3}{\partial z} dz = dx^3 \qquad (A.6c)$$

where $d\vec{r} = \hat{i}\, dx + \hat{j}\, dy + \hat{k}\, dz$. Alternately, however, we may write for $d\vec{r}$ in the x^i system,

$$d\vec{r} = \frac{\partial \vec{r}}{\partial x^1} dx^1 + \frac{\partial \vec{r}}{\partial x^2} dx^2 + \frac{\partial \vec{r}}{\partial x^3} = \vec{a}_i \cdot dx^i \qquad (A.7)$$

where the Einstein summation convention has been used, in which an index appearing once as a superscript and once as a subscript in a given expression is assumed to be summed over, unless otherwise stated. Thus, from Eqs. (A.6a) and (A.7) we have

$$\vec{\nabla} x^1 \cdot d\vec{r} = \vec{a}^1 \cdot \vec{a}_1\, dx^1 + \vec{a}^1 \cdot \vec{a}_2\, dx^2 + \vec{a}^1 \cdot \vec{a}_3\, dx^3 \qquad (A.8)$$

However, since also from Eq. (A.6a)

$$\vec{a}^1 \cdot d\vec{r} = dx^1 \qquad (A.9)$$

the coefficient of dx^1 in Eq. (A.7) must equal unity, or

$$\vec{a}^1 \cdot \vec{a}_1 = 1 \qquad (A.10a)$$

and thus also

$$\vec{a}^1 \cdot \vec{a}_2 = \vec{a}^1 \cdot \vec{a}_3 = 0 \qquad (A.10b)$$

Similarly, from Eqs. (A.6b), and (A.7)

$$\vec{\nabla} x^2 \cdot d\vec{r} = \vec{a}^2 \cdot d\vec{r} = \vec{a}^2 \cdot \vec{a}_1\, dx^1 + \vec{a}^2 \cdot \vec{a}_2\, dx^2 + \vec{a}^3 \cdot \vec{a}_3\, dx^3 \qquad (A.11)$$

while also from Eq. (A.6b)

$$dx^2 = \vec{a}^2 \cdot d\vec{r} \tag{A.12}$$

so that from comparison of Eqs. (A.11) and (A.12) we have

$$\vec{a}^2 \cdot \vec{a}_2 = 1 \tag{A.13a}$$

$$\vec{a}^2 \cdot \vec{a}_1 = \vec{a}^2 \cdot \vec{a}_3 = 0 \tag{A.13b}$$

Finally, from Eqs. (A.6c) and (A.7) we have

$$\vec{\nabla} x^3 \cdot d\vec{r} = \vec{a}^3 \cdot \vec{a}_1 \, dx^1 + \vec{a}^3 \cdot \vec{a}_2 \, dx^2 + \vec{a}^3 \cdot \vec{a}_3 \, dx^3 \tag{A.14}$$

while from Eq. (A.6c) we see that

$$dx^3 = \vec{a}^3 \cdot d\vec{r} \tag{A.15}$$

so that comparing Eqs. (A.14) and (A.15) yields

$$\vec{a}^3 \cdot \vec{a}_3 = 1 \tag{A.16a}$$

$$\vec{a}^3 \cdot \vec{a}_1 = \vec{a}^3 \cdot \vec{a}_2 = 0 \tag{A.16b}$$

Equations (A.10a), (A.10b), (A.13a), (A.13b), (A.16a), and (A.16b) when collected yield Eq. (A.4). Either set of unitary base vectors may be used to expand a given vector as follows.

$$\vec{A} = A^i \vec{a}_i = A_i \vec{a}^i \tag{A.17}$$

where the A^i are called the *contravariant components* of $\vec{A}$, the A_i are called the *covariant components* of $\vec{A}$ and are computed as

$$A^i = \vec{A} \cdot \vec{a}^i \tag{A.18a}$$

$$A_i = \vec{A} \cdot \vec{a}_i \tag{A.18b}$$

respectively. The second-rank symmetric tensor containing the information that describes the coordinate system, g_{ij}, is called the *metric tensor*, and its elements are computed as follows.

$$g_{ij} = \vec{a}_i \cdot \vec{a}_j \tag{A.19}$$

Similarly, we compute

$$g^{ij} = \vec{a}^i \cdot \vec{a}^j \tag{A.20}$$

It can be shown that g^{ij} is the inverse of g_{ij}, so that

$$g^{il}g_{lj} = \delta_j^i \tag{A.21}$$

where l is a dummy index of summation.

We note in passing that the coefficients of the differentials appearing in the expression for the square of the line element, ds^2, are the elements of the metric tensor and are often introduced in this way. Thus,

$$ds^2 = \vec{r} \cdot \vec{r} = \vec{a}_i \cdot \vec{a}_j \, dx^i \, dx^j = g_{ij} \, dx^i \, dx^j \tag{A.22}$$

With the help of the metric tensor and its inverse we can compute the covariant components of a vector or tensor given its contravariant components and vice versa, as follows. Beginning with a vector $\vec{A}$ expanded in terms of its contravariant components, we have $\vec{A} = A^i \vec{a}_i$. However, from Eq. (A.18b) $A_j = \vec{A} \cdot \vec{a}_j$, so that

$$A_j = \vec{A} \cdot \vec{a}_j = A^i \vec{a}_i \cdot \vec{a}_j = A^i g_{ij} \tag{A.23a}$$

Similarly, from Eqs. (A.17) and (A.18) we have

$$A^j = \vec{A} \cdot \vec{a}^j = A_i \vec{a}^i \cdot \vec{a}^j = A_i g^{ij} \tag{A.23b}$$

The process of computing contravariant components of a vector or tensor from covariant components and vice versa is called raising and/or lowering indices and can be applied to vectors and tensors of any rank as well as indexed, but nontensorial objects, such as the Christoffel symbols, which we shall discuss shortly, and the differentials of the x^i.

We define the physical components of a vector or tensor, the $\tilde{A}^i$, for example, as those lying along the lines of intersection of the coordinate surfaces of Figure A.1 and expanded in terms of unit, rather than unitary, base vectors as follows.

$$\vec{A} = A^i \vec{a}_i = A^i |\vec{a}_i| \frac{\vec{a}_i}{|\vec{a}_i|} = A^i \sqrt{g_{ii}}\, \hat{e}_i = \tilde{A}^i \hat{e}_i \tag{A.24a}$$

where

$$|\vec{a}_i| = (\vec{a}_i \cdot \vec{a}_i)^{1/2} = \sqrt{g_{ii}} \tag{A.24b}$$

is the absolute value of $\vec{a}_i$ (no summation in this case) and $\hat{e}_i$ is the unit vector along the direction of a_i, namely

$$\hat{e}_i = \frac{\vec{a}_i}{|\vec{a}_i|} \tag{A.25a}$$

and, from eqs. (A.24b) and (A.25a)

$$\vec{a}_i = \sqrt{g_{ii}}\,\hat{e}_i \tag{A.25b}$$

Writing $\tilde{A}^i$ for the physical components of $\vec{A}$ as above yields

$$\tilde{A}^i = A^i \sqrt{g_{ii}} \tag{A.25c}$$

with no summation over i in this case. Beginning with a vector expressed in terms of its covariant components, we have

$$\vec{A} = A_i \vec{a}^i = A_i g^{ij} \tilde{A}^i g_{ij} = A_i g^{ij} \vec{a}_j = A_i g^{ij} \sqrt{g_{jj}}\,\hat{e}_j = A^j \sqrt{g_{jj}}\,\hat{e}_j \tag{A.26a}$$

so that

$$\tilde{A}^j = A^j \sqrt{g_{jj}} \tag{A.26b}$$

$$\vec{A} = \tilde{A}^j \hat{e}_j \tag{A.26c}$$

Note in passing that we could write for a set of unit vectors normal to the coordinate surfaces

$$\hat{e}^i = \frac{\vec{a}^i}{|\vec{a}^i|} \tag{A.27a}$$

or

$$\vec{a}^i = \sqrt{g^{ii}}\,\hat{e}^i \tag{A.27b}$$

so that

$$\vec{A} = A_i \sqrt{g^{ii}}\,\hat{e}^i \tag{A.28}$$

We now consider the differentiation of scalars, vectors, and tensors. Actually, of course, a scalar is a tensor of rank zero and a vector is a tensor of rank one, so no distinction need be made, but we shall adhere to the customary terminology. We begin with what is called the covariant derivative of $\vec{A}$, expressed in terms of its contravariant components as follows:

$$\frac{\partial \vec{A}}{\partial x^k} = \frac{\partial A^j \vec{a}_j}{\partial x^k} = A^j \frac{\partial \vec{a}_i}{\partial x^k} + \vec{a}_j \frac{\partial A^j}{\partial x^k} \tag{A.29}$$

We now write $\partial \vec{A}/\partial x^k$ in terms of its contravariant components to yield

$$\frac{\partial \vec{A}}{\partial x^k} \cdot \vec{a}^i = A^j \frac{\partial \vec{a}_j}{\partial x^k} \cdot \vec{a}_j + \vec{a}_j \cdot \vec{a}^i \frac{\partial A^j}{\partial x^k} = A^i_{,k} \tag{A.30}$$

Using Eq. (A.4) with Eq. (A.30) yields

$$A^k_{,k} = A^j \frac{\partial \vec{a}_j}{\partial x^k} \cdot \vec{a}^i + \frac{\partial A^j}{\partial x^k} \delta^i_j = \frac{\partial A^i}{\partial x^x} + A^j \frac{\partial \vec{a}_j}{\partial x^k} \cdot \vec{a}^i \tag{A.31}$$

We now set

$$\frac{\partial \vec{a}_j}{\partial x^k} \cdot \vec{a}^i = \Gamma^i_{jk} \tag{A.32}$$

where Γ^i_{jk} is called the Christoffel symbol of the second kind. Thus, we have finally for the covariant derivative of a vector written in terms of its contravariant components

$$A^i_{,k} = \frac{\partial A^i}{\partial x^k} + \Gamma^i_{jk} A^j \tag{A.33}$$

We now consider the covariant derivative of a vector expressed in terms of its covariant components as follows:

$$\frac{\partial \vec{A}}{\partial x^k} = \frac{\partial A_j \vec{a}^j}{\partial x^k} = A_j \frac{\partial \vec{a}^j}{\partial x^k} + \vec{a}^j \frac{\partial A_j}{\partial x^k} \tag{A.34}$$

We now express $\partial \vec{A}/\partial x^k$ in terms of its covariant components to obtain

$$\frac{\partial \vec{A}}{\partial x^k} \cdot \vec{a}_i = A_{i,k} = \frac{\partial \vec{a}^j}{\partial x^k} \cdot \vec{a}_i + \vec{a}^j \cdot \vec{a}_i \frac{\partial A_j}{\partial x^k} = \frac{\partial A_i}{\partial x^k} + \frac{\partial \vec{a}^j}{\partial x^k} \cdot \vec{a}_i \tag{A.35}$$

We now note that

$$\frac{\partial (\vec{a}_j \cdot \vec{a}^i)}{\partial x^k} = 0 \tag{A.36a}$$

so that

$$\vec{a}_j \cdot \frac{\partial \vec{a}^i}{\partial x^k} = -\vec{a}^i \cdot \frac{\partial \vec{a}_j}{\partial x^k} = -\Gamma^i_{jk} \tag{A.36b}$$

or, on interchanging i and j in Eq. (A.36b)

$$-\Gamma^j_{ik} = \vec{a}_i \cdot \frac{\partial \vec{a}^j}{\partial x^k} \tag{A.37}$$

Substituting Eq. (A.37) into Eq. (A.35) yields for the covariant derivative of a vector expressed in its covariant components

$$A_{i,k} = \frac{\partial A_i}{\partial x^k} - \Gamma^j_{ik} A_j \tag{A.38}$$

These results can be generalized to tensors of arbitrary order in both covariant and contravariant indices as follows:

$$T^{i \cdots l}_{j \cdots m, k} = \frac{\partial T^{i \cdots l}_{j \cdots m}}{\partial x^k} + \Gamma^i_{kn} T^{n \cdots l}_{j \cdots m} + \cdots \Gamma^l_{kn} T^{i \cdots n}_{j \cdots m}$$

$$- \Gamma^n_{kj} T^{i \cdots l}_{n \cdots m} - \Gamma^n_{km} T^{i \cdots l}_{j \cdots n} \tag{A.39}$$

In Eq. (A.39) there is one additive Christoffel symbol term for each contravariant index of the tensor being differentiated. Each contravariant index of the tensor being differentiated appears as the superscript of each successive Christoffel symbol in each additive term and in its place in the original tensor is a running index of summation, n, in Eq. (A.39), which also appears as one of the subscripts of each successive Christoffel symbol. The other subscript in each Christoffel symbol is the index of the covariant coordinate with respect to which the original tensor is being differentiated. Terms each containing a Christoffel symbol are subtracted as shown in Eq. (A.39), where the running index, n, is the superscript in the Christoffel symbol and replaces, subtractive term by subtractive term, each covariant index in the original tensor. Each successive index being so replaced appears as one of the subscripts of the Christoffel symbol in that subtractive term while the other subscript is the index of the coordinate with respect to which the original tensor is being differentiated. Note that covariant differentiation results in a tensor of one covariant rank higher than that of the of the tensor being differentiated. As an example of the use of Eq. (A.39) we consider the covariant differentiation of a mixed tensor (one that has both covariant and contravariant indices) of rank four.

$$T_{jl,k}^{im} = \frac{\partial T_{jl}^{im}}{\partial x^k} + \Gamma_{kn}^i T_{jl}^{nm} + \Gamma_{kn}^m T_{jl}^{in} - \Gamma_{kj}^n T_{nl}^{im} - \Gamma_{kl}^n T_{jn}^{im} \qquad (A.40)$$

The covariant differentiation of a scalar, a tensor of rank zero, results in a covariant vector, a tensor of rank one, the gradient of the scalar, as follows:

$$\vec{\nabla}\Phi(x^1, x^2, x^3) = \frac{\partial \Phi}{\partial x^k}\vec{a}^k = \Phi_{,k}\vec{a}^i = g_k\vec{a}^k \qquad (A.41)$$

where only the first term in Eq. (A.39) survives since Φ, a scaler, has no indices. Using Eq. (A.23b) we can express $\Phi_{,k}$ in terms of its contravariant components as follows:

$$g^i = g_k g^{kj} = \Phi_{,k} g^{kj} \qquad (A.42)$$

The physical components of the gradient, $\tilde{g}^j$, can now be expressed as

$$\tilde{g}^j = g^j\sqrt{g_{jj}} = \Phi_{,k} g^{kj}\sqrt{g_{jj}} \qquad (A.43)$$

We now observe that the covariant derivative of a tensor is the generalization of the usual gradient operation and as such gives explicit meaning to expressions such as $\vec{\nabla}\vec{v}$, the gradient of the velocity vector, which tensor may be defined as the usual covariant derivative of $\vec{v}$.

If the index of the coordinate with respect to which differentiation has been carried out is subsequently raised, we call the result the *contravariant derivative*. For example, A^{ij}, the contravariant derivative of A^i, is given by

$$A^{ij} = A^i_{,k} g^{kj} \tag{A.44}$$

We now express the Christoffel symbols in terms of derivatives of the elements of the metric tensor. We set

$$\vec{a}_i \cdot \frac{\partial \vec{a}_j}{\partial x^k} = [jk, i] \tag{A.45}$$

where the $[jk, i]$ are the covariant components of $\partial \vec{a}_j / \partial x^k$ and the quantity $[jk, i]$ is called the Christoffel symbol of the first kind. Since the Christoffel symbol of the second kind represents the contravariant components of $\partial \vec{a}_j / \partial x^k$ [cf. Eq. A.32], we have on using Eq. (A.23a)

$$[jk, i] = g_{il} \Gamma^l_{jk} \tag{A.46a}$$

$$\Gamma^i_{jk} = g^{il} [jk, l] \tag{A.46b}$$

Further,

$$\frac{\partial \vec{a}_j}{\partial x^k} = \frac{\partial}{\partial x^k} \frac{\partial \vec{r}}{\partial x^j} = \frac{\partial}{\partial x^j} \frac{\partial \vec{r}}{\partial x^k} = \frac{\partial \vec{a}_k}{\partial x^j} \tag{A.47}$$

where

$$\vec{r} = x^i \vec{a}_i \tag{A.48}$$

as usual and the order of partial differentiation has been interchanged. Thus the Christoffel symbols of the first and second kind are symmetric in j and k so that

$$[jk, i] = [kj, i] \tag{A.49a}$$

$$\Gamma^i_{jk} = \Gamma^i_{kj} \tag{A.49b}$$

We can now write for the Christoffel symbol of the first kind

$$\begin{aligned}
[jk, i] &= \vec{a}_i \cdot \frac{\partial \vec{a}_j}{\partial x^k} = \frac{1}{2} \vec{a}_i \cdot \left(\frac{\partial \vec{a}_j}{\partial x^k} + \frac{\partial \vec{a}_k}{\partial x^j} \right) \\
&= \frac{1}{2} \left(\frac{\partial (\vec{a}_i \cdot \vec{a}_j)}{\partial x^k} - \vec{a}_j \cdot \frac{\partial \vec{a}_i}{\partial x^k} + \frac{\partial (\vec{a}_i \cdot \vec{a}_k)}{\partial x^j} - \vec{a}_k \cdot \frac{\partial \vec{a}_i}{\partial x^j} \right) \\
&= \frac{1}{2} \left(\frac{\partial g_{ij}}{\partial x^k} + \frac{\partial g_{ik}}{\partial x^j} - \vec{a}_j \cdot \frac{\partial \vec{a}_k}{\partial x^i} - \vec{a}_k \cdot \frac{\partial \vec{a}_j}{\partial a^i} \right) \\
&= \frac{1}{2} \left(\frac{\partial g_{ij}}{\partial x^k} + \frac{\partial g_{ik}}{\partial x^j} - \frac{\partial (\vec{a}_j \cdot \vec{a}_k)}{\partial x^i} \right) \tag{A.50a}
\end{aligned}$$

Thus

$$[jk, i] = \frac{1}{2} \left(\frac{\partial g_{ij}}{\partial x^k} + \frac{\partial g_{ik}}{\partial x^j} - \frac{g_{ik}}{\partial x^i} \right) \qquad (A.50b)$$

or

$$[jk, i] = \frac{1}{2} \left(\frac{\partial g_{ij}}{\partial x^k} + \frac{\partial g_{ki}}{\partial x^j} - \frac{\partial g_{jk}}{\partial x^i} \right) \qquad (A.51a)$$

and for the Christoffel symbol of the second kind

$$\Gamma^i_{jk} = g^{il}[jk, l] = \frac{1}{2} g^{il} \left(\frac{\partial g_{lj}}{\partial x^k} + \frac{\partial g_{kl}}{\partial x^j} - \frac{\partial g_{jk}}{\partial x^l} \right) \qquad (A.51b)$$

Equations (A.51a) and (A.51b) allow the computation of the Christoffel symbols for any coordinate system.

We now consider the total derivative of a vector or tensor with respect to a scalar parameter, t, on which the vector or tensors can depend explicitly, and on which the x^k depend. We have for the derivative, with respect to t, of a vector $\vec{A}$ expressed in terms of its contravariant components

$$\frac{d\vec{A}}{dt} = \frac{d(A^i \vec{a}_i)}{dt} = \frac{dA^i}{dt} \vec{a}_i + A^i \frac{d\vec{a}_i}{dt}$$

$$= \left(\frac{\partial A^i}{\partial t} + \frac{\partial A^i}{\partial x^k} \frac{dx^k}{dt} \right) \vec{a}_i + A^i \frac{\partial a_i}{\partial x^k} \frac{dx^k}{dt} \qquad (A.52)$$

where we note from Eq. (A.2) that $\vec{a}_i$ depends on time only through the x^k–no explicit time dependence–so that $\partial a_i / \partial t = 0$. Since the Γ^i_{jk} are the contravariant components of $\partial \vec{a}_j / \partial x^k$ (from Eq. (A.32)), we have

$$\frac{\partial \vec{a}_j}{\partial x^k} = \Gamma^i_{jk} \vec{a}_i \qquad (A.53)$$

Setting $i = j$ in the second term of Eq. (A.52) and then substituting Eq. (A.53) yields

$$\frac{d\vec{A}}{dt} = \left(\frac{\partial A^i}{\partial t} + \frac{\partial A^i}{\partial x^k} \frac{dx^k}{dt} \right) \vec{a}_i + A^j \Gamma^i_{jk} \vec{a}_i \frac{dx^k}{dt}$$

$$= \left[\frac{\partial A^i}{\partial t} + \left(\frac{\partial A^i}{\partial x^i} + A^j \Gamma^i_{jk} \right) \frac{dx^k}{dt} \right] \vec{a}_i$$

$$= \left(\frac{\partial A^i}{\partial t} + A^i_{,k} \frac{dx^k}{dt} \right) \vec{a}_i \qquad (A.54)$$

where

$$\frac{\delta A^i}{\delta t} = \frac{\partial A^i}{\partial t} + A^i_{,k} \frac{dx^k}{dt} \qquad (A.55)$$

is called the intrinsic, or absolute, derivative, of A^i with respect to t.

APPENDIX B

Review of Elements of Bessel Functions

The basic form of Bessel's equation is

$$\frac{d^2 R(r)}{dr^2} + \frac{1}{r}\frac{dR(r)}{dr} + \left(k^2 - \frac{n^2}{r^2}\right) R(r) = 0 \tag{B.1}$$

which has solutions

$$R(r) = A J_n(kr) + B Y_n(kr) \tag{B.2}$$

where the $J_n(kr)$ are Bessel functions of the first kind of order n, $Y_n(kr)$ are Bessel functions of the second kind of order n, and A and B are constants. Plots of $J_n(x)$ and $Y_n(x)$ for $n = 0$, 1, and 2, are given in Figures B.1 and B.2, respectively. Explicit representations, though not the only ones available, for $J_n(x)$ and $Y_n(x)$ are given in Eqs. (B.3) and (B.4), respectively.

$$J_n(x) = \left(\frac{x}{2}\right)^n \sum_{\ell=0}^{\infty} \frac{(-x^2/4)^\ell}{\ell!\,\Gamma(\ell + n + 1)} \tag{B.3}$$

and

$$Y_n(x) = -\frac{(x/2)^{-n}}{\pi} \sum_{\ell=0}^{n-1} \frac{(n-\ell-1)!}{\ell!}\left(\frac{x^2}{4}\right)^\ell + \frac{2}{\pi}\ln\left(\frac{x}{2}\right) J_n(x)$$

$$-\frac{(x/2)^n}{\pi} \sum_{\ell=0}^{\infty} [\psi(\ell+1) + \psi(n+\ell+1)] \frac{(-x^2/4)^\ell}{\ell!\,(n+\ell)!} \tag{B.4}$$

where $\psi(m)$, the Psi, or digamma, function is given for integer arguments by

$$\psi(m) = -\gamma + \sum_{q=1}^{m-1} \frac{1}{q}, \qquad \text{for } m \geq 2 \tag{B.5}$$

where $\gamma = 0.5772156649 \cdots$ is Euler's constant and

$$\psi(1) = -\gamma \tag{B.6}$$

More generally,

$$\psi(x) = \frac{d\ln[\Gamma(x)]}{dx} = \frac{\Gamma'(x)}{\Gamma(x)} \tag{B.7}$$

where $\Gamma(x)$ is the usual gamma function. If we now make the following changes of variable in Eq. (B.1), namely

$$\beta = k \tag{B.8a}$$

347

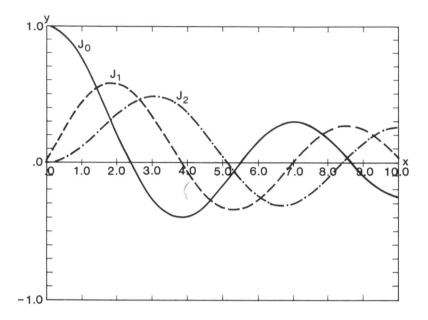

Figure B.1: Ordinary Bessel functions of the first kind of orders 0–2.

$$r = x^\delta \tag{B.8b}$$

$$R(r) = y(r)x^\alpha \tag{B.8c}$$

it becomes

$$\frac{d^2y(x)}{dx^2} - \frac{2\alpha - 1}{x}\frac{dy(x)}{dx} + \left(\beta^2\delta^2 x^{2\delta-2}\frac{dy(x)}{dx} + \frac{\alpha^2 - n^2\delta^2}{x^2}\right)y(x) = 0 \tag{B.9a}$$

with the solution

$$y(x) = Ax^\alpha J_n(\beta x^\delta) + Bx^\alpha Y_n(\beta x^\delta) \tag{B.9b}$$

Equation (B.9a) is called the generalized Bessel equation and is used to solve a second-order differential equation (not all second-order ordinary differential equations can be so solved, of course, but when it can be used, this technique is excellent) by choosing α, β, δ, and n such that Eq. (B.9a) is brought into the same form as the equation to be solved and writing down the solution to within two arbitrary constants, A and B, using Eq. (B.9b).

A change of sign in Eq. (B.9a) yields

$$\frac{d^2R(r)}{dr^2} + \frac{1}{r}\frac{dR(r)}{dr} - \left(k^2 + \frac{n^2}{r^2}\right)R(r) = 0 \tag{B.10}$$

which, to within two arbitrary constants, has the solution

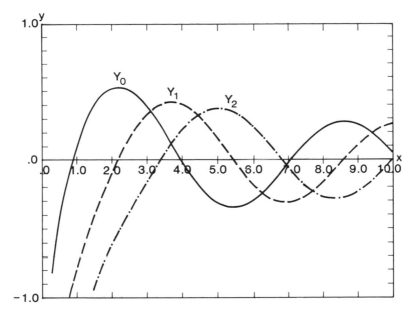

Figure B.2: Ordinary Bessel functions of the second kind of orders 0–2.

$$R(r) = AI_n(kr) + BK_n(kr) \tag{B.11}$$

where $I_n(kr)$ is the modified Bessel function (Bessel function of imaginary argument) of the first kind of order n, $K_n(kr)$ is the modified Bessel function of the second kind of order n, and A and B are constants fitted from the boundary conditions of the problem being solved. Making the same changes of variable in Eq. (B.10) that led to Eq. (B.9a) yields

$$\frac{d^2y(x)}{dx^2} - \frac{2\alpha - 1}{x}\frac{dy(x)}{dx} - \left(\beta^2\delta^2 x^{2\delta-2} + \frac{\alpha^2 - n^2\delta^2}{x^2}\right)y(x) = 0 \tag{B.12}$$

the solution of which, to within two arbitrary constants, is

$$y(x) = Ax^\alpha I_n(\beta x^\delta) + Bx^\alpha K_n(\beta x^\delta) \tag{B.13}$$

Equation (B.12) is the generalized Bessel equation for modified Bessel functions and is used in the same way as Eq. (B.10). Plots of $I_n(x)$ and $K_n(x)$ for $n = 0$, 1, and 2, are given in Figures B.3 and B.4, respectively. The modified Bessel functions have a number of explicit representations. For example,

$$I_n = \left(\frac{x}{2}\right)^n \sum_{\ell=0}^{\infty} \frac{(x^2/4)^\ell}{\ell!\,\Gamma(n+\ell+1)} \tag{B.14}$$

and

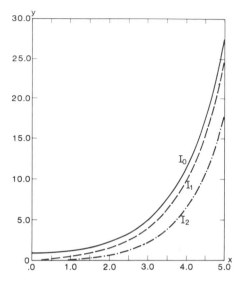

Figure B.3: Modified Bessel functions of the first kind of orders 0–2.

$$K_n(x) = \frac{1}{2}\left(\frac{x}{2}\right)^{-n}\sum_{\ell=0}^{n-1}\frac{(n-\ell-1)!}{\ell!}\left(-\frac{x^2}{4}\right)^{\ell} + (-1)^{n+1}\ln\left(\frac{x}{2}\right)I_n(x)$$

$$+\frac{1}{2}(-1)^n\left(\frac{x}{2}\right)^n\sum_{\ell=0}^{\infty}[\psi(n+1)+\psi(n+\ell+1)]\frac{(x^2/4)^{\ell}}{\ell^2(n+\ell)!} \quad \text{(B.15)}$$

where $\psi(m)$ has been defined in Eq. (B.5). Also,

$$K_0(x) = -\left[\ln(x/2)+\gamma\right]I_0(x) + 2\sum_{\ell=1}^{\infty}\frac{I_{2\ell}(x)}{\ell} \quad \text{(B.16)}$$

We now consider the generating function for ordinary Bessel functions of the first kind. We begin with the product of two exponentials, which we express in terms of their MacLaurin series.

$$e^{xt/2}\,e^{-x/2t} = \sum_{k=0}^{\infty}\left(\frac{xt}{2}\right)^k\frac{1}{k!}\sum_{\ell=0}^{\infty}\left(\frac{-x}{2t}\right)^{\ell}\frac{1}{\ell!} \quad \text{(B.17)}$$

We now set $n+\ell = k$ to yield

$$e^{x(t-1/t)/2} = \sum_{n+\ell=0}^{\infty}\sum_{\ell=0}^{\infty}\frac{1}{(n+\ell)!\ell!}\left(\frac{x}{2}\right)^{n+2\ell}(-1)^{\ell}t^n$$

$$= \sum_{n+\ell=0}^{\infty}t^n J_n(x) \quad \text{(B.18)}$$

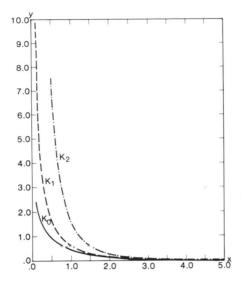

Figure B.4: Modified Bessel functions of the second kind of orders 0–2.

where Eq. (B.3) has been used. Since $0 \leq \ell \leq \infty$, the only way in which $0 \leq n + \ell \leq \infty$ is for the following inequalities to hold: $-\ell \leq n \leq \infty$, or $-\infty \leq n \leq \infty$. Thus,

$$e^{x(t-1/t)/2} = \sum_{n=-\infty}^{\infty} t^n J_n(x) \tag{B.19}$$

Various useful series can be derived from Eq. (B.19) by assigning suitable values to t. For example, if $t = 1$, we have

$$1 = \sum_{n=-\infty}^{\infty} J_n(x) = \sum_{n=-\infty}^{-1} J_n(x) + J_0(x) + \sum_{n=1}^{\infty} J_n(x) \tag{B.20}$$

Recalling that $J_{-n}(x) = (-1)^n J_n(x)$ for integer n, we see that the first sum on the right of Eq. (B.20) becomes, on setting $n = -n$,

$$\sum_{n=-\infty}^{-1} J_n(x) = \sum_{n=\infty}^{1} (-1)^n J_n(x) = \sum_{n=1}^{\infty} (-1)^n J_n(x) \tag{B.21}$$

so that Eq. (B.20) becomes

$$1 = J_0(x) + \sum_{n=1}^{\infty} (-1)^n J_n(x) + \sum_{n=1}^{\infty} J_n(x)$$

$$= J_0 + \sum_{n=1}^{\infty} [1 + (-1)^n] J_n(x) \tag{B.22}$$

in which the indicated sums have been added to each other term by term. The quantity $1 + (-1)^n$ equals zero for odd n; that is, $n = 1, 3, 5, \ldots$ and equals 2 for even n; that is, $n = 2, 4, 6, \ldots$. Thus,

$$1 = J_0(x) + 2J_2(x) + 2J_4(x) + 2J_6(x) + \cdots = J_0(x) + 2\sum_{k=1}^{\infty} J_{2k}(x) \quad \text{(B.23)}$$

Two somewhat more elaborate series are available if we put

$$t = ie^{i\theta} = e^{i(\theta + \pi/2)} \quad \text{(B.24)}$$

as follows.
$$e^{x(t-1/t)/2} = e^{x(ie^{i\theta} + ie^{-i\theta})/2}$$

$$= \cos[x\cos(\theta)] + i\sin[x\cos(\theta)] = \sum_{n=-\infty}^{\infty} e^{i(\theta + \pi/2)n} J_n(x)$$

$$= \sum_{n=-\infty}^{\infty} \{\cos[n(x + \pi/2)] + i\sin[n(x + \pi/2)]\} J_n(x) \quad \text{(B.25)}$$

Equating real and imaginary parts on both sides of Eq. (B.25) yields

$$\cos[x\cos(\theta)] = \sum_{n=-\infty}^{\infty} \cos[n(x + \pi/2)]J_n(x) \quad \text{(B.26)}$$

$$\sin[x\cos(x)] = \sum_{n=-\infty}^{\infty} \sin[n(x + \pi/2)]J_n(x) \quad \text{(B.27)}$$

It can further be shown that

$$\cos[x\cos(\theta)] = J_0 + 2\sum_{k=1}^{\infty} (-1)^k J_{2k}(x)\cos(2k\theta) \quad \text{(B.28)}$$

$$\sin[x\cos(\theta)] = 2\sum_{k=0}^{\infty} (-1)^k J_{2k+1}(x)\cos[(2k+1)\theta] \quad \text{(B.29)}$$

By setting

$$t = e^{i\theta} \quad \text{(B.30)}$$

in Eq. (B.19) and carrying out similar manipulations, it can be shown that

$$\cos[x\sin(\theta)] = J_0 + 2\sum_{k=1}^{\infty} J_{2k}(x)\cos(2k\theta) \quad \text{(B.31)}$$

and

$$\sin[x\sin(\theta)] = 2\sum_{k=0}^{\infty} J_{2k+1}(x)\sin[(2k+1)\theta] \tag{B.32}$$

We now note selected integrals involving Bessel functions as follows.

$$\int_0^x I_{n-1}(u)u^n \, du = x^n I_n(x) \tag{B.33}$$

$$\int_0^x K_{\nu-1}(u)u^\nu \, du = -xK_\nu(x) + 2^{\nu-1}\Gamma(\nu) \tag{B.34}$$

$$\int_0^\infty K_0(x) \, dx = \frac{\pi}{2} \tag{B.35}$$

$$\int_0^\infty K_0(x)x \, dx = 1 \tag{B.36}$$

$$\int_0^\infty K_1(x)x \, dx = \frac{\pi}{2} \tag{B.37}$$

$$\int_0^\infty K_\nu(x)x^\nu \, dx = 2^{\nu-1}\sqrt{\pi}\,\Gamma\left(\nu + \frac{1}{2}\right) \tag{B.38}$$

We note the derivative relations for z a complex variable

$$I_0'(z) = I_1(z) \tag{B.39}$$

$$K_0'(z) = -K_1(z) \tag{B.40}$$

and the asymptotic relations valid as $z \to 0$

$$I_\nu \sim \frac{\left(\frac{z}{2}\right)^\nu}{\Gamma(\nu+1)} \tag{B.41}$$

for ν not equal to a negative integer, and

$$K_\nu(z) \sim \frac{\Gamma(\nu)}{2\left(\frac{z}{2}\right)^\nu} \tag{B.42}$$

Recursion relations for $I_\nu(z)$ and $e^{\nu\pi i}K_\nu(z)$, where $i = \sqrt{-1}$, are

$$\frac{2\nu}{z}F_\nu(z) = F_{\nu-1}(z) - F_{\nu+1}(z) \tag{B.43}$$

$$F_\nu'(z) = F_{\nu-1}(z) - \frac{\nu}{z}F_\nu(z) \tag{B.44}$$

$$2F_\nu'(z) = F_{\nu-1}(z) + F_{\nu+1}(z) \tag{B.45}$$

$$F_\nu'(z) = F_{\nu+1}(z) + \frac{\nu}{z}F_\nu(z) \tag{B.46}$$

APPENDIX C
Lagrange Multipliers

We now consider the computation of the stationary value (minimum or maximum) of a function subject to given conditions, which we will cast into the form of equations of constraint, that is, the method of Lagrange multipliers. Lagrange multipliers arise in many areas in the physical sciences as an aid in solving the problem of finding stationary values (local maxima or minima as the case may be) of functions of one or more independent variables subject to one or more conditions cast as equations of the form $g = $ const., which are called *equations of constraint*. The number of unknown quantities to be solved for in a given problem is equal to the number of independent variables of the original function being maximized or minimized plus the number of Lagrange multipliers in the problem, one for each equation of constraint. In general, then, if a stationary value of the function $f(x, y, z, u, v, w, \ldots)$ is to be found subject to the constraints $g_1(x, y, z, u, v, w) = c_1$, $g_2(x, y, z, u, v, w) = c_2$, etc., we write

$$d(f + \lambda_1 g_1 + \lambda_2 g_2 + \cdots) = 0 \qquad \text{(C.1)}$$

where the λ_i are the (undetermined) Lagrange multipliers. Equation (C.1) may be expanded to yield

$$\frac{\partial f}{\partial x} + \lambda_1 \frac{\partial g_1}{\partial x} + \lambda_2 \frac{\partial g_2}{\partial x} + \cdots = 0 \qquad \text{(C.2a)}$$

$$\frac{\partial f}{\partial y} + \lambda_1 \frac{\partial g_1}{\partial y} + \lambda_2 \frac{\partial g_2}{\partial y} + \cdots = 0 \qquad \text{(C.2b)}$$

$$\frac{\partial f}{\partial z} + \lambda_1 \frac{\partial g_1}{\partial z} + \lambda_2 \frac{\partial g_2}{\partial z} + \cdots = 0 \qquad \text{(C.2c)}$$

$$\frac{\partial f}{\partial u} + \lambda_1 \frac{\partial g_1}{\partial u} + \lambda_2 \frac{\partial g_2}{\partial u} + \cdots = 0 \qquad \text{(C.2d)}$$

$$\frac{\partial f}{\partial v} + \lambda_1 \frac{\partial g_1}{\partial v} + \lambda_2 \frac{\partial g_2}{\partial v} + \cdots = 0 \qquad \text{(C.2e)}$$

$$\frac{\partial f}{\partial w} + \lambda_1 \frac{\partial g_1}{\partial w} + \lambda_2 \frac{\partial g_2}{\partial w} + \cdots = 0 \qquad \text{(C.2f)}$$

$$\vdots$$

etc.

The equations (C.2a)–(C.2f) together with the equations of constraint, the $g_i = c_i$, form the equation set to be solved for the values of the independent variables yielding a stationary value for f. The λ_i are found only if doing so increases the convenience of arriving at the solution for f.

APPENDIX D

In this Appendix statistical tables for the Mann–Whitney statistic, Table D.1, the probability tables for the runs test, Table D.2, a table of the Gaussian probability distribution, Table D.3, critical values of the t distribution, Table D.4, critical values of the χ^2 distribution, Table D.5, and values of the F distribution, Table D.6 are presented.

Table D.1 contains the probabilities that the Mann–Whitney statistic is less than or equal to the value computed by considering all possible occurrences of one variate having ranks larger than the other variate.

Table D.1a: $n = 2$

u_i/m	1	2
0	0.33333	0.16667
1	0.66667	0.33333
2		0.66667
3		0.83333

Table D.1b: $n = 3$

u_i/m	1	2	3
0	0.25000	0.10000	0.05000
1	0.50000	0.20000	0.10000
2	0.75000	0.40000	0.20000
3		0.60000	0.35000
4		0.80000	0.50000
5		0.90000	0.65000
6			0.80000
7			0.90000
8			0.95000

Table D.1c: $n = 4$

u_i/m	1	2	3	4
0	0.20000	0.06667	0.02857	0.01429
1	0.40000	0.13333	0.05714	0.02857
2	0.60000	0.26667	0.11429	0.57143
3	0.80000	0.40000	0.20000	0.10000
4		0.60000	0.31429	0.17143
5		0.73333	0.42857	0.24286
6		0.86667	0.57143	0.34286
7		0.93333	0.68571	0.44286
8			0.80000	0.55714
9			0.88571	0.65714
10			0.94286	0.75714
11			0.97143	0.82857
12				0.90000
13				0.94286
14				0.97143

Table D.1d: $n = 5$

u_i/m	1	2	3	4	5
0	0.16667	0.04762	0.01786	0.00794	0.00397
1	0.33333	0.09524	0.03571	0.01587	0.00794
2	0.50000	0.19048	0.07143	0.03175	0.01587
3	0.66667	0.28571	0.12500	0.05556	0.02778
4	0.83333	0.42857	0.19643	0.09524	0.04762
5		0.57143	0.28571	0.14286	0.0754
6		0.71429	0.39286	0.20635	0.11111
7		0.80952	0.50000	0.27778	0.15476
8		0.90476	0.60714	0.36508	0.21032
9		0.95238	0.71429	0.45238	0.27381
10			0.80357	0.54762	0.34524
11			0.87500	0.63492	0.42063
12			0.92857	0.72222	0.50000
13			0.96429	0.79365	0.57937
14			0.98214	0.85714	0.65476
15				0.90476	0.72619
16				0.94444	0.78968
17				0.96825	0.84524
18				0.98413	0.88889
19				0.99206	0.92460
20					0.95238

Table D.1e: $n = 6$

u_i/m	1	2	3	4	5	6
0	0.14286	0.03571	0.01191	0.004762	0.00217	0.00108
1	0.28571	0.07143	0.02381	0.009524	0.00433	0.00217
2	0.42857	0.14286	0.04762	0.01905	0.00866	0.00433
3	0.57143	0.21429	0.08333	0.03333	0.01515	0.00758
4	0.71429	0.32143	0.13095	0.05714	0.02597	0.01299
5	0.85714	0.42857	0.19048	0.08571	0.04113	0.02056
6		0.57143	0.27381	0.12857	0.06277	0.03247
7		0.67857	0.35714	0.17619	0.08875	0.04654
8		0.78571	0.45238	0.23810	0.12338	0.06602
9		0.85714	0.54762	0.30476	0.16450	0.08983
10		0.92857	0.64286	0.38095	0.21429	0.12103
11		0.96429	0.72619	0.45714	0.26840	0.15476
12			0.80952	0.54286	0.33117	0.19697
13			0.86905	0.61905	0.39610	0.24242
14			0.91667	0.69524	0.46537	0.29437
15			0.95238	0.76190	0.53463	0.34957
16			0.97619	0.82381	0.60390	0.40909
17			0.98810	0.87143	0.66883	0.46861
18				0.91429	0.73160	0.53139
19				0.94286	0.78571	0.59091
20				0.96667	0.83550	0.65043
21				0.98095	0.87662	0.70563
22				0.99048	0.91126	0.75758
23				0.99524	0.93723	0.80303
24					0.95887	0.84524
25					0.97403	0.87987
26					0.98485	0.91017
27					0.99134	0.93398
28					0.99567	0.95346
29					0.99784	0.96753
30						0.97944
31						0.98701
32						0.99242
33						0.99567
34						0.99784
35						0.99892

Table D.1f: $n = 7$

u_i/m	1	2	3	4	5	6	7
0	0.12500	0.02778	0.00833	0.00303	0.00126	0.00058	0.00029
1	0.25000	0.05556	0.01667	0.00606	0.00253	0.00117	0.00058
2	0.37500	0.11111	0.03333	0.01212	0.00505	0.00233	0.00117
3	0.50000	0.16667	0.05833	0.02121	0.00884	0.00408	0.00204
4	0.62500	0.20500	0.09167	0.03636	0.01515	0.00699	0.0035
5	0.75000	0.33333	0.13333	0.05455	0.02399	0.01107	0.00554
6	0.87500	0.44444	0.19167	0.08182	0.03662	0.01748	0.00874
7		0.55556	0.25833	0.11515	0.05303	0.02564	0.01311
8		0.66667	0.33333	0.15758	0.0745	0.03671	0.01894
9		0.75000	0.41667	0.20606	0.10101	0.0507	0.02652
10		0.83333	0.50000	0.26364	0.13384	0.06877	0.03642
11		0.88889	0.58333	0.32424	0.17172	0.09033	0.04866
12		0.94444	0.66667	0.39394	0.21591	0.11713	0.06410
13		0.97222	0.74167	0.46364	0.26515	0.14744	0.08246
14			0.80833	0.53636	0.31944	0.18298	0.10431
15			0.86667	0.60606	0.37753	0.22261	0.12966
16			0.90833	0.67576	0.43813	0.26690	0.15880
17			0.94167	0.73636	0.50000	0.31410	0.19143
18			0.96667	0.79394	0.56187	0.36538	0.22786
19			0.98333	0.84242	0.62247	0.41783	0.26748
20			0.99167	0.88485	0.68056	0.47261	0.31002
21				0.91818	0.73485	0.52739	0.35519
22				0.94545	0.78409	0.58217	0.40239
23				0.96364	0.82828	0.63462	0.45076
24				0.97879	0.86616	0.68590	0.50000
25				0.98788	0.89899	0.73310	.54924
26				0.99394	0.92551	0.77739	0.59761
27				0.99697	0.94697	0.81702	0.64481
28					0.96338	0.85256	0.69000
29					0.97601	0.88287	0.73252
30					0.98485	0.90967	0.77214
32					0.99495	0.94930	0.84120
34					0.99874	0.97436	0.89569
36						0.98893	0.93590
38						0.99592	0.96358
40						0.99883	0.98106

Table D.1g: $n = 8$

u_i/m	3	4	5	6	7	8
0	0.00606	0.00202	0.00078	0.00033	0.00016	0.00008
1	0.01212	0.00404	0.00155	0.00067	0.00031	0.00016
2	0.02424	0.00808	0.00311	0.00133	0.00062	0.00031
3	0.04242	0.01414	0.00544	0.00233	0.00199	0.00054
4	0.06667	0.02424	0.00932	0.00400	0.00186	0.00093
5	0.09697	0.03636	0.01476	0.00633	0.00295	0.00148
6	0.13939	0.05455	0.02253	0.00999	0.00466	0.00233
7	0.18788	0.07677	0.03263	0.01465	0.00699	0.00350
8	0.24848	0.10707	0.04662	0.02131	0.01026	0.00521
9	0.31515	0.14141	0.06371	0.02964	0.01445	0.00738
10	0.38788	0.18384	0.08547	0.04063	0.02005	0.01033
11	0.46061	0.23030	0.11111	0.05395	0.02704	0.01406
12	0.53939	0.28485	0.14219	0.07093	0.03605	0.01896
13	0.61212	0.34141	0.17716	0.09058	0.04693	0.02494
14	0.68485	0.40404	0.21756	0.11422	0.06030	0.03248
15	0.75152	0.46667	0.26185	0.14119	0.07599	0.04149
16	0.81212	0.53333	0.31080	0.17249	0.09464	0.05245
17	0.86061	0.59596	0.36208	0.20679	0.11593	0.06519
18	0.90303	0.65859	0.41647	0.24542	0.14048	0.08026
19	0.93333	0.71515	0.47164	0.28638	0.16783	0.09744
20	0.95758	0.76970	0.52836	0.33100	0.19845	0.11725
21	0.97576	0.81616	0.58353	0.37729	0.23170	0.13932
22	0.98788	0.85859	0.63792	0.42591	0.26791	0.16410
23	0.99394	0.89293	0.68920	0.47486	0.30629	0.19114
24		0.92323	0.73815	0.52514	0.34716	0.22090
25		0.94545	0.78244	0.57409	0.38943	0.25268
26		0.96364	0.82284	0.62271	0.43326	0.28687
27		0.97576	0.85781	0.66900	0.47754	0.32269
28		0.98586	0.88889	0.71362	0.52246	0.36045
29		0.99192	0.91453	0.75458	0.56674	0.39922
30		0.99596	0.93629	0.79321	0.61057	0.43924
32			0.96737	0.85881	0.69371	0.52044
34			0.98524	0.90942	0.76830	0.60078
36			0.99456	0.94605	0.83217	0.67731
38			0.99845	0.97036	0.88407	0.74733
40				0.98535	0.92401	0.80886

Table D.1h: $n = 9$

u_i/m	4	5	6	7	8	9
0	0.00140	0.00050	0.00020	0.000087	0.00004	0.00002
1	0.00280	0.00100	0.00040	0.00017	0.00008	0.00004
2	0.005594	0.00200	0.00080	0.00035	0.00016	0.00008
3	0.00980	0.00350	0.00140	0.00060	0.00029	0.00014
4	0.01678	0.00599	0.00240	0.00105	0.00049	0.00025
5	0.02518	0.00949	0.00380	0.00166	0.00078	0.00039
6	0.03776	0.01449	0.00599	0.00262	0.00123	0.00062
7	0.05315	0.02098	0.00879	0.00393	0.00185	0.00093
8	0.07413	0.02997	0.01279	0.00577	0.00276	0.00138
9	0.09930	0.04146	0.01798	0.00822	0.00395	0.00200
10	0.13007	0.05594	0.02478	0.01145	0.00555	0.00282
11	0.16503	0.07343	0.03317	0.01556	0.00761	0.00389
12	0.20699	0.09491	0.04396	0.02089	0.01033	0.00531
13	0.25175	0.11988	0.05674	0.02745	0.01370	0.00710
14	0.30210	0.14885	0.07233	0.03558	0.01798	0.00938
15	0.35524	0.18182	0.09051	0.04537	0.02320	0.01222
16	0.41259	0.21878	0.11189	0.05708	0.02962	0.01573
17	0.46993	0.25924	0.13606	0.07080	0.03723	0.02000
18	0.53007	0.30320	0.16384	0.08689	0.04636	0.02515
19	0.58741	0.34965	0.19421	0.10524	0.05697	0.03126
20	0.64476	0.39860	0.22797	0.12614	0.06940	0.03850
21	0.69790	0.44905	0.26434	0.14956	0.08359	0.04696
22	0.74825	0.50000	0.30350	0.17552	0.09979	0.05675
23	0.79301	0.55095	0.34446	0.20393	0.11794	0.06796
24	0.83497	0.60140	0.38781	0.23488	0.13830	0.08075
25	0.65035	0.86993	0.43197	0.26801	0.16063	0.09513
26	0.90070	0.69680	0.47732	0.30323	0.18519	0.11121
27	0.92587	0.74076	0.52268	0.34030	0.21172	0.12904
28	0.94685	0.78122	0.56803	0.37885	0.24035	0.14866
29	0.96224	0.81818	0.61219	0.41853	0.27071	0.17005
30	0.97483	0.85115	0.65554	0.45909	0.30292	0.19325
32	0.99021	0.90509	0.73566	0.54091	0.37149	0.24471
34	0.99720	0.94406	0.80579	0.62115	0.44418	0.30241
36		0.97003	0.86394	0.69677	0.51872	0.36522
38		0.98551	0.90949	0.76512	0.59260	0.43165
40		0.99401	0.94326	0.82448	0.66351	0.50000

Table D.1i: $n = 10$

u_i/m	4	5	6	7	8	9	10
0	0.00010	0.00033	0.00013	0.00005	0.00002	0.00001	0.00001
1	0.00200	0.00067	0.00025	0.00100	0.00005	0.00002	0.00001
2	0.00400	0.00133	0.00050	0.00021	0.00009	0.00004	0.00002
3	0.00699	0.00233	0.00087	0.00036	0.00016	0.00008	0.00004
4	0.01199	0.00233	0.00087	0.00036	0.00016	0.00008	0.00004
5	0.01798	0.00633	0.00237	0.00098	0.00043	0.00021	0.00010
6	0.02697	0.00966	0.00375	0.00154	0.00069	0.00032	0.00016
7	0.03796	0.01399	0.00549	0.00231	0.00103	0.00049	0.00024
8	0.05295	0.01998	0.00799	0.00339	0.00153	0.00073	0.00036
9	0.07093	0.02764	0.01124	0.00483	0.00219	0.00105	0.00053
10	0.09391	0.03763	0.01561	0.00679	0.00311	0.00149	0.00075
11	0.11988	0.04962	0.02098	0.00926	0.00427	0.00207	0.00104
12	0.15185	0.06460	0.02797	0.01250	0.00583	0.00284	0.00144
13	0.18681	0.08225	0.03634	0.01651	0.00777	0.00381	0.00194
14	0.22677	0.10323	0.04670	0.02155	0.01026	0.00507	0.00260
15	0.26973	0.12721	0.05894	0.02766	0.01332	0.00664	0.00342
16	0.31768	0.15485	0.07355	0.03512	0.01714	0.00861	0.00447
17	0.36663	0.18548	0.09029	0.04391	0.02171	0.01101	0.00575
18	0.41958	0.21978	0.10989	0.05440	0.02726	0.01396	0.00735
19	0.47253	0.25674	0.13174	0.06654	0.03380	0.01749	0.00927
20	0.52747	0.29704	0.15659	0.08063	0.04157	0.02174	0.01162
21	0.58042	0.33933	0.18382	0.09662	0.05055	0.02674	0.01162
22	0.63337	0.38395	0.21391	0.11477	0.06100	0.03263	0.01773
23	0.68232	0.42957	0.24613	0.13492	0.07286	0.03945	0.02163
24	0.73027	0.47652	0.28109	0.15739	0.08641	0.04736	0.02621
25	0.77323	0.52348	0.31768	0.18192	0.10154	0.05638	0.03151
26	0.81319	0.57043	0.35639	0.20866	0.11849	0.06665	0.03763
27	0.84815	0.61605	0.39623	0.23735	0.13714	0.07820	0.04461
28	0.88012	0.66067	0.43744	0.26810	0.15771	0.09116	0.05256
29	0.90609	0.70296	0.47890	0.30044	0.17997	0.10551	0.06150
30	0.92907	0.74326	0.52110	0.33453	0.20412	0.12140	0.07157
32	0.96204	0.81452	0.60377	0.40626	0.25738	0.15769	0.09516
34	0.98202	0.87279	0.68232	0.48113	0.31672	0.20009	0.12373
36	0.99301	0.91775	0.75387	0.55661	0.38091	0.24835	0.15750
38	0.99800	0.95038	0.81618	0.63019	0.44838	0.30189	0.19652
40		0.97236	0.86826	0.69956	0.51730	0.35985	0.24063

The cumulative probability, $P(n_1, n_2, R)$, from Eq. (8.23) of the text, that the total number of runs, R, for a set of data containing n_1 results of outcome one and n_2 results of outcome two, is less than or equal to a given number of runs, r is contained in Table D.2.

Table D.2a: $P(n_1, n_2, R \le r)$

n_1	n_2	r						
		2	3	4	5	6	7	8
3	3	0.10000	0.30000	0.70000	0.90000	1.00000		
4	3	0.05714	0.20000	0.54286	0.80000	0.97143	1.00000	
5	3	0.03571	0.14286	0.42857	0.71429	0.92857	1.00000	
6	3	0.02381	0.10714	0.34524	0.64286	0.88095	1.00000	
7	3	0.01667	0.08333	0.28333	0.58333	0.83333	1.00000	
8	3	0.01212	0.06667	0.23636	0.53333	0.78788	1.00000	
9	3	0.00909	0.05455	0.20000	0.49091	0.74545	1.00000	
10	3	0.00699	0.04546	0.17133	0.45455	0.70629	1.00000	
4	4	0.02857	0.11429	0.37143	0.62857	0.88571	0.97143	1.00000
5	4	0.01587	0.07143	0.26190	0.50000	0.78571	0.92857	0.99206
6	4	0.00952	0.04762	0.19048	0.40476	0.69048	0.88095	0.97619
7	4	0.00606	0.03333	0.14242	0.33333	0.60606	0.83333	0.95455
8	4	0.00404	0.02424	0.10909	0.27879	0.53333	0.78788	0.92929
9	4	0.00280	0.01818	0.08532	0.23636	0.47133	0.74545	0.90210
10	4	0.00200	0.01399	0.06793	0.20280	0.41858	0.70629	0.87413
5	5	0.00794	0.03968	0.16667	0.35714	0.64286	0.83333	0.96032
6	5	0.00433	0.02381	0.11039	0.26190	0.52165	0.73810	0.91113
7	5	0.00253	0.01515	0.07576	0.19697	0.42424	0.65152	0.85354
8	5	0.00155	0.01010	0.05361	0.15152	0.34732	0.57576	0.79332
9	5	0.00100	0.00699	0.03896	0.11888	0.28671	0.51049	0.73427
10	5	0.00067	0.00500	0.02897	0.09491	0.23876	0.45455	0.67832
6	6	0.00216	0.01299	0.06710	0.17532	0.39177	0.60823	0.82468
7	6	0.00117	0.00758	0.04254	0.12121	0.29604	0.50000	0.73310
8	6	0.00067	0.00466	0.02797	0.08625	0.22611	0.41259	0.64569
9	6	0.00400	0.00300	0.01898	0.06294	0.17483	0.34266	0.56643
10	6	0.00025	0.00200	0.01324	0.04695	0.13686	0.28671	0.49650

Table D.2b: $P(n_1, n_2, R \leq r)$

n_1	n_2	r						
		11	12	13	14	15	16	17
6	5	1.0000						
7	5	1.0000						
8	5	1.0000						
9	5	1.0000						
10	5	1.0000						
6	6	0.99784	1.0000					
7	6	0.99242	0.99942	1.0000				
8	6	0.98368	0.99767	1.0000				
9	6	0.97203	0.99441	1.0000				
10	6	0.95804	0.98951	1.0000				
7	7	0.97494	0.99592	0.99942	1.0000			
8	7	0.94872	0.98788	0.99767	0.99984	1.0000		
9	7	0.91608	0.97483	0.99441	0.99930	1.0000		
10	7	0.87937	0.95712	0.98951	0.99815	1.0000		
8	8	0.89977	0.96830	0.99114	0.99876	0.99984	1.0000	
9	8	0.84266	0.93941	0.97932	0.99585	0.99930	0.99996	1.0000
10	8	0.78219	0.90313	0.96360	0.99047	0.99815	0.99979	1.0000
9	9	0.76203	0.89103	0.95553	0.98778	0.99700	0.99963	0.99996
10	9	0.68141	0.83417	0.92328	0.97420	0.99239	0.99863	0.99979
10	10	0.58593	0.75779	0.87236	0.94874	0.98148	0.99551	0.99901

Table D.3a contains critical values of the Gaussian probability distribution computed in double precision using a bisection algorithm.

Table D.3a: Critical Values of the Gaussian Distribution

Pr	z	Pr	z	Pr	z
0.02	0.05015	0.22	0.58284	0.4200	1.40507
0.04	0.10043	0.24	0.64335	0.4400	1.55477
0.06	0.15097	0.26	0.70630	0.4600	1.75069
0.08	0.20189	0.28	0.77219	0.4750	1.95996
0.10	0.25335	0.30	0.84162	0.4800	2.05375
0.12	0.30548	0.32	0.91537	0.4900	2.32635
0.14	0.35846	0.34	0.99446	0.4950	2.57583
0.16	0.41246	0.36	1.08032	0.4990	3.09023
0.18	0.46770	0.38	1.17499	0.4995	3.29053
0.20	0.52440	0.40	1.28155	0.4999	3.71902

Table D.3b contains values of the Gaussian probability distribution, $P(z)$, where z is the variate measured in units of the standard deviation of the Gaussian distribution, or z score, and in this table has a range between 0 and 3.00 standard deviations.

Table D.3b: $\Pr(0 \le z = 3.00)$

z	0.00	0.01	0.02	0.03	0.05	0.07	0.09
0.00	0.00	0.00399	0.00798	0.01197	0.01994	0.02790	0.03586
0.10	0.03989	0.04380	0.04776	0.05172	0.05962	0.06750	0.07535
0.20	0.07926	0.08317	0.08706	0.09095	0.09871	0.10642	0.11409
0.30	0.11791	0.12172	0.12552	0.12930	0.13683	0.14431	0.15173
0.40	0.15542	0.15910	0.16276	0.16640	0.17364	0.18082	0.18793
0.50	0.19146	0.19497	0.19847	0.20194	0.20884	0.21566	0.22240
0.60	0.22575	0.22907	0.23237	0.23565	0.24215	0.24857	0.25490
0.70	0.25804	0.26115	0.26424	0.26730	0.27337	0.27935	0.28524
0.80	0.28814	0.29103	0.29389	0.29673	0.30234	0.30785	0.31327
0.90	0.31594	0.31859	0.32121	0.32381	0.32894	0.33398	0.33891
1.00	0.34134	0.34375	0.34614	0.34849	0.35314	0.35769	0.36214
1.10	0.36433	0.36650	0.36864	0.37076	0.37493	0.37900	0.38298
1.20	0.38493	0.38686	0.38877	0.39065	0.39435	0.39796	0.40147
1.30	0.40320	0.40490	0.40658	0.40824	0.41149	0.41466	0.41774
1.40	0.41924	0.42073	0.42220	0.42364	0.42647	0.42922	0.43189
1.50	0.43319	0.43448	0.43574	0.43699	0.43943	0.44179	0.44408
1.60	0.44520	0.44630	0.44738	0.44845	0.45053	0.45254	0.45449
1.70	0.45543	0.45637	0.45728	0.45818	0.45994	0.46164	0.46327
1.80	0.46407	0.46485	0.46562	0.46638	0.46784	0.46926	0.47062
1.90	0.47128	0.47193	0.47257	0.47320	0.47441	0.47558	0.47670
2.00	0.47725	0.47778	0.47831	0.47882	0.47982	0.48077	0.48169
2.10	0.48214	0.48257	0.48300	0.48341	0.48422	0.48500	0.48574
2.20	0.48610	0.48645	0.48679	0.48713	0.48778	0.48840	0.48899
2.30	0.48928	0.48956	0.48983	0.49010	0.49061	0.49111	0.49158
2.40	0.49180	0.49202	0.49224	0.49254	0.49286	0.49324	0.49361
2.50	0.49375	0.49396	0.49413	0.49430	0.49461	0.49492	0.49520
2.60	0.49534	0.49547	0.49560	0.49573	0.49598	0.49621	0.49643
2.70	0.49653	0.49664	0.49674	0.49683	0.49702	0.49720	0.49736
2.80	0.49744	0.49752	0.49760	0.49767	0.49781	0.49795	0.49807
2.90	0.49813	0.49819	0.49825	0.49831	0.49841	0.49851	0.49861
3.00	0.49865	0.49869	0.49874	0.49878	0.49886	0.49893	0.49900

Table D.4 contains critical values of the t distribution in which the probability values range from 0.20 through 0.99 and the values in the table are the corresponding upper limits of the probability integral.

Table D.4: Critical values of the t distribution

$\nu \backslash 2\alpha$	0.20	0.50	0.80	0.90	0.95	0.99
1	0.324919	0.999995	3.07855	6.31901	12.7210	63.7406
2	0.288675	0.816491	1.88568	2.92090	4.30736	9.96690
3	0.276671	0.764888	1.63775	2.35364	3.18424	5.86937
4	0.270722	0.740693	1.53320	2.13197	2.77736	4.62216
5	0.267180	0.726683	1.47587	2.01511	2.57114	4.04445
6	0.264835	0.717554	1.43074	1.94321	2.44729	3.71649
7	0.263167	0.711138	1.41491	1.89459	2.36491	3.50659
8	0.261921	0.706383	1.39680	1.85955	2.30622	3.36122
9	0.260955	0.702719	1.38301	1.83311	2.26233	3.25480
10	0.260184	0.699809	1.37216	1.81245	2.22828	3.17361
11	0.259556	0.697442	1.36341	1.79587	2.20111	3.10967
12	0.259033	0.695480	1.35620	1.78227	2.17892	3.05804
13	0.258591	0.693826	1.35015	1.77092	2.16046	3.01549
14	0.258212	0.692414	1.34501	1.76129	2.14487	2.97982
15	0.257885	0.691194	1.34059	1.75303	2.13152	2.94950
16	0.257599	0.690129	1.33674	1.74586	2.11997	2.92341
17	0.257347	0.689192	1.33336	1.73958	2.10987	2.90072
18	0.257123	0.688361	1.33037	1.73404	2.10097	2.88081
19	0.256923	0.687619	1.32771	1.72911	2.09307	2.86321
20	0.256742	0.686952	1.32532	1.72469	2.08600	2.84752
21	0.256588	0.686349	1.32317	1.72072	2.07965	2.83346
22	0.256432	0.685802	1.32122	1.71712	2.07391	2.82070
23	0.256297	0.685303	1.31944	1.71384	2.06869	2.80931
24	0.256173	0.684847	1.31782	2.06393	2.49268	2.79886
25	0.256060	0.684427	1.31633	1.70811	2.05956	2.78930
26	0.255954	0.684040	1.31495	1.70559	2.05555	2.78054
27	0.255857	0.683682	1.31368	1.70326	2.05185	2.77247
28	0.255768	0.683350	1.31251	1.70110	2.04843	2.76500
29	0.255683	0.683041	1.31141	1.69910	2.04525	2.75810
30	0.255605	0.682753	1.31040	1.69723	2.04229	2.75167

Table D.5 contains critical values of the probability integral, P, for the χ^2 distribution. P takes on values ranging from 0.005 through 0.999, and the numbers in the table represent the upper limits of the probability integral corresponding to the given values of P.

Table D.5a: Critical values of the χ^2 distribution

$\nu\backslash P$	0.010	0.025	0.050
1	1.5708786×10^{-4}	9.8026912×10^{-4}	3.9321400×10^{-3}
2	0.020100672	0.050635616	0.10258659
3	0.11483180	0.21579528	0.35184632
4	0.29710948	0.48441856	0.71072302
5	0.55429808	0.83121161	1.1454762
6	0.87209033	1.2373442	1.6353829
7	1.2390423	1.6898692	2.1673499
8	1.6464974	2.1797307	2.7326368
9	2.0879007	2.7003895	3.3251128
10	2.5582122	3.2469728	3.9402991
11	3.0538441	3.8157483	4.5748131
12	3.5705690	4.4037885	5.2260295
13	4.1069155	5.0087505	5.8918643
14	4.6604251	5.6287261	6.5706314

Table D.5b: Critical values of the χ^2 distribution

$\nu\backslash P$	.100	.250	.500	.750	.900
1	0.0157908	0.1015310	0.45493642	1.3233037	2.7055435
2	0.2107210	0.5753641	1.3862944	2.7725887	4.6051702
3	0.5843744	1.2125329	2.3659739	4.1083449	6.2513886
4	1.0636232	1.9225575	3.3566940	5.3852691	7.7794403
5	1.6103080	2.6746028	4.3514602	6.6256798	9.2363569
6	2.2041307	3.4545988	5.3481206	7.8408041	10.644641
7	2.8331069	4.2548522	6.3458112	9.0371475	12.017037
8	3.4895391	5.0706404	7.3441215	10.218854	13.361566
9	4.1681590	5.8988259	8.3428327	11.388751	14.683657
10	4.8651821	6.7372008	9.3418178	12.548861	15.987179
11	5.5777848	7.5841428	10.340998	13.700693	17.275009
12	6.3037961	8.4384188	11.340322	14.845404	18.549348
13	7.0415046	9.2990655	12.339756	15.983906	19.811930
14	7.7895336	10.165314	13.339274	17.116934	21.064144

Table D.5c: Critical values of the χ^2 distribution

$\nu \backslash P$	.950	.975	.990	.995	.999
1	3.8414588	5.0238862	6.6348966	7.8794386	10.827566
2	5.9914645	7.3777589	9.2103404	10.596635	13.815510
3	7.8147279	9.3484036	11.344867	12.838156	16.266236
4	9.4877290	11.143287	13.276704	14.860259	18.466827
5	11.070498	12.832502	15.086272	16.749602	20.515005
6	12.591587	14.449375	16.811894	18.547584	22.457743
7	14.067140	16.012764	18.475307	20.277740	24.321899
8	15.507313	17.534546	20.090235	21.954954	26.124205
9	16.918978	19.022768	21.665992	23.589328	27.876761
10	18.307038	20.483177	23.209248	25.188185	29.592384
11	19.675137	21.920046	24.724809	26.757905	31.265481
12	21.026069	23.336663	26.216905	28.300462	33.034556
13	22.362030	24.735614	27.689000	29.836797	33.805202
14	23.684788	26.118842	29.138402	31.316828	35.018655

Table D.6 contains a few critical values of the F distribution. The values in the table are the upper limits of the probability integral, which has the value $1 - \alpha$, as functions of the two degrees of freedom, ν_1, and ν_2, of the distribution.

Table D.6: Critical values of the F distribution for $\alpha = .05$

$\nu_2 \backslash \nu_1$	2	3	4	5	6	7	8
5	5.7912	5.4148	5.1981	5.0564	4.9566	4.8823	4.8249
6	5.1475	4.7614	4.5383	4.3922	4.2888	4.2116	4.1518
7	4.7410	4.3505	4.1241	3.9754	3.8699	3.7909	3.7296
8	4.4622	4.0694	3.8411	3.6907	3.5837	3.5035	3.4411
9	4.2594	3.8653	3.6359	3.4843	3.3763	3.2952	3.2319
10	4.1055	3.7108	3.4805	3.3281	3.2193	3.1374	3.0735
15	3.6844	3.2891	3.0571	2.9025	2.7914	2.7073	2.6412
20	3.4946	3.0998	2.8672	2.7116	3.5994	2.5141	2.4469
25	3.3868	2.9925	2.7596	2.6035	2.4906	2.4046	2.3366

INDEX